Membranes for Gas Separations

World Scientific Series in Membrane Science and Technology: Biological and Biomimetic Applications, Energy and the Environment

Series Editor: Joel R Fried *(University of Louisville, USA)*

Published

Vol. 1 Membranes for Gas Separations
by Moises A Carreon

Forthcoming

Biomedical Membranes and (Bio)Artificial Organs
edited by Dimitrios Stamatialis

World Scientific Series in Membrane Science and Technology:
Biological and Biomimetic Applications, Energy and the Environment Vol. 1

Membranes for Gas Separations

Moises A Carreon
Colorado School of Mines, USA

World Scientific
NEW JERSEY • LONDON • SINGAPORE • BEIJING • SHANGHAI • HONG KONG • TAIPEI • CHENNAI • TOKYO

Published by

World Scientific Publishing Co. Pte. Ltd.

5 Toh Tuck Link, Singapore 596224

USA office: 27 Warren Street, Suite 401-402, Hackensack, NJ 07601

UK office: 57 Shelton Street, Covent Garden, London WC2H 9HE

British Library Cataloguing-in-Publication Data
A catalogue record for this book is available from the British Library.

World Scientific Series in Membrane Science and Technology:
Biological and Biomimetic Applications, Energy and the Environment — Vol. 1
MEMBRANES FOR GAS SEPARATIONS

ISBN 978-981-3207-70-7

Printed in Singapore

Preface

Membrane technology has been demonstrated to be a viable energy-saving method to effectively separate challenging molecular gas mixtures, and potentially to leapfrog the energy-intensive and expensive benchmark technologies employed to separate gases. This book consists of a series of chapters devoted to illustrate distinctive examples of different membrane compositions ranging from inorganic, polymeric, metallic, metal organic framework, and composite which have been successfully employed to separate industrially relevant gas mixtures including hydrogen, nitrogen, methane, carbon dioxide, olefins/parafins among others. Each book chapter highlights some of the current and key fundamental and technological challenges for these membranes that must be overcome in order to envision its effective application at industrial level.

Contents

Chapter 1

Mixed Matrix Membranes for Gas Separation Applications

Maria Carreon*, Ganpat Dahe†, Jie Feng† and Surendar R Venna†,‡

*Department of Chemical Engineering, University of Tulsa, Tulsa, OK

†National Energy Technology Laboratory, Pittsburgh, PA

‡surendar.venna@netl.doe.gov.

Mixed matrix membranes (MMMs) have gained lot of interest in the research community over the last few years in order to overcome the performance trade-off shown by the polymeric membranes. An MMM utilizes the benefits of both polymeric and inorganic materials and has the potential to show very high performance without much increase in the cost of fabrication. But the fabrication of an MMM without conceding the mechanical properties is very challenging. In this chapter, the materials selection and fabrication techniques of MMM in both flat sheet and hollow fiber configurations are discussed. Formation of defects at the interface is one of the biggest challenges in fabricating the MMM. Different techniques used in the literature to compatibilize the polymer and particle were summarized. In addition, the recent progress of MMM for gas separation applications and performance prediction models were provided in detail.

1. Introduction

Chemical separations using membrane technology has attracted an increasing interest in recent years for several applications such as gas separation, water purification, food processing, pervaporation, membrane contactor and reactors, etc. Gas separation using membranes is one of the most important areas in membrane technology and is growing rapidly from early 1960s after the invention of asymmetric membranes by Loeb and Sourirajan.[1] Membrane technology has several advantages such as easy operation, small foot print, easy scale up, low capital cost, energy efficient, reliability and high product purity without any phase changes compared to conventional technologies like sorbents and scrubbing solutions which are generally energy-intensive. The three main types of membrane materials that were used for gas separation are polymeric, inorganic and polymer–inorganic mixed matrix membranes (MMMs).

The separation of gas mixture in the conventional polymeric membranes is based on the difference in the permeation rates of each gas species through membrane materials. Robeson developed a plot showing the performance limitation (called "Upper Bound") of polymeric membranes which was later supported by theoretical framework and provided the empirical correlation.[2–4] On the other hand, porous inorganic membranes such as zeolite and metal organic framework membranes showed high potential for gas separation applications.[5–13] Apart from high gas transport properties, the inorganic membranes also have very good mechanical, chemical and thermal properties. But these inorganic materials are very difficult and have expensive process into different membrane configurations and cost of the materials is very high and therefore, there are very few commercial inorganic membranes. Therefore, polymer membranes are still attractive for industrial applications, but alternative materials are needed to enhance the gas transport properties.

In order to overcome the "upper bound" limitation of the polymers and processing problems of inorganic materials, researchers focused on the development of MMMs. MMMs have been recently introduced as a new alternative approach that combines the benefits of both polymeric and inorganic materials. A typical MMM consists of inorganic fillers dispersed at a nanometer scale in a polymer matrix. The incorporation of inorganic fillers in the polymer matrix allows these membranes inherit the superior separation performance from this material whereas the presence of a polymeric phase solves the inherent brittleness related to these fillers. In hybrid membranes such as MMMs, inorganic fillers act to create preferential permeation pathways that allow selective permeability imposing a barrier to undesired infiltrates in order to improve the membrane performance.

MMMs were fabricated by dispersing the porous inorganic materials in the continuous polymeric matrix. There have been several key issues related to fabrication that have prevented the commercialization of these membranes. The fabrication of MMMs presented several practical challenges at the polymer–filler interface due to the incompatibility of organic–inorganic interfaces.[14] There are several factors that are important in the design of an MMM. The compatibility of the polymer and filler particles in terms of their gas transport properties is very important. Substantial improvement relative to the polymer occurs only when the permeabilities of the faster permeating component through the polymer and filler are approximately matched. If one phase has high gas transport resistance, it dictates the overall performance of the MMM. The other factor that influences the selection of the materials is the processing capability of the polymers after the addition of filler particles. Particle distribution within the matrix is another important factor to obtain the defect-free membranes. Wide variety of filler particles were used to fabricate the MMMs such as zeolites, metal organic frameworks, carbon nanotubes, graphene sheets, activated carbon, carbon molecular sieves, metal oxides, etc. These filler particles were also modified in order to improve polymer–filler interface and increased the gas transport properties. As stated above, initial evaluation of the component selection for the MMM can be done using the Maxwell equation with careful matching of the transport properties of each phase. Several other modified Maxwell models were developed to predict the gas separation performance of MMMs.[15] For the mixed matrix approach to achieve its full potential, the polymer matrix should have properties such that it lies close to the "upper bound" on the Robeson plot.

This chapter is organized as follows. Section 2 provides the brief overview of the mixed matrix fabrication techniques in different configurations using various filler particles. Section 3 provides the gas separation properties of the MMMs. The change in the gas transport properties of MMMs by the addition of filler particles compared to their polymer counterparts will be discussed. Section 4 focuses on the models for the prediction of gas transport properties in MMMs and different models are compared for their validity. Section 5 provides the summary and future directions needed for MMMs.

2. Fabrication and Characterization of MMMs

In 1991, L. L. Robeson introduced the concept of "upper bound" in gas separation for membranes which depicts the trade-off between permeability and selectivity of a polymeric membrane.[2] According to him, the permeability of the polymeric membranes increases as the selectivity decreases. It was also proposed that irrespective of the materials the polymeric membranes could not surpass the selectivity for any instantaneous permeability which is the so-called "upper bound". He proposed revised limits for the same relationships in the year 2008.[3] Figure 1 depicts the proposed "Upper Bound" in 1991 and 2008 which shows the improvement in the gas separation properties of newly developed polymers.

As seen from Figure 1, MMMs are capable of overcoming the Robeson "upper bound" through providing better transport properties.[13,16] Porous fillers act as molecular sieving agents separating gas

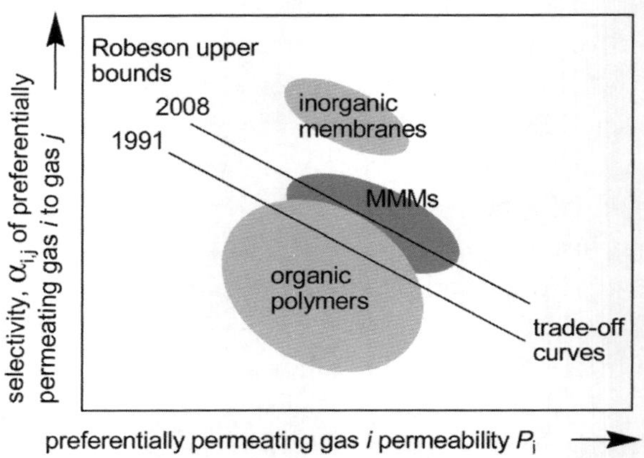

Fig. 1. Schematic representation of 1991 and 2008 Robeson upper bound.[19]
Source: Reproduced from Ref. 19 with permission from Royal Society of Chemistry.

molecules by their size or shape.[11,13,17,18] These porous inorganic particles exhibit high permeability and selectivity above the Robesons upper bound due to their small precise apertures. These highly selective porous fillers allow passing the desired components making the permselectivity higher compared with a polymeric membrane. Thus, the addition of porous fillers to a polymer matrix results in an efficient way to overcome the traditional permeability–selectivity trade-off of the polymeric membrane.

In the last two decades, MMMs have shown improvement in gas separation performance in comparison to their pure polymer membranes. MMMs are made by incorporation of inorganic filler (solid) phase in a polymer matrix. These hybrid membranes have the high permeability of the polymer membranes and results in high flux and selectivity as well, which is a property of inorganic membranes. It not only solves the trade-off between selectivity and permeability but also improves the thermal, chemical and mechanical stability along with filter morphology.[19–21]

MMMs could have two types of morphology: symmetric and asymmetric, as shown in Fig. 2. Symmetric dense membranes have been widely studied due to their simple synthesis process. In a typical synthesis process, the polymer solutions casted onto a glass plate using a knife and evaporate the solvent in the membranes under solvent saturated environment to form the dense membranes. Another disadvantage of this type of membranes is the thickness required for mechanical stability ($>50\,\mu$m) which reduces the gas flux through the membrane.

Contrary to symmetric membranes, asymmetric membranes have a selective thin ($<1\,\mu$m) skin used for gas separation and a non-selective porous support for mechanical stability. These membranes are very thin compared to the symmetric. Typically, it is the selective skin that contains the dispersed inorganic filler.[20,21] Asymmetric membranes could be fabricated as flat sheet and hollow fiber configurations.[20] The asymmetric membranes can be synthesized using several methods such as phase separation, interfacial polymerization, solution coating, plasma polymerization, etc.[22] As mentioned earlier, first fabrication of asymmetric membrane with ultra-thin dense

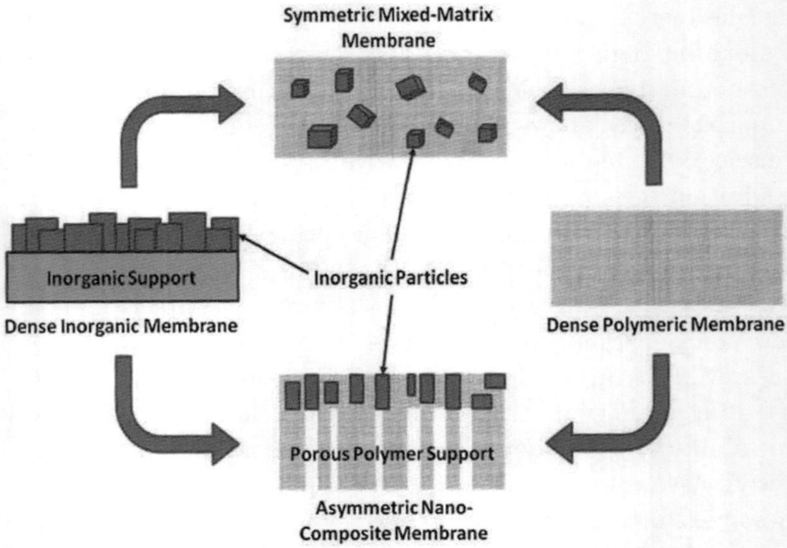

Fig. 2. Schematic representation of different MMM morphologies.
Source: Reproduced from Ref. 21 with permission from Royal Society of Chemistry.

selective layer on a porous support was reported by Loeb and Sourirajan.[1] All industrial applications prefer asymmetric membranes because of their high gas fluxes. Asymmetric hollow fibers are the most industrially practical geometry for gas separation membranes in terms of productivity and membrane packing efficiency. Although MMMs show potential to be novel materials for high performance gas separation applications, several challenges have to be overcome in the design and fabrication of these membranes for industrial applications. The following sections discuss the different factors that affect the fabrication and performance of MMMs.

3. Polymer-filler Compatibility

In an ideal case, theoretically, MMM shows improved gas transport properties. But in the real world, the dispersed phase may affect the surrounding polymer matrix at the interface between the two phases. Moore and Koros summarized several different non-idealities that

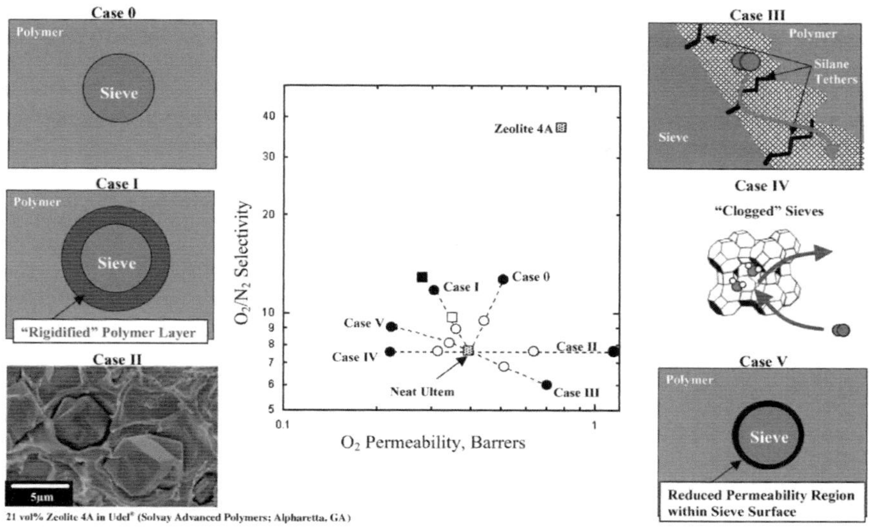

Fig. 3. Summary of the relationship between MMM morphologies and transport properties.

Notes: Circles represent calculated values; squares represent experimental Ultem data. Solid markers are 35 vol% zeolite 4A; open markers are 15 vol% 4A.

Source: Reprinted from Ref. 14. Copyright (2005) with permission from Elsevier Science.

arises during the formation of MMM as shown in Fig. 3. As shown in Fig. 3, the interfacial defects can be categorized into five cases as presented by Moore and Koros:[16] (i) rigidified polymer layers wrapped around the fillers, (ii) large voids (>∼10–100 nm) at the polymer–filler interface, which could be an unfavorable result after the MMM processing, (iii) small voids at the interface mainly because of the rigidity of polymer chains and less attractive interaction between polymers and fillers, (iv) clogged fillers by gas molecules because of strong holding from the inside surface of fillers, and (v) bottlenecked at filler side of the interfacial region.

In order to avoid these non-ideal behavior, the polymer and filler particle needs to be compatible both in terms of chemical properties and gas transport properties. One way to achieve ideal interface is by choosing polymer and inorganic materials with the same polarity.[23]

The interaction forces between the polymer and filler should be higher than both polymer–solvent interaction and filler–solvent interaction in order to achieve good interface. The additional requirement of polymer flexibility during formation would be harder to achieve because polymers close to the "upper bound" are inherently rigid, but this can still be achieved by the use of suitable plasticizing mechanisms. In particular, it is desirable to prepare membranes in such a manner that the stress at the organic–inorganic interface is minimized to suppress the tendency to form undesirable voids at that interface. Other methods include the casting of the membranes at a temperature above or at glass transition temperature,[24,25] annealing of membranes above glass transition temperature,[26] adding surface coupling agents,[27,28] preparation of membranes using melt processing,[14,29] using soft–hard block copolymer,[30] etc. Furthermore, processing steps and solvents that can "clog" the zeolites should be avoided.[31] As mentioned in the introduction, it is very important to match not only chemical properties of polymer and filler but also the gas transport properties. Moore and Koros made a great conclusion that the substantial improvement in MMM relative to the neat polymer can be achieved only when the permeabilities of the faster permeating component through the polymer and sieve are approximately matched.[14] The careful selection of the polymer and the filler materials greatly impacts the resulting morphology and performance of the MMM. The appropriate choice of both of these materials requires a clear understanding of the transport mechanisms. Pore size and surface functionality of the filler particles is also very important in selecting the filler materials to achieve high gas transport properties in MMM. Most of the fillers used in the formation of MMM are microporous materials such as zeolites, carbon molecular sieves and metal organic frameworks. Other fillers that were used in the MMMs as fillers are silica, activated carbon, metal nanoparticles, graphene, carbon nanotubes, porous polymers, covalent organic frameworks, etc.[19–21,31–39] The microporous materials show very high selectivity owing to their molecular sieving capability. Another factor that affects the properties are their surface functionality which improves adsorption of the gases on their surfaces which in turn results in improved molecular diffusion through its pores. Rigid materials with pore sizes bigger than the penetrants have also been studied as well. In this case, the most adsorption capacity or condensable component will adsorb and diffuse more effectively through the particles while the less adsorption capacity will permeate slowly.[40] When using these large pore size rigid materials, it is the gas feed mixture that determines surface flow or molecular sieving as the dominant mechanism for transport. An example of this kind of filler is activated carbon. Typically, the mean pore size of activated carbons is bigger than the molecular diameter of the penetrants; consider a common mixture such as CO_2 and CH_4. In this case, the adsorption capacity of the penetrants is a key aspect of activated carbons when used as fillers.[34] The other type of fillers used in MMMs is the non-porous particles. These fillers improve the separation by increasing the matrix tortuous pattern and decreasing the diffusion of larger molecules.[36]

3.1. *Filler Size and Morphology*

The typical membrane thickness for asymmetric membranes should be much less than 1 μm in order to be applicable in the industry. The thickness of the membrane layer in MMMs largely depends on the particle size. In order to achieve very thin selective layer, the particle must be much less than the layer thickness. Also, the smaller particles have higher surface area–volume ratios that normally equate to enhanced mass transfer between the phases. Nanoscale inorganic materials may alter the packing of the polymer chain increasing the free volume between chains resulting in an increase in the gas diffusion causing an increase in the permeability of the penetrants.[33] The effect of particle size on the gas separation properties of MMM was studied by several groups. There is one study where they systematically changed the particle size of silicalite in polydimethylsiloxane (PDMS). They found that the permeability decreased with a decrease in particle size which was contradictory to the theoretical prediction. They hypothesized that this result was due to enhanced solid–polymer interfacial contact or densification of the polymer around the particle because of the rubbery nature of PDMS. Su *et al.* also studied the particle size effect by using different size silica nanoparticles and dispersing them in cross linked polyethylene oxide polymer.[41] They show that the gas separation performance of MMM deviates from the theoretical values as the particle size decreases.[42]

A critical characteristic in the fabrication of MMM is the limit of particle loading which should not exceed 50% due to agglomeration issues. resulting in the gas transport dominated by the polymer phase.[21,43] The amount of particle loading also depends on the particle size. As the particle size goes down, the weight percent that can be added into the polymer decreases. Furthermore, the filler loading is an important aspect in fabrication. In most cases, the increase in particle loading above the critical weight percentage results in agglomeration of particles and reduction in the mechanical properties.[26] It has been observed that at low loading of zeolite, the viscosity of the solution is low, hence the dispersion of the particles in the polymer phase is not uniform.[35] While high loadings increase the viscosity, sedimentation occurs resulting in defects, then the loading must be optimized in order to have appropriate MMM. Agglomeration is very common at high temperatures which could cause wide gaps and voids between the filler and polymer. This could also occur due to the surface tension and buoyancy of the processing fluid.[35]

In the case of MMM, the polymer–filler solution needs to be homogenous in order to disperse the inorganic filler in the membrane structure. This can be achieved by dispersing the inorganic particles in the solvent using ultrasonication horn before adding the polymer. Koros *et al.* developed a technique called "Priming" in order to keep the filler particles dispersed in the polymer solution.[44] It was observed that particles should be added into the solvent in order to disperse them well. A small amount of polymer addition to the particle dispersed solution avoids the agglomeration of particles. Large particle size results in the sedimentation or setting of particle at the bottom of the film. This can be avoided by using the nanosize filler particle which stays in suspension for longer periods. The settling can also be reduced by increasing the viscosity of the solution. A recent review by Dong *et al.* clearly explained several other methods used for the dispersion of particles in the MMMs.[21]

Most of the research reports were focused on improving the interface between the particle and polymer. These reports used several techniques as mentioned above such as surface modification of filler, using of ionic liquids, melt processing, using of plasticizers, etc. Most of these techniques showed good improvement compared to the MMM that were made without any modification.

3.2. *Modification of the Fillers and Casting Conditions*

The filler and polymer matrix compatibility and the polymer–sieve–solvent interaction are the major intrinsic factors affecting the structure of the MMMs.[45] The surface functionalization of the particles and controlling the casting conditions were the two most used methods to achieve the defect-free interface and tune the filler surface chemistry compared to the other methods mentioned in Section 2.1. Other methods used were summarized by Aroon *et al.*[46]

Surface modification of particles using coupling agents have been extensively used especially in the case of zeolites.[20,27,35,37,46–50] One end of the silane agent can be reacted to the hydroxyl groups on the surface of zeolites while the other end attaches to the polymeric chain. In this way, the silanes act as a bonding agent between the polymer and the particles improving the interface between the polymer and zeolite. General procedures and mechanisms for silanization of a zeolite have been described elsewhere.[47,49] Furthermore, the addition of hydroxyl and other functional groups on the surface of these nanomaterials may lead to interactions with polar gases such as CO_2 improving the penetrant solubility in the MMMs.[51] The simplest approach to enhance the link between the polymer and the inorganic particles is the introduction of a third component. For this purpose, low molecular additives are used such as 2,4,6-triaminopyrimide (TAP) and p-nitroaniline (PNA).[52,53] Figure 4 shows cross-section single electron microscope (SEM) images of zeolites before and after being modified with 3-aminopropyl-diethoxymethyl silane as coupling agent. This approach led to a better interface and higher selectivity and permeability compared to unmodified particles. It is clear from the SEM

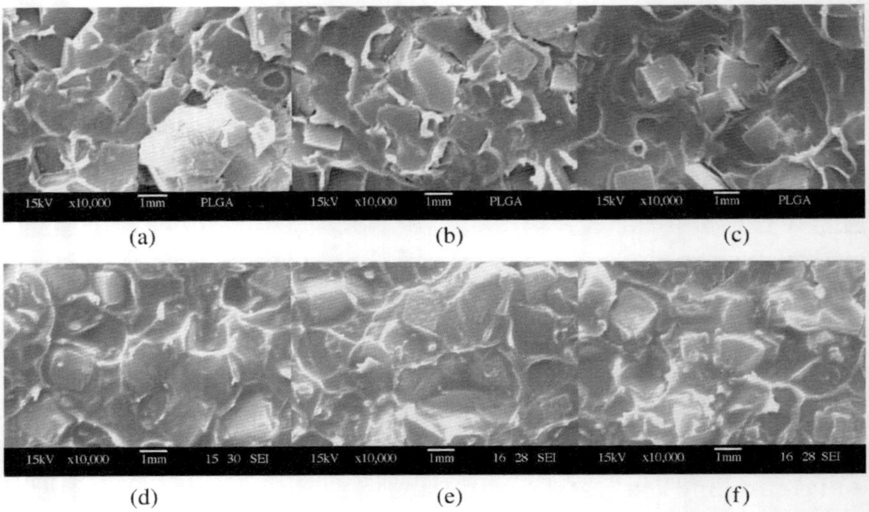

Fig. 4. Comparison of cross-section SEM images of MMMs before and after the chemical modification of zeolite surface (a)–(c): PES–zeolite3A,4A and 5A MMMs, respectively; (d)–(f): PES–zeolite 3A-NH_2, 4A-NH_2 and 5A-NH_2 MMMs, respectively).

Source: Reprinted from Ref. 47. Copyright (2006) with permission from Elsevier Science.

images that the contact between the zeolite and the polymer is better for the modified zeolites.

Due to the incompatibility between the polymer and inorganic phases that can lead to non-ideal morphologies of the synthetized MMMs, there are several casting conditions that have been extensively followed. One of these widely used approaches is to perform the casting at temperature above the polymer glass transition or either use a polymer with low glass transition temperature and perform the membrane fabrication close to the polymer glass transition temperature.[26,47,49,54,55] To maximize stress relaxation in this method, the matrix flexibility is maintained during the membrane formation. Thus, a low volatile solvent is the choice, however, such a solvent is difficult to find since the typical solvents used for film casting boil below this temperature.[14,26] An alternative is annealing above the glass transition temperature, however, the defective membranes formed using this method do not lead to significant improvements.[26] To maintain polymer chain flexibility and mobility and reduce the glass transition temperature during the membrane synthesis, the incorporation of a plasticizer into the polymer solution such as RDP Fyroflex, 4-hydroxy benzophenone and di-butyl phthalate is common.[26,52]

Grignard treatment is another technique to minimize the interface defects. The zeolite 4A particles shown in Fig. 5 were treated with thionyl chloride followed by subsequent reaction with Gringnard reagent. Grignard reagent grows whiskers of $Mg(OH)_2$) around the inorganic filler and roughens the surface of the particles which helps create a defect-free interface.[56]

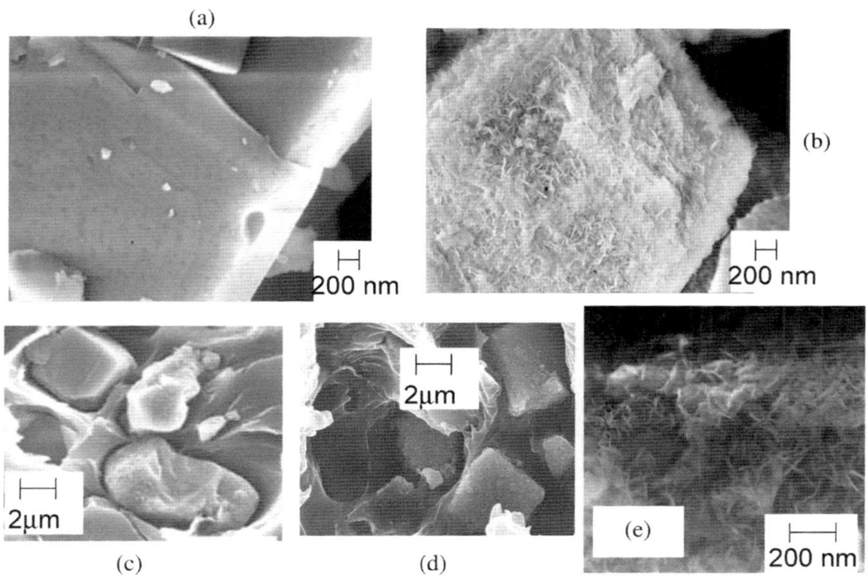

Fig. 5. Representative SEM micrographs: (a) unmodified 4A surface, (b) modified 4A surface with the inorganic whisker morphology, (c) Ultem composite film containing unmodified 4A particles, (d) Ultem composite film containing modified 4A particles, (e) high-magnification image of the surface whisker structure as shown in (b). In all the composite films, the zeolite particles are evenly distributed throughout the polymer matrixes without apparent aggregation.

Source: Reprinted with permission from Ref. 56. Copyright (2007) American Chemical Society.

Although metal organic frameworks are intrinsically more compatible with polymer because of the organic nature of the structure, several researchers modified them with different functional groups to tune its compatibility with polymer. UiO-66 metal organic framework (MOF) was modified with several organic functional groups such as amine, phenylacetyl chloride, decanoyl chloride, succinic anhydride and improved the interface properties.[16,57,58] Ti-exchanged UiO-66 MOFs were used to increase the interface and performance of the MMMs using PIM-1 as the polymer matrix.[59] MIL-based MOFs were also modified with several chemical surface groups and metal exchange methods which resulted in the improved interaction between the MOF and the polymer.[60–64]

MMMs have also been prepared by combining a small amount of highly absorbing non-toxic ionic liquid and MOFs in order to avoid the void structure while improving the gas separation performance by utilizing the high CO_2 absorption of capacity of ionic liquids.[65] The ionic liquid, being charged and non-volatile, can remain within a polymer matrix prepared using polymerizable ionic liquids. Addition of sufficient ionic liquid to wet the zeolite keeps the particles dispersed within the matrix. An ionic liquid has been used to wet both the polymer phase and a zeolite (SAPO-34) for selective gas separations.[66–68] Another method is to perform the casting on a dense liquid surface with high surface tension in order to avoid strong membrane adhesion to the casting substrate surface.[14] Herein, the polymer solution can contract freely after casting and the stresses distribute uniformly in every direction and forms void-free membranes. Melt processing is another method used to maintain the polymer flexibility during membrane preparation; however, it cannot lead to feasible industrial relevant membranes.[29] Furthermore, low molecular weight materials or bridging agents could be used to fill the voids between the inorganic filler and the polymer phase. These materials must be able to form hydrogen bond, hydroxyl or carbonyl bond besides being soluble in the solvent while making the polymer dope solution. Long aliphatic and polyatomic-based compounds with polar atoms, rigid and planar structure serves this purpose; p-nitroaniline (pNA),

2,4,6-triaminopyrimidine (TAP) and 2-hydroxy 5-methylaniline are a few examples.[35,37,69]

Another way is cross-linking of the copolymers through acetal linkages which could provide the flexible component of rubbery polymer and good contact for filler and rigidity of the glassy polymer. No additional substances must be added in this method.[35] The use of some copolymers to enhance the interfacial contacts polymer–particle has also been explored. This method allows obtaining void-free membranes, however, finding an industry suitable copolymer or furthermore synthetized increases the number of preparation steps.[30]

The minimization of zeolite-solvent–non-solvent interaction method is another way to obtain interfacial void-free MMMs. Hydrophobization of the zeolite surface by replacing the hydroxyl groups with methyl groups avoid the zeolite particles which behave as nucleating agents to improve the adhesion.[48] Alternatively, one can use a hydrophobic material for both phases, inorganic and organic. An example of this method is the zeolite Nu-6(2) and polysulfone. MMMs synthetized using these latest methods can result in free void interfaces.[70] Another method called "sizing technique" which uses the sizing agents (typically polyvinylpyrrolidone) to increase the strength of the interface by increasing the chemical reactive site and the surface area for adhesion without using complicated chemical reactions.[71–73]

3.3. *Flat Sheet and Hollow Fiber Configurations*

MMMs can be fabricated as flat sheet membranes and hollow fiber membranes. Initial studies on MMMs were solely focused on the fabrication of flat sheet membranes. Flat sheet membranes have been used widely in gas separation industry and have been exhaustively studied for the past 20 years despite the disadvantage of dense selective layer and lower gas permeation flux. For symmetric flat dense membranes, the thickness would be more than 50 μm and for asymmetric, it could go from 5 to 25 μm.[20] Recently, flat sheet membranes have attracted increasing attention for potential application in water desalination,.[74,75] and waste water treatment.[76–78] Most of the flat sheet MMMs were fabricated as dense membranes. There are very few research reports that focused on the fabrication of asymmetric flat sheet membranes.[79–81] Basu *et al.* prepared asymmetric Matrimid®/[$Cu_3(BTC)_2$] MMMs using phase inversion technique to improve the membrane gas separation performance.[79] The selectivity and permeance increased with increasing filler loading compared to pure polymer membranes.

Recently, researchers were focused on the fabrication of hollow fiber MMMs because of their intrinsic advantages such as high surface area to volume ratio. Furthermore, high selectivity, good mechanical strength, good chemical resistance, low fouling, thermal stability, cost-effective, defect-free production and good compatibility make them more flexible during operation and handling in the module fabrication.[24,35,76] Despite these advantages, there are several challenging factors related to fabrication of hollow MMMs such as the thickness of the membrane (should be as thin as possible), low defects in the selective thin skin, thin dense separating layer which lead to faster production (but must be able to hold the pressure during operation) and sub-microsized particles which could fit inside the ultrathin layer and suppressed polymer and polymer–particle defects are important features of hollow fiber asymmetric MMMs. It has been demonstrated that in hollow fiber membrane, the particle size distribution should be controlled by manipulating the air gap or draw ratio instead of high loading in order to avoid processing problems.[20]

Typically, hollow fiber MMMs are produced by single-layer spinning where phase inversion method is used and the inorganic filler is introduced in the spinning polymer.[48] Figure 6 shows a hollow fiber with no observable interface between the sheath and core layers which indicates good adhesion. Furthermore, the skin region exposed here shows the sieve-in-a-cage morphology with no particle agglomerates.[48] Another common method used is double-layer spinning method. Dai *et al.* fabricated mixed matrix asymmetric dual-layer hollow fiber membranes using dry jet-wet quench process containing the metal organic framework sieve ZIF-8 in an Ultem® polymer matrix.[82] Dual-layer fibers were produced via co-extrusion of the

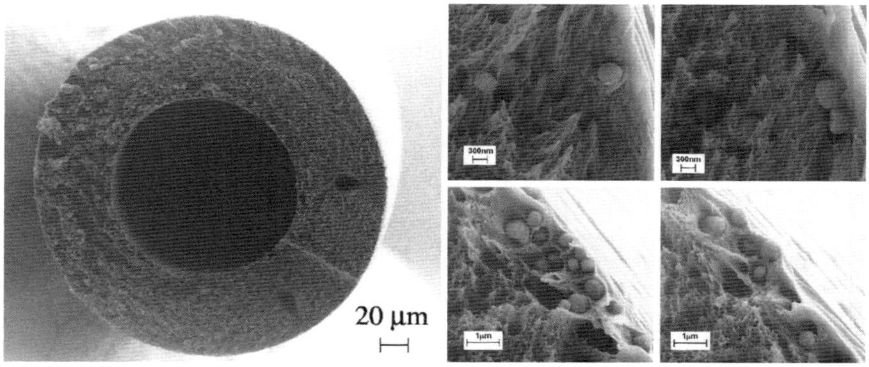

Fig. 6. SEM microphotographs of dual layer mixed matrix hollow fiber membrane incorporating Ultem® "sized" HSSZ-13 zeolites and of the skin region showing sieve-in-a-cage morphology in mixed matrix hollow fiber membranes incorporating Ultem® "sized" HSSZ-13 zeolites.[48]

Source: Reprinted from Ref. 48. Copyright (2007) with permission from Elsevier Science.

core and sheath dopes with a bore fluid solution through a composite spinneret. Other advantages of this method are high performance but brittle layer by choosing two different material in layers, control over the porosity of the membrane, easy fabrication and higher fluxes using single-step production.

4. Gas Separation Performance of MMM

Porous solids with higher permeability and selectivity than polymers are highly desirable as additives in MMM. As mentioned above, many different type of particles such as zeolite, MOF, carbon molecular sieves, porous organic polymer (POP), covalent organic framework (COF) and nanoparticles were used for the improvement of gas separation performance of MMMs. Polymer and porous particle interfacial interactions such as pore blockage, chain rigidification, free volume increase, interfacial void formation dictate the gas transport of MMMs.[14]

In general, the permeability is the product of diffusivity (D) and solubility (S). Variation in permeability of MMM can be explained using diffusion and solubility coefficient. The effect of polymer-free volume on penetrants diffusion coefficient can be described using statistical mechanical model of hard spheres diffusion in liquid, which is proposed by Cohen and Turnbull.[83] The penetrant diffusion coefficient (D) can be described using the following equation:

$$D = Ae^{\left(\frac{-\gamma V^*}{V_f}\right)}, \tag{1}$$

where A is a pre-exponential factor weakly dependent on temperature, γ is an overlap factor introduced to avoid double counting free volume, V^* is minimum free volume element size that can accommodate penetrant molecule, and V_f is the average free volume in the media accessible to penetrants transport. According to this equation, increase in free volume is expected to enhance penetrant diffusion. Free volume measured using positron annihilation lifetime spectroscopy (PALS) in poly(1-trimethylsilyl-1-propyne) (PTMSP)/fumed silica (FS) MMM exhibited increase in free volume with FS loading, correspondingly significant increase in N_2 permeability was observed with increase in FS content.[84]

The solubility coefficient of the penetrant depends on the interaction with polymer and filler. Functional groups of polymer and filler such as hydroxyl amine interacts with polar gases such as CO_2, SO_2, which results in increased penetrants solubility in the MMM, hence increase in the gas permeability. The solubility (S) of penetrants dependence with temperature (T) and enthalpy of sorption (ΔH_s) is described in terms of van 't Hoff relationship.[85]

$$S = S_0 e^{\left(-\frac{\Delta H_s}{RT}\right)}, \tag{2}$$

where S_0 is a constant, R is ideal gas constant and T is absolute temperature. Increase in the interaction between functional groups and polar penetrants molecules decreases ΔH_s, leading to increase in gas solubility. Solubility coefficient of CO_2 is significantly increased with increase in silica content in poly(amide-6-b-ethylene oxide) (PEBAX®) and silica MMMs attributed to strong interaction of CO_2 molecules with SiO_2 and polyamide block in PEBAX®.[86]

5. Zeolite-based MMM

Zeolites are microporous crystalline solid structures formed from silicon, aluminum and oxygen. They are available in a wide range of structures with different chemical compositions and physical properties and are extensively used as additives in MMMs.[87] Zeolites have significantly higher diffusivity and selectivity than polymeric materials due to their molecular sieving properties.[88] A list of some of the MMMs synthesized using zeolites and their gas transport properties are shown in Table 1. Bastani *et al.* extensively discussed about polymer zeolite MMMs.[35] Percentage loading, particle size, pore size, shape and modification of zeolite surface significantly affect the separation performance. Increase in zeolite loading enhanced the permeability of MMM in some cases,[53,89] while permeability is decreased in some other cases.[49,90] Possible reasons for reduction in permeability are inhibition of polymer chain mobility at zeolite–polymer interface and pore blockage of zeolites by polymer chains.[49] Poor polymer–zeolite interface resulted in voids and led to increase in permeability.[91,92] The interactions between zeolite and polymer are determined using glass transition temperature (T_g) of MMM. Increase in T_g of polyethersulfone/zeolite A MMM is observed with increase in zeolite A loading. The inhibition of polymer chain mobility near the polymer–zeolite interface may be responsible for increase in T_g.[49]

Zeolite particle sizes found contradictory effect on permeability of MMM. Increase in the particle size of zeolite increases the permeability of silicalite–PDMS MMMs,[41] while nanosized zeolite 4A into the polyethersulfone showed enhanced permeability and selectivity than microsized zeolite 4A in polyethersulfone.[93] Another important factor in zeolite-based MMM is pore size. Permeability of MMM increases with increase in pore size of zeolites. This phenomenon indicates that increase in pore size zeolite would potentially decrease the partial pore blockage and polymer chain rigidification.[47,49,53] Zeolite particles are available in different shapes and sizes, however, very limited data is available to better understand the shape dependence on gas permeability and selectivity. Barrier films employ

Table 1. Performance of polymer–zeolite MMMs.

	Pure polymer					Mixed matrix membranes					
	Permeability (Barrer)			Selectivity		Permeability (Barrer)			Selectivity		
MMMs	N_2	CO_2	CH_4	CO_2/N_2	CO_2/CH_4	N_2	CO_2	CH_4	CO_2/N_2	CO_2/CH_4	Refs.
Polyimide/13X (1/0.43 by wt.)	0.218	8.34	6.86	38.3	1.22	1.35	33.4	4.87	24.7	6.86	53
Polyimide/4A (1/0.43 by wt.)	0.218	8.34	6.86	38.3	1.22	0.45	9.36	4.19	20.6	2.23	53
PEEK-WC/NaA (70/30 wt.)	0.12	3.96	—	33	—	90	75.6	—	0.87	—	90
Polyimide/ZSM-5 (15 wt.%)	74.4	1156	62.1	15.5	18.6	164	1492	203	9.1	7.3	97
Matrimid®5218/NaY (80/20 wt.)	—	8.34	0.23	—	36.3	—	22	0.8	—	27.6	98
Pebax/Zeolite NaX (80/20 wt.)	—	—	—	—	—	0.33	32.3	—	98	—	99
Matrimid/SAPO-34 (80/20 wt.)	—	4.5	0.13	—	34.0	6.9	—	0.10	—	66.7	100

impermeable flakes oriented perpendicular to gas diffusion path to enhance gas diffusion resistance for larger kinetic diameter gases and achieve higher selectivity towards smaller gas molecules. Polyimide/layered aluminophosphate MMM showed significant improvement in ideal gas selectivity of O_2/N_2 and CO_2/CH_4.[94] Zornoza *et al.* studied the effect of the hollow zeolites sphere (HZS) in polysulfone and polyimide MMM. The separation performances of H_2/CH_4, CO_2/N_2 and O_2/N_2 are significantly improved due to the spherical shape.[95]

Several researchers studied zeolite surface modification using amino silane coupling agents as mentioned in Section 2.3 to improve the interaction between polymer and filler. The gas selectivity is significantly improved with increasing amino silane concentration of zeolite particles. Improvement in the selectivity was attributed to enhanced interaction between zeolite and polymers.[28,96]

6. MOF-based MMM

MOFs are porous crystalline compounds forming one-, two- or three-dimensional structures consisting of metal ions or clusters coordinated to organic ligands. Properties of MOFs such as surface area and pore size can be easily tuned using appropriate building block.[107] The most important feature of MOF that makes them more attractive for MMM application is its intrinsic compatibility with the polymers because of its organic linkers and their functionalities. Another important feature is its porosity and the surface area is much higher than the zeolites which makes them the most suitable additives for MMMs. The availability of different MOFs with outstanding properties leads to rapid growth of MOF in MMMs. Some of the examples of MOF-based MMMs are shown in Table 2. First fabrication of MOF-based MMM incorporated copper (II) biphenyl dicarboxylate–triethylenediamine MOF in the poly(3-acetoxyethylthiophene matrix for gas separation.[108] The permeability of CH_4 is enhanced with MOF loading. The authors claimed that the increased adsorption of methane in the copper MOF is responsible for increase in CH_4 permeability.[108] Several MOF types based

Table 2. Performance of polymer–MOF MMMs.

	Pure polymer					MMM					
	Permeability (Barrer)			Selectivity		Permeability (Barrer)			Selectivity		
MMMs	N_2	CO_2	CH_4	CO_2/ N_2	CO_2/ CH_4	N_2	CO_2	CH_4	CO_2/ N_2	CO_2/ CH_4	Refs.
Polyvinylacetate/ Cu-BDC (15 w/w)	0.078	2.44	0.068	31	35.0	0.09	3.26	0.80	35.7	40.4	101
Matrimid/ZIF-8 (60% w/w)	0.31	9.52	0.24	30.6	39.8	0.44	8.08	0.10	18.3	80.7	43
SPEEK/MIL-101(cr) (40% w/w)	0.48	15.8	0.69	33.0	27.2	0.83	31	1.0	37.0	31.0	102
Matrimid/NH_2-MIL 53(Al) (8% w/w)	—	8.01	0.20	—	37.3	—	9.03	0.25	—	36.5	103
Matrimid/NH_2-MIL-125 (30% w/w)	—	6.00	0.20	—	30.0	—	50.0	1.35	—	37.0	104
Matrimind/UiO-66-NH_2 (23% w/w)	—	8.5	0.29	29	—	—	29.0	0.78	37.0	—	57
Polysulfone/NH_2-MIL-125(Ti) (30 wt.%)	—	9.5	0.43	—	22	—	40	1.36	—	29.2	105
Matrimid/MIL-53 (37 wt.%)	0.25	8.4	0.22	33.6	38.1	1.8	51.0	1.1	28.3	46.4	106

on Cu, Zn, Mn, Al, Cr metal ion and benzene-1,3,4-tricarboxylate (BTC), benzene-1,4-dicarboxylate (BDC), 2-methylimidazolate, 2-carboxyaldehydeimidazolate, 2-amino-terephthalate linkers are developed.[19]

Cu-BDC MOF synthesized from copper nitrate trihydrate and terephthalic acid is added to polyvinyl acetate (PVAc) matrix. 15 wt.% Cu-BDC-MOF loading PVAc/Cu–BDC MOF MMM showed improvements in the pure gas permeabilities of He, CO_2, O_2, N_2 and CH_4 along with enhancements in He/CH_4, O_2/N_2, N_2/CH_4, CO_2/N_2 and CO_2/CH_4 selectivity. The permeation time lag behavior of the MMMs suggested good PVAc-Cu-BDC MOF adhesion as well as easy penetrant access to the channels of Cu-BDC.[101] [36]. Car *et al.* incorporated $Cu_3(BTC)_2$ (copper(II)–benzene-1,3,5-tricarboxylate) and $Mn(HCOO)_2$ (manganese (II) formate) in polysulfone (PSf) polymer matrix.[109] MMM containing MOF showed more or less higher permeabilities than the pure PSf, while improvement in selectivities as a consequence of increased diffusivity factor.[109]

Zeolite imidazolate frameworks (ZIFs) are another MOF type synthesized from the linker containing imidazolate and Zinc (Zn), named as ZIF due to their resemblance to zeolite topologies and the metal–imidazolate–metal bond angles to the zeolite Si–O–Si angles. ZIF-8 (Zn(2-methylimidazole)2) prepared from Zn metal and 2-methylimidazole linker is most widely studied MOF in MMMs. ZIF-8 embedded into PSf matrix showed improved gas adsorption in the MMM with increase in ZIF-8 loading.[110] ZIF-8/Matrimid® MMMs are fabricated with loading up to 80 % (w/w) and observed loss of mechanical strength and decrease in flexibility.[43] The permeabilities of H_2, CO_2, O_2, N_2, CH_4, C_3H_8 are increased with increase in ZIF-8 loading up to 40% (w/w) and decreased for higher loading of 50% and 60% (w/w). This result indicates a transition from a polymer-driven to a ZIF-8-controlled gas transport process, where loadings higher than 40% (w/w) evidenced the sieving effect. The ideal selectivities for a gas pair containing small molecules, such as H_2/CH_4, CO_2/CH_4, H_2/C_3H_8, and CO_2/C_3H_8, are increased consistently with increase in ZIF-8 loading suggesting molecular sieving.[43]

MIL-53 MOF type synthesized from Al^{3+} or Cr^{3+} metal ions and benzene-1,4-dicarboxylate (BDC) linker is a special type of flexible MOF that can reversibly contract or open the pores upon the adsorption of molecules like CO_2 and H_2O.[111] Basu *et al.* incorporated MIL-53 (Al) MOF in Matrimid® and observed 218% increase in CO_2 permeance compared to pure Matrimid® membranes at 30% (w/w) loading. Selectivity of CO_2/CH_4 increased up to 15% at 30% (w/w) loading indicating good interaction between MIL-53(Al) and Matrimid®.[112] NH_2-MIL-53(Al) MOF synthesized from Al^{3+} ion and 2-amino-terephthalate are embedded in polysulfone and exhibited moderate increase in CO_2 permeability up to 40% (w/w) loading and CO_2/CH_4 selectivity increased two-fold up to 25% (w/w). The significant enhancement in the selectivity can be related to good interaction between amino functionalized framework and polysulfone due to hydrogen bonding between amine and sulfone groups at their interface.[113]

7. Carbon-based MMM

Carbon molecular sieves (CMS), activated carbon, fullerenes, carbon nanotubes and graphene oxides, are carbon-based materials and are used as additives for MMMs and their gas separation performances sre shown in Table 3. Carbon molecular sieves have high porosity and fine pore size distribution with pore opening in the same order as the size of gas molecule. Vu *et al.* incorporated CMS into polyimide (Matrimid®) and polyetherimide (Ultem® 1000) for gas separations.[38] CMS are formed from polyimide precursor via pyrolysis and CMS films are transformed into fine particles (1–10 μm) using a ball mill/mixer. Matrimid® and Ultem® 1000 MMM exhibited significant improvement in CO_2/CH_4 and O_2/N_2 selectivities and minor improvement in permeability. The author claimed that CMS particles offer better adhesion and polymer–CMS contacts than zeolites.[38] Activated carbon (AC) is another carbon-based porous material used for MMMs study in acrylonitrile–butadiene–styrene (ABS) polymer.[34] MMM of ABS-AC exhibited increase in CO_2 gas permeability up to 600% and CO_2/CH_4 selectivity up to 100% over

Table 3. Performance of polymer-carbon based MMM.

	Pure polymer					Mixed matrix membranes					
	Permeability (Barrer)			Selectivity		Permeability (Barrer)			Selectivity		
MMMs	N_2	CO_2	CH_4	CO_2/N_2	CO_2/CH_4	N_2	CO_2	CH_4	CO_2/N_2	CO_2/CH_4	Refs.
Matrimid® 5218/ CMS (36% w/v)	0.32	10	0.28	26.3	35.3	0.38	12.6	0.24	33.1	51.7	38
Ultem® 1000/CMS (35% v)	0.05	1.45	0.037	27.8	38.8	0.14	4.48	0.08	32.9	53.7	38
Acrylonitrile-butadiene-styrene/AC (10% w/v)	—	2.8	0.33	—	8.48	—	18.4	0.92	—	19.9	34
Polystyrene/fullerene (10/1 wt.)	0.38	9.3	—	24.4	—	0.56	9.6	—	17.1	—	114
Polysulfone/SWNT (15% wt.)	0.17	3.9	0.17	22.9	22.9	0.22	4.52	0.28	20.4	16.1	115
Polyether block amide/GO (0.1 wt.%)	1.26	48	—	38.0	—	—	95.0	1.14	83.0	—	116
Pebax 1657/ MWNTs–NH_2 (23 wt.%)	2.74	133	8.0	48.6	16.6	6.0	268	18.0	44.6	14.8	117

intrinsic ABS membranes with increase in AC loading up to 40% (w/v). The increase in perselectivity of CO_2 was attributed to the selective adsorption of CO_2 and tight interfacial contact between ABS and AC.

Fullerenes are the carbon-based large spheroidal molecule consisting of a hollow cage of atoms called as buckminsterfullerene. Higuchi *et al.* introduced fullerenes into polystyrene matrix for gas separation.[114] Fullerene/polystyrene showed significant enhancement in the permeation of pure N_2, O_2, CO_2, C_2H_6, and C_2H_4 and slight decrease in the selectivity of O_2/N_2 and C_2H_4/C_2H_6 compared to pure polystyrene. The increase in permeability is due to increase in diffusion coefficients in the fullerene/polystyrene membranes.[114] Carbon nanotubes (CNTs) are graphite sheets rolled into circular bundles of several angstroms in diameter and several micrometers in length. These tubes can be single-walled or series of different shells spaced with common axis called multi-walled nanotubes and ends of the tubes can be either opened or closed. CNTs showed attractive diffusivities of gases using simulation studies make more attractive filler for MMMs. Kim *et al.* incorporated single-walled nanotubes in polyimide-siloxane copolymer for gas separation. Permeability of He is dropped with addition of close-ended CNTs due to good adhesion of copolymer to CNTs. In case of open-ended CNTs, the permeabilities of He, H_2, CO_2, O_2, N_2 and CH_4 are increased with CNTs proportions in polymer matrix.[115] The enhancement in permeability might be due to the presence of CNT tunnels in the polymer matrix leading to increase in diffusion coefficients of gases.[115]

Graphene is another carbon-based single atomic sheet with extraordinary electrical, thermal and mechanical properties. As high van der Walls attraction between graphene sheets makes it difficult to use, graphene is chemically or thermally modified for their applications.[118] Graphene oxide (GO) is a stable material used to create GO by chemical reduction. GO sheets are impermeable to most of the gases and make the most suitable additives for gas barrier applications.[119] Yang *et al.* prepared polyethyleneimine (PEI)/GO thin films using layer by layer (LbL) technique for super gas barrier application. Depositing alternating layers of oppositely charged materials forms

LbL films.[120] Oxygen transmission rates (OTR) of PEI/GO films deposited on poly(ethylene terephthalate) (PET) film significantly decreases with increase in number of bilayers depositions. OTR is dropped from 8.48 for uncoated PET films to 1.23 cc m^{-2} atm^{-1} day^{-1} for coated PEI/GO PET films. The authors claim that the reduction in OTR is due to the overlapping of GO platelets limiting the oxygen diffusion pathways.[120] Recently, the various lateral GO sizes are used to fabricate polyether block amide (PEBA)-GO MMMs for CO_2 separation.[116] Three different types of GO with later sizes 100–200 nm (GO-S), 1–2 μm (GO-M) and 5–10 μm (GO-L) are used to fabricate MMMs. CO_2 permeability and CO_2/N_2 selectivity are significantly improved for 0.1 wt.% GO-M MMMs crossing Robeson "upper bound" as shown in Fig. 7. The increase in CO_2/N_2 permselectivity is due to good dispersion of GO-M resulting in longer N_2 path length compared to CO_2. However, CO_2/N_2 permselectivity dropped sharply with GO-L additive due to reduction in its dispersion ability in MMMs.[116]

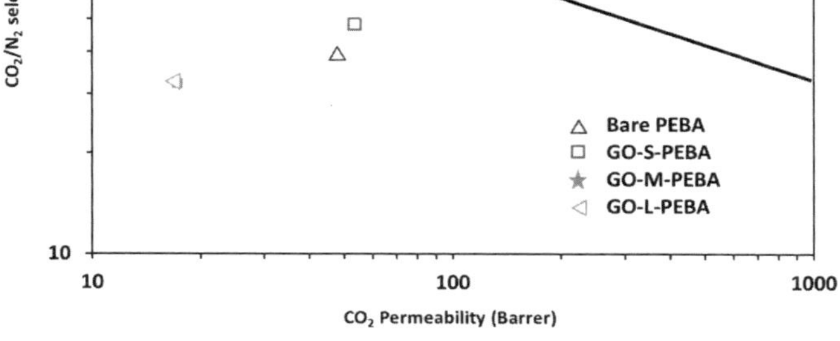

Fig. 7. Polyether block amide/GO-M MMM containing 0.1 wt.% GO-M showing the CO_2/N_2 separation performance above 2008 Robeson upper bound.

Source: Reprinted with permission from Ref. 116. Copyright (2016) American Chemical Society.

8. Other Novel Materials

Table 4 shows the gas transport properties of the MMM that were fabricated using several novel porous inorganic materials as fillers. Porous organic polymers (POPs) and COFs are the class of micro and nanoporous polymers. POPs are disordered amorphous polymers with wider or narrow pore size distribution, while COFs are well ordered crystalline polymers with uniform pore size.[121] POPs and COFs are excellent additives for CO_2 separation MMM due to their extremely high CO_2 sorption capacity.[122] Polymer of intrinsic microporosity (PIM-1) is an example of porous organic polymer and owns extremely high permeance and good selectivity for CO_2/N_2 separation.[123] PIM-1 ages fast resulting in decrease of permeability due to its ability to pack into denser phase over a time. Lau *et al.* first time developed an anti-ageing PIM-1 nanocomposite membrane using porous aromatic framework (PAF-1) particles as an additive.[124] PIM-1/PAF-1 MMM exhibited no change in CO_2 permeability over 240 days and significant improvement in CO_2/N_2 selectivity. This effect is achieved using an intimate mixing of PIM-1 polymer chains within the porous PAF-1 matrix, which are evidenced by tethering bulky chemical moieties within PAF-1 pores. PIM-1 has H_2/CO_2 selectivity of 0.8, which is dramatically improved to 9.6 using chemical modification of PIM-1/Matrimid membranes. PIM-1/Matrimid (90/10) dense films are modified by immersing into different diamine solutions.[125] The modified MMM showed exceptional separation performance surpassing the Robeson "Upper Bound" for H_2/CO_2, H_2/N_2, H_2/CH_4 and O_2/N_2. The X-ray diffraction (XRD) analysis and sorption data confirmed that the modified PIM-1/Matrimid MMM has a smaller d-spacing and a decrease in the diffusivity coefficient. Kang *et al.* incorporated two-dimensional COFs in the polyetherimide (Ultem®) and polybenzimidazole (PBI) MMM for CO_2 separation.[126] PBI MMMs exhibited significant improvement in H_2/CO_2 permselectivity due to excellent compatibility between COF and PBI, which surpassed 2008 Robeson upper bound as shown in Fig. 8. However, author reported that separation performance is slightly deteriorated due to competitive adsorption and diffusion between CO_2 and H_2 in MMMs.

Table 4. Performance of polymer–POP, COF, nanomaterials-based MMMs.

	Pure polymer					Mixed matrix membranes					
	Permeability (Barrer)			Selectivity		Permeability (Barrer)			Selectivity		
MMMs	N_2	CO_2	CH_4	CO_2/N_2	CO_2/CH_4	N_2	CO_2	CH_4	CO_2/N_2	CO_2/CH_4	Refs.
PIM-1/PAF-1 (10 wt.%) at start	200	4000	—	20	—	1200	14000	—	11.6	—	124
PIM-1/PAF-1 (10 wt.%) at 240 days	60	1520	—	25.3	—	720	13300	—	18.5	—	124
PIM-1/Matrimid (90/10)	192	3825	268	19.9	14.3	144	2855	173	19.8	16.5	125
PIM-1/Matrimid (90/10) wt., 2h pXDA treated	—	—	—	—	—	20.8	388	17.2	18.6	22.5	125
Ultem/COF NUS-2 (20 wt.%) at 2 bar	—	2.2	0.11	—	20.2	—	4.92	0.22	—	22.4	126
Polysulfone/silica NP (20% vol.)	0.24	6.3	0.22	26.2	28.6	1.12	19.7	1.1	17.6	19.1	33
Matrimind/TiO_2 NP (25% vol.)	0.22	4.3	0.21	19.5	20.5	1.35	12.0	1.84	8.8	6.5	129
Matrimid® 5218/ MgO NP (20 wt.%)	—	—	—	—	—	0.36	8.27	0.27	23.0	29.8	130
Matrimid®5218/ MgO NP (20 wt.%), 350°C HT for 1 h	—	—	—	—	—	0.52	11.64	0.41	22.4	28.7	130
PIM-1/BILP-101 (40 wt.%)	240	4700	—	19.3	—	5100	290	—	17.4	—	128

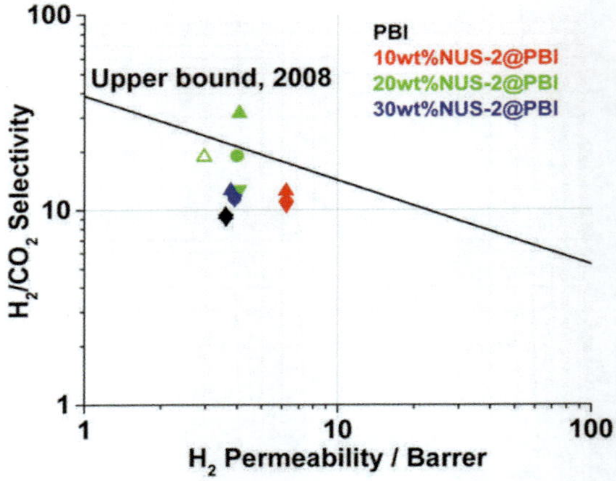

Fig. 8. Ultem/COF (NUS-2) MMMs exhibiting H_2/CO_2 separation performance above Robeson "upper" bound for 20 wt.% loading MMM. Solid and hollow symbols represent single gas data and mixed gas data, respectively (down triangle, 2 bar; circle, 3.5 bar; up triangle, 5 bar).

Source: Reprinted with permission from Ref. 126. Copyright (2016) American Chemical Society.

Recently, a new class of imidazole linkage-based porous organic polymers called as benzimidazole-linked polymers (BILPs) was developed. BILPs exhibited outstanding chemical and thermal properties and excellent CO_2 and H_2 uptakes with high CO_2/N_2 and CO_2/CH_4 selectivities.[127] Therefore, it is considered as the most suitable additives for CO_2 separation MMMs. Ali *et al.* incorporated BILP in PIM-1 MMMs and observed significant improvement in permeability of CO_2 with slight decrease in CO_2/N_2 selectivity up to 30 wt.% loading.[128] Further increase in loading to 40 wt.% resulted in decrease in CO_2 permeability and enhancement in CO_2/N_2 selectivity. The authors claimed that initial improvements in the permeability are due to increase in interfacial void formation with BILP loading. PIM-1/BILP membrane tested in real flue gas conditions showed stable separation performance.[128]

The nanoparticles are used to fabricate MMMs and separation performance is strongly affected by polymer chains–nanoparticles interactions as well as nanoparticles–penetrants interactions. These interactions may induce the free volume between the polymer chains resulting in enhancement in gas diffusion. The functionalized nanoparticles with hydroxyl or amine groups interacts with polar gases such as CO_2 and SO2 resulting in enhancement in solubility of the gases.[51] The inorganic silica nanoparticles in polymer matrix are the widely studied MMMs for gas separations. Merkel *et al.* introduced nonporous FS nanoparticle into poly(1-trimethylsilyl-1-propyne) (PTMSP) matrix and observed systematic increases in gas permeability with increasing FS content due to increase in free volume of MMMs.[131] Enhancement in pure gas H_2/CH_4 selectivity with FS content is observed, while 64% reduction in mixed gas n-C_4H_{10}/CH_4 selectivity at 50 wt.% FS content is observed compared to pure PTMSP. The reduction in mixed gas selectivity due to increase in phase transport mechanisms such as Knudsen diffusion with increase FS content in the polymer.[131]

Overall, MMM possesses great potential for tremendous improvements in both permeability and selectivity, especially for metal organic frameworks and porous organic frameworks as an additive. The separation factor in MMM depends on polymer–filler interface and nature of filler dispersion. The next generation MMM can be produced using nanosize fillers without agglomeration. The nanosized particles would have more uniform particle distribution in polymer matrix than microsize particles and allow forming thin films for high throughput membrane development. The transformation of separation performance of film-based MMMs to the industry-relevant hollow fiber-based MMMs is highly critical and difficult due to different preparation conditions used for both flat and hollow membrane forms. Further research and development are needed for new additives synthesis, polymer selection and membrane preparation techniques.

8.1. *Comparison of MMM with Inorganic Membranes*

Pure MOF membranes showed very high gas permeabilities, around 1–3 orders of magnitude, compared to pure polymeric membrane.[13] Interestingly, dispersing these highly permeable inorganic particles in the polymers did not show very high improvement in gas transport properties. The overall gas separation performance is still close to polymer membranes even after 30–40 wt.% loadings of inorganic particles. Moore *et al.* found that the gas permeability of MMM increases significantly compared to pure polymer membranes if the permeabilities of the faster permeating component through the polymer and sieve are approximately matched.[14] If polymer permeability is much less than the filler permeability, the resultant MMM does not show great improvement in gas permeability. For example, the CO_2 permeability of SAPO-34 zeolite membranes is 6865 Barrer with CO_2/CH_4 selectivity of 158, which is very high compared to the pure polymers.[12] But addition of these SAPO-34 particles in the Matrimid resulted in very minimal improvement in CO_2 permeability. The CO_2 permeability only increased from 4.45 Barrer to 6.9 Barrer with the addition of SAPO-34 in Matrimid polymer.[100] Similar results were obtained with the addition of carbon molecular sieves in Matrimid. CO_2 permeability increased from 10 Barrer to 12.6 Barrer with the addition of 36 wt.% of CMS in Matrimid polymer while intrinsic CO_2 permeability of CMS is 44 Barrer.[132] Adding these CMS particles in Ultem polymer also showed similar trend in CO_2 permeability. In both these cases, CO_2/CH_4 selectivity is increased significantly. CO_2/CH_4 selectivity increased from 34 to 67 with the addition of SAPO-34 which has an intrinsic selectivity of 158. In the case of CMS with intrinsic selectivity around 200, the selectivity increased from 35 to 51.7.[132] Recent years, MOF-based MMM gained a lot of interest and promise. MOF-5, one of the well-known MOFs showed intrinsic CO_2 permeability of 4354 Barrer with very low selectivity.[133] Addition of MOF-5 in Matrimid showed improvement in both CO_2 permeability and CO_2/N_2 selectivity.[134] Similar to SAPO-34 and CMS, the increase in the gas separation performance is very minimal. There is a need to develop the techniques to fabricate these MMMs with higher volume loadings of fillers in order to achieve the gas separation performance close to that of inorganic fillers. It is a challenge to increase the volume loading above optimum loading because of the formation of defects and weak mechanical properties of MMM.

9. Predictive Models for Gas Separation Performance

The gas separation performance of the MMMs has been influenced by many factors such as gas species, percentage of filler loading, pore size of fillers, flexibility of the filler structure, polymer glass transition temperature (T_g), fraction of free volume (FFV) of polymers, dispersion of the fillers, and polymer–filler interface morphology.[15,20,21,26,32,37,69,135–138] Essentially, these properties are determined by the interactions among gas molecules, polymers and fillers, which can be computed using molecular simulation with verified force fields accurately. Molecular modeling can reach tens of nanoseconds and nanometers at maximum, which is large enough to compute the gas transportation in the homogenous phases like pure polymer membranes and inorganic porous fillers. However, the size of fillers in the MMMs is larger than 100 nm mostly.[21] A representative "unit" of hybrid MMMs is significantly beyond the scope of molecular modeling and cannot be simulated by molecular simulation alone.

Permeation models such as Maxwell model, Bruggeman model, Lewis–Nielson model and Pal model have been applied to describe systems at larger scales and predict the performance of the MMMs.[15] To make the prediction, the gas permeabilities in dispersed phases and continuous phases need to be known as inputs for permeation model. The experimental data of gas permeation in polymer membranes have been reported comprehensively.[2,3] However, it is still hard to measure the gas permeability in inorganic particles using experiments. Although the gas permeability of inorganic films have been studied in some experimental research groups,[10–13] the gas permeabilities in the particles used in MMMs are much different from inorganic membranes because of the grain boundary defects between the crystals and other defects in the membrane which was formed by many crystals.[32] Molecular simulation is an important tool to investigate the gas transportation in the single phase, especially in filler crystals.[139] The interfacial structure between fillers and polymers influences the performance of composites significantly. It is hard to explore the filler–polymer interface at nanoscale using current experimental instruments or permeation models, while molecular modeling is an ideal method to fill this gap. Therefore, it is necessary to combine the molecular models with permeation models in order to predict the gas separation performance of MMMs.

10. Permeation Models

10.1. *Ideal MMMs*

Most of the permeation models have been developed with the assumption of ideal interfacial morphology between the particle and polymer. In an ideal interfacial morphology, there are no defects, densifications and distortions in the membranes (Case I in Fig. 3). Under such consideration, the properties of polymers and fillers at interfacial region are the same as those in their bulks. A well-known permeation model is Maxwell–Wagner–Sillars equation that was originally developed from the Maxwell's theory for predicting the electrical conductivity of hybrid materials.[140–142] It has been applied to calculate permeability of the MMMs due to the similarity between electrical conduction through composite materials and gas permeation through MMMs. The full version of Maxwell–Wagner–Sillars equation is

$$P_{\mathrm{MMM}} = P_c \left[\frac{nP_d + (1-n)\,P_c - (1-n)\,\emptyset_d\,(P_c - P_d)}{nP_d + (1-n)\,P_c + n\emptyset_d(P_c - P_d)} \right], \tag{3}$$

where P_{MMM} is the mixed-matrix membrane permeability, P_d is the permeability of the dispersed filler phase and P_c is the permeability of the continuous polymer phase, and $\emptyset_d$ is the volume fraction of the fillers. $0 < n < 1$ is the shape factor of fillers, at the limits of $n = 0$ for the parallel platelets, which represents an arithmetic average of gas permeation of fillers and polymers and $n = 1$ for gas transportation through the laminates. When the equation becomes Hamilton-Crosser model for cylindrical fillers (e.g., carbon nanotubes). When $n = 1/3$ for the spherical fillers, the Maxwell–Wagner–Sillars equation can be simplified into most well-known and commonly used Maxwell equation:

$$P_{\mathrm{MMM}} = P_c \left[\frac{P_d + 2P_c - 2\emptyset_d(P_c - P_d)}{P_d + 2P_c + \emptyset_d(P_c - P_d)} \right]. \tag{4}$$

Although the maximum packing volume fraction can reach 0.64 for random close packing of uniform spherical fillers, the Maxwell–type equations can only be applied to the systems for the low volume loading of fillers in MMMs because the model was derived on the condition of streamlining morphology around fillers, which can be interfered and even broken by the close presence of fillers. In practice, over 0.3 volume loading of fillers can lead to significant errors using Maxwell models to predict gas permeation through the MMMs.

For the systems of higher volume loading of fillers, other models have been developed.[143] adopted from the model of dielectric constant in composite materials, has been applied to calculate the gas permeation through MMMs as shown:

$$\frac{P_{\mathrm{MMM}}}{P_c} = \frac{1}{(1-\emptyset_d)^3}\left(\frac{\frac{P_{\mathrm{MMM}}}{P_c} - \frac{P_d}{P_c}}{1-\frac{P_d}{P_c}}\right)^3. \tag{5}$$

However, the prediction capability of Bruggeman model cannot reach to the maximum volume loading. The exclusion of factors of filler shape, size distribution and aggregation of fillers decrease the accuracy of the prediction. Bruggeman equation is not feasible to use for the prediction of gas separation performance because of implicit functional form.

The Lewis–Nielsen model is the one can be applied to the system with loading at maximum packing volume fraction of fillers.[144,145] It has been adopted from the field of material mechanics to calculate the elastic modulus of composites, as expressed as follows:

$$P_{\mathrm{MMM}} = P_c\left[\frac{1+2\emptyset_d\frac{\left(\frac{P_d}{P_c}\right)-1}{\left(\frac{P_d}{P_c}\right)+2}}{1-\psi\emptyset_d\frac{\left(\frac{P_d}{P_c}\right)-1}{\left(\frac{P_d}{P_c}\right)+2}}\right], \tag{6}$$

where

$$\psi = 1+\phi_d\left(\frac{1-\phi_m}{\phi_m^2}\right) \tag{7}$$

and ϕ_m is the maximum packing volume fraction of fillers, which is a function of filler size, filler distribution and aggregation. The Lewis–Nielsen model includes the effects of morphology partially on permeability via setting up ϕ_m. With $\phi_m = 0.64$, the random close packing of uniform spheres has been assumed in many systems. When $\phi_m \longrightarrow 1$, the Lewis–Nielsen model reduces to Maxwell model.

Pal model is another permeation model that can be applied to the system with filler loading at maximum packing volume fraction as shown.:[138,146]

$$\left(\frac{P_{\mathrm{MMM}}}{P_c}\right)^{\frac{1}{3}}\left(\frac{\frac{P_d}{P_c}-1}{\frac{P_d}{P_c}-\frac{P_{\mathrm{MMM}}}{P_c}}\right)=\left(1-\frac{\phi_d}{\phi_m}\right)^{-\phi_m}. \tag{8}$$

The Pal model is an implicit function, and reduces to Bruggeman equation when $\phi_m \longrightarrow 1$.

Among above-mentioned models, Maxwell model is the simplest one. In comparison to the others, Maxwell model cannot be applied in the high filler loading systems. As shown in Fig. 9,[21] under low fractional filler loading (<0.2), the simulation results of Maxwell

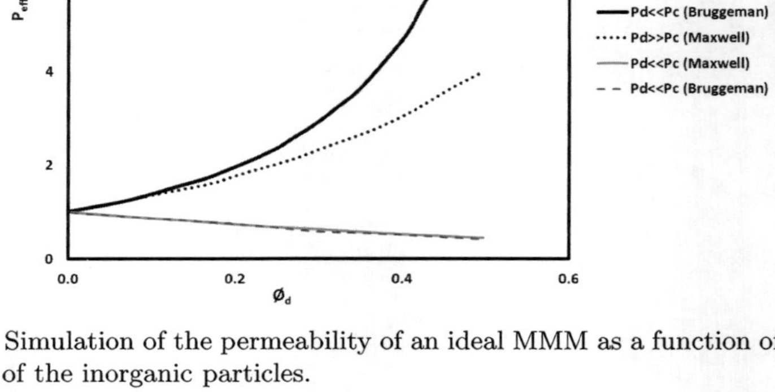

Fig. 9. Simulation of the permeability of an ideal MMM as a function of volume fraction of the inorganic particles.

Source: Reproduced from Ref. 21 with permission from Royal Society of Chemistry.

and Bruggeman equations are almost identical since chain conformation of polymers around fillers is not affected by the neighboring fillers and kept as that in polymer bulk. At a higher fractional particle loading (>0.2), there is a substantial deviation in the simulation results between the Maxwell and Bruggeman models, and Maxwell model is no longer valid at this loading since polymer chains between the neighboring fillers distort seriously. Nevertheless, Maxwell equation is still the most commonly used permeation model for the following reasons: (1) it is very straightforward and convenient to use, and (2) in practice, the filler volume loading is usually lower than 0.3 in most of the MMMs since it is difficult to fabricate defect-free membranes at high filler loading systems. Above discussion is based on the assumption of ideal interface between fillers and polymers. Once defects are formed in the MMMs (as described in Fig. 3), a modification for the original equations has to be applied.

10.1.1. *Non-ideal MMMs*

It is helpful to avoid the complexity in building permeation models for MMMs with the assumption of ideal interface. However, these "ideal" models could not describe the real MMMs accurately if non-ideal interface exists. As shown in Fig. 3, the interfacial defects can be categorized into five cases as presented by Moore and Koros:[14] (I) rigidified polymer layers wrapped around the fillers, (II) large voids (>∼10–100 nm) at the polymer–filler interface, which could be an unfavorable result after the MMM processing, (III) small voids at the interface mainly because of the rigidity of polymer chains and less attractive interaction between polymers and fillers, (IV) clogged fillers by gas molecules because of strong holding from the inside surface of fillers, and (V) bottlenecked at filler side of the interfacial region.

Moore and Koros attributed the formation of interface cases I–III to the stresses encountered during the MMM fabrication. In the case I, gas permeability is decreased at the rigidified polymers wrapping around the filler surface. The rigidified polymer layer has been described to be a denser and less mobile phase because of rigid structure of fillers and contraction of polymers with long time relaxation after solvent evaporation. The rigidified polymers are similar to semi-crystal polymers, in which the gas permeability is 1/3 to 1/4 of it in amorphous polymers. As shown in Fig. 3, when the MMMs have been fabricated as films on substrates, the contraction of polymers is asymmetric and results in void structures at the filler–polymer interface. Moore and Koros classified case III as a special situation with small interfacial gap of case II. However, even if the interface stress has been relaxed, the interfacial defects cannot be eliminated completely. The interaction between polymers and fillers also contributes to the formation of case I and case III. When the interaction between polymers and fillers is attractive, it drives the polymers into a dense rigidified layer. In contrast, the repulsive interaction between polymers and fillers makes nanosize voids at interface. Since the molecular interaction is nanodistance force, the preference between polymers and fillers has no effects on void formation in case II with large size (e.g., >10–100 nm). In some contexts, rigidified layers have been described as densified layers, immobilized structure, "pancake" morphology, etc. Actually, the "pancake" morphology of polymer chains always exists in the region close to the filler surface whether the surface is attractive, neutral or repulsive due to the entropy effects. Also, these chains lose mobility because of the deformation and blockage by filler surface. At attractive filler surface, the density at rigidified layer is higher than in the bulk and the polymer density at repulsive filler surface is lower than in the bulk.

Moore and Koros modified the basic Maxwell equation to be applicable for the MMM systems involving non-ideal interfaces as shown in Table 5.[14] The Maxwell equation has been first applied to the dispersed filler and interface layer of voids assumed to involve Knudsen diffusion, rigidification with permeation reduction by "chain immobilization factor" β or "permeability reduction factor" β'. Then, the interface layer and fillers together have been joined as a pseudodispersed phase. The Maxwell equation has been applied second time to the pseudodispersed phase and matrix. The modified Maxwell

Table 5. Comparison of models for cases I–IV based on nested applications of the Maxwell model.[14]

	Case I, matrix rigidification	Cases II–III, sieve-in-cage	Cases IV–V, "clogged" or "bottle-bottlenecked" zeolites
Parameters	β, "Chain immobilization factor" $P_I = P_c/\beta$	P_I, Void permeability $P_I = DS$ $D^a = D_{\text{Knudsen}}\left(1 - \frac{\sigma_p}{2l_I}\right)$ $S^a = \frac{1}{RT}\left(1 - \frac{\sigma_p}{2l_I}\right)^2$	β', "Permeability reduction factor" $P_I = P_c/\beta'$
	l_I, Thickness of the rigidified region	l_I, Void thickness	l'_I, Thickness of the reduction permeability region
Model for the pseudo-dispersed phase permeability	$P_{\text{eff}} = P_I\left[\frac{P_d + 2P_I - 2\emptyset_s(P_I - P_d)}{P_d + 2P_I + \emptyset_s(P_I - P_d)}\right]$ where $\emptyset_s = \frac{\emptyset_d}{\emptyset_d + \emptyset_I} = \frac{r_d^3}{(r_d + l_I)^3}$ or $\frac{(r_d - l'_I)^3}{r_d^3}$		
Pseudo-dispersed phase components	Zeolite + surrounding polymer of reduced permeability	Zeolite + surrounding void	Zeolite (including bulk zeolite and region of reduced permeability)
Model for the overall mixed matrix material	$P_{3\text{mm}} = P_c\left[\frac{P_{\text{eff}} + 2P_c - 2(\emptyset_d + \emptyset_I)(P_c - P_{\text{eff}})}{P_{\text{eff}} + 2P_c + (\emptyset_d + \emptyset_I)(P_c - P_{\text{eff}})}\right]$		

[a]The term $(1 - \frac{\sigma_p}{2l_I})$ must be added to account for the finite size of the penetrants when the void dimension is of the same order of magnitude as the penetrants. D_{Knudsen} is given by the expression: $D_{\text{Knudsen}} = (d_{\text{Pore}}/3)(8RT/\pi M_p)^{1/2}$, which is based on the kinetic theory of gases. We used the hydraulic diameter of the pore ($\sim 2l_I$) in the place of the diameter of the pore, d_{pore}. P_I, P_c and P_d are gas permeabilities at interface, polymer and particle separately. P_{eff} and $P_{3\text{mm}}$ are gas permeabilities at the pseudo-dispersed phase and the whole membrane individually. D and S are gas diffusivity and solubility in the voids separately. D_{Knudsen} is the Knudsen diffusivity. σ_P and M_p are dynamical dimeter and mass weight of the penetrate (gas) molecule. $\emptyset_d$ and $\emptyset_I$ are volume fractions of dispersed phase and interface phase individually. r_d is the radius of particle.

equation has included interface defects quantitatively, but still has been subjected to the same limitation as original Maxwell equation in filler loading, particle size distribution, particle shape, and particle aggregation. There are some more complicated situations in MMMs involving more than one interface, e.g., polymer rigidification happens around the outside of the fillers with pore blockage occurring inside of the particle, then the basic Maxwell equation has to be applied three times to describe the overall MMM performance.[14] Similar strategy of "nested application" has also been employed in other models.

Felske developed a model originally for thermal conductivity of composites of core–shell fillers (core filler covered with an interfacial layer), which is also valid for permeability prediction. The Felske model can be expressed as follows.:[147]

$$P_r = \frac{P_{\text{eff}}}{P_c} = \frac{2(1-\emptyset_d) + (1+2\emptyset_d)\beta/\gamma}{(2+\emptyset_d) + (1-\emptyset_d)\beta/\gamma}, \tag{9}$$

where

$$\begin{aligned} \beta &= (2+\delta^3)\lambda_{dm} - 2(1-\delta^3)\lambda_{Im}, \\ \gamma &= (1+\delta^3) - (1-\delta^3)\lambda_{dI}, \end{aligned} \tag{10}$$

where δ is the ratio of outer radius of interfacial shell to a core radius. $\lambda_{dm} = \frac{P_d}{P_m}$, $\lambda_{Im} = \frac{P_I}{P_m}$ and $\lambda_{dI} = \frac{P_d}{P_I}$ are ratios of permeabilities in dispersed fillers, polymer matrix, and interphase, respectively. When $\delta = 1$, the interfacial layer is absent and the Felske model reduces to the original Maxwell model.

Felske model provides almost the same predictions as the (modified) Maxwell model can do, and suffers same limitations (e.g., $\emptyset_d < 0.2$) the Maxwell model encounters. Pal further modified the Felske model with introducing composite morphology parameter ψ a function of ϕ_m (maximum packing volume fraction of core–shell filler) and ϕ_d as defined in Eq. (5).[138] The loading range of modified Felske model has been extended to $\emptyset_d \to \emptyset_m$. When $\phi_m = 1$ the modified Felske model reduces to original one. When $\delta = 1$, the modified Felske model reduces to Lewis–Nielsen model. And when both ϕ_m

and δ are equal to 1, the modified Felske model reduces to original Maxwell model.

Shimekit *et al.* modified Pal model including the effects of interfacial rigidification on the permeability of MMMs as follows:[148]

$$P_r = f(\delta, \lambda_{Im}, \lambda_{dI}, \phi_d, \phi_m, \phi_s), \tag{11}$$

where δ, λ_{Im}, $\lambda_{dI}\phi_d$, ϕ_m and ϕ_s are defined as same as in other models discussed in the context.

The Pal model for relative permeability P_r can be rewritten as follows:

$$(P_r)^{1/3}\left(\frac{\lambda_{dI}-1}{\lambda_{dI}-P_r}\right) = \left(1-\frac{\phi_d}{\phi_m}\right)^{-\phi_m}. \tag{12a}$$

By replacing P_r with $\frac{P_{\text{eff}}}{P_I}$, the effective permeability for pseudodispersed phase (the combination of filler and interphase) can be calculated:

$$\left(\frac{P_{\text{eff}}}{P_I}\right)^{1/3}\left(\frac{\lambda_{dI}-1}{\lambda_{dI}-\frac{P_{\text{eff}}}{P_I}}\right) = \left(1-\frac{\phi_d}{\phi_m}\right)^{-\phi_m}. \tag{12b}$$

Same as in modified Maxwell model, the permeability at rigidified interphase is defined as $P_I = \frac{P_m}{\beta}$, and β, the chain immobilization factor, is usually set to 3.

As shown in Fig. 10, the predictions of different permeation models on the Matrimid®-CMS mixed matrix membrane have been compared with experimental data.

In all the models for non-ideal MMMs, δ, the ratio of outer radius of interfacial shell to the core radius or the thickness of interphase has to be determined. However, it cannot be measured directly using experimental ways. Alternative, a fitting program with the least-square method has been used to estimate the value of δ. In this procedure, experimental permeation data need to be used for the fitting. This actually fails the capability of permeation models to predict the performance. Therefore, independent tools to determine interphase thickness are desired.

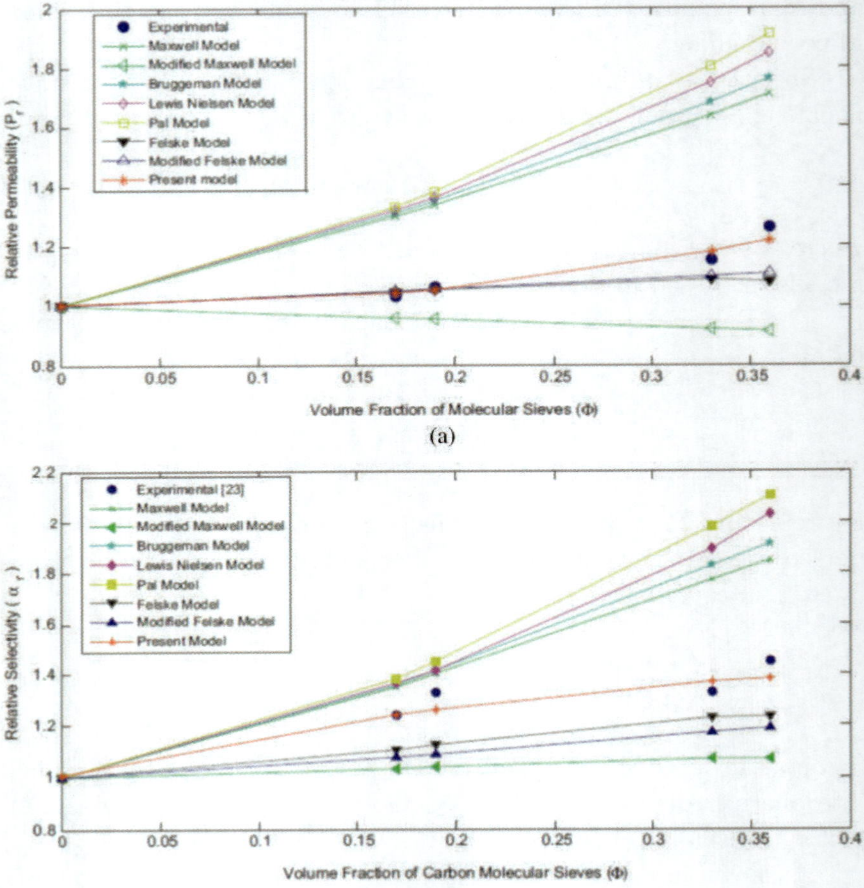

Fig. 10. (a) Comparison of the relative permeability predicting ability of different models using experimental data.[132] (b) Comparison of the relative selectivity predicting ability of different models using experimental data.[132] Here, the present model is the modified Pal model.[148]

10.2. *Molecular Models*

Non-ideal interface happens in almost all MMMs and is critical to affecting the gas separation performance of MMMs, but interfacial information of filler–polymer is hard to be measured using experimental tools. A representative unit of MMMs involving three phases (bulk polymer, interface, and a whole filler) is too large to be simulated using molecular modeling. With current computing resources, using molecular modeling to investigate the local interface is a feasible way and the most important part to understand MMMs.

Molecular simulation can be divided into three main categories by length–time scales: atomistic model, coarse-grained off-lattice model, and coarse-grained lattice model. Atomistic simulation is the most accurate but most time consuming; coarse-grained lattice model is the least time consuming and provides general trends but lose chemical details; in comparison, the off-lattice model is a compromise solution between the atomistic model and coarse-grained lattice model. It takes a very long time to relax polymer chains, and the relaxation time (τ) increases nonlinearly with the chain length in condensed phase; e.g., τ of a chain with 10 segments is about 10 ns and $\tau \propto$ L3.3.4.[145]

To complete a simulation in a reasonable time, most work using atomistic model starts from a desired equilibrium state that is determined by molecular mechanics.[149] However, it is difficult to apply this strategy to the systems of condensed polymers on inhomogeneous surfaces like filler surfaces in MMMs. In contrast, many polymer microstructures confined by surface have been predicted using coarse-grained lattice models and verified by experiments.[150–155] Employing lattice Monte Carlo simulation, Termonia predicted the stretching and widening of polymer chains near a nanoparticle surface and the formation of a 1–2 nm interfacial region.[156] When the two small particles approach in a distance smaller than the radius of gyration of the polymer chain, an overall swelling of polymer chains between the particles has been observed. Gao *et al.* observed the formation of the pancake-like morphology of the polymer in the region close to a nanosheet surface using coarse-grained molecular dynamics simulations.[157]

It is interesting that this phenomena is independent of the interfacial strength since configurational entropy rather than interaction energy dominates the interfacial morphology. However, the interfacial interaction strength still influences the glass-transition temperature. It has been found that the glass-transition temperature for repulsive interface interaction is lower, and for attractive interface interaction it is higher than the pure polymer system. Furthermore, they predicted the rigidification of the polymer layer on a strongly attractive surface. Very recently, Feng *et al.* studied the polymer morphology and free volume at polymer–particle interface using lattice coarse-grained Monte Carlo simulations.[158] It was found that the interfacial properties are dominated by the interaction of the polymer with the filler particle and the geometry of the filler particle surface. Filler particles that have brush-like surfaces tend to result in larger free volume at the interface than that occuring with smooth surfaced filler particles.

Until now, the research on MMMs using molecular simulation is still rare and incomplete. In the future, it is desired to develop: (1) Accurate force fields for coarse-grained model in order to simulate the polymer–filler interface quantitatively, and (2) novel method for predicting the gas permeability at the interface region.

11. Conclusions and Outlook

MMMs possess great potential to significantly improve the gas transport properties compared to their parent polymer matrix without much increase in the cost of processing. MMMs have been used to separate several gases such as CO_2, N_2, CH_4, H_2, O_2, He, H_2O, Olefins, paraffins, etc. The selection of polymers and inorganic fillers plays a major role in the fabrication of MMMs and dictates the overall properties of MMMs for specific gas pair separation. Zeolites, silica, and carbon molecular sieves were used as fillers in the initial phases of MMM research. But it gained more interest among the researchers after the invention of other porous materials such as metal organic frameworks, CNTs, covalent organic frameworks, and porous polymers. Overall, MMMs have been studied widely in the literature in order to optimize the interface structure, size of particles, compatibility between polymer and filler, mechanical properties, etc. using a variety of inorganic materials. There are very few reports on studies on the effect of morphology and size of the above-mentioned filler particles on the gas transport properties. Further research on understanding the size and morphology effects including the alignment of the pores will be useful for further development of these MMMs. Agglomeration and sedimentation of the particles also pose a challenge to process the MMM especially using nanomaterials. Synthesizing the particles during the fabrication of membranes also results in good dispersion and formation of smaller particles to improve gas separation performance. Future research should also extend on developing novel techniques to keep these particles in the dispersed phase. The research so far focused on the polymers with very low gas permeability. But the research must focus on the polymers that are close to Robeson "upper bound" to obtain the industrially viable gas separation performance with the MMM.

However, successful industrial implementation of these MMMs for commercial separations has not yet been achieved because of several problems related to their processability into thin films on the porous supports. The challenging factor for the fabrication and understanding of these membranes includes the synthesis of smaller particles, compatibilization of the polymer particle, characterization of the interface, etc. Several reports were published on the tuning of the interface using ionic liquids, plasticizers, surface functionalization and by applying several processing modifications as mentioned in Sec. 2. But the complete interface structure and its involvement in MMM are still unknown. The research must be focused on understanding the complex structure of the MMM especially at the interface of polymer and filler in order to develop them further. Several researchers used conventional characterization techniques such as SEM, TEM, CT scan, and PALS to study the limited interface volume, but all those methods give the qualitative information. Experimental techniques needs to be developed to study the interface structure and its importance in improving the gas transport properties. The research should also focus on developing molecular simulation models to analyze the polymer–filler interface quantitatively. It is also important to study novel methods for predicting the gas transport properties of MMM including the non-ideal effects. Several mathematical models were developed for the prediction of the gas transport properties of the MMM, but these methods were based on several assumptions and were difficult to predict the properties accurately. This also helps in proper selection of materials using a consistent approach

Although MMM shows gas separation performance well above the Robeson "Upper bound", economic processing of the MMMs at industrial scale is the main hurdle for applying these membranes in the industry. The transformation of separation performance of film-based MMMs to the industry-relevant hollow fiber-based MMMs is highly critical and difficult due to challenging hollow fiber preparation conditions. Hollow fiber MMMs are advantageous at industrial scale compared to flat sheet MMMs because of their high packing density. Research on the fabrication of mixed matrix hollow fiber membranes is very sporadic. The research needs more focus on studying the single layer and dual layer hollow fiber mixed matrix membranes. Fabrication of thin MMMs on the hollow fiber supports is also a challenge. Synthesis of smaller porous particles is advantageous and essential to prepare thin MMMs and the research should also concentrate on the development of techniques to keep these particles dispersed. These smaller particles also provide more interfacial area and disrupt the polymer chain packing which enhances the gas transport properties of MMM. Most importantly, MMMs should be studied at industrial conditions in order to analyze their stability to routinely apply MMMs in the industry and focus the research in improving their stability.

References

1. Loeb, S., Sourirajan, S. *Adv. Chem.* 1963, 38, 117–132
2. Robeson, L. M. Correlation of separation factor versus permeability for polymeric membranes. *J. Memb. Sci.* 1991, 62, 165–185.

3. Robeson, L. M. The upper bound revisited. *J. Memb. Sci.* 2008, 320, 390–400.
4. Freeman, B. D. Basis of permeability/selectivity tradeoff relations in polymeric gas separation membranes. *Macromolecules* 1999, 32, 375–380.
5. Carreon, M. A., Li, S., Falconer, J. L., Noble, R. D. SAPO-34 seeds and membranes prepared using multiple structure directing agents. *Adv. Mater.* 2008, 20, 729–732.
6. Carreon, M. L., Li, S., Carreon, M. A. AlPO-18 membranes for CO_2/CH_4 separation. *Chem. Commun. (Camb).* 2012, 48, 2310–2312.
7. Carreon, M. A., Li, S., Falconer, J. L., Noble, R. D. Alumina-supported SAPO-34 membranes for CO_2/CH_4 separation. *J. Am. Chem. Soc.* 2008, 130, 5412–5413.
8. Li, S., Fan, C. Q. High-Flux SAPO-34 Membrane for CO_2/N_2 Separation. *Ind. Eng. Chem. Res.* 2010, 49, 4399–4404.
9. Li, S., Carreon, M. A., Zhang, Y., Funke, H. H., Noble, R. D., Falconer, J. L. Scale-up of SAPO-34 membranes for CO_2/CH_4 separation. *J. Memb. Sci.* 2010, 352, 7–13.
10. Venna, S. R., Zhu, M., Li, S., Carreon, M. A. Knudsen diffusion through ZIF-8 membranes synthesized by secondary seeded growth. *J. Porous Mater.* 2013, 21, 235–240.
11. Venna, S. R., Carreon, M. A. Highly permeable zeolite imidazolate framework-8 membranes for CO2/CH4 separation. *J. Am. Chem. Soc.* 2010, 132, 76–78.
12. Venna, S. R., Carreon, M. A. Amino-functionalized SAPO-34 membranes for CO_2/CH_4 and CO_2/N_2 separation. *Langmuir.* 2011, 27, 2888–2894.
13. Venna, S. R., Carreon, M. A. Metal organic framework membranes for carbon dioxide separation. *Chem. Eng. Sci.* 2015, 124, 3–19.
14. Moore, T. T., Koros, W. J. Non-ideal effects in organic–inorganic materials for gas separation membranes. *J. Mol. Struct.* 2005, 739, 87–98.
15. Vinh-Thang, H., Kaliaguine, S. Predictive models for mixed-matrix membrane performance: A review. *Chem. Rev.* 2013, 113, 4980–5028.
16. Nik, O. G., Chen, X. Y., Kaliaguine, S. Functionalized metal organic framework-polyimide mixed matrix membranes for CO_2/CH_4 separation. *J. Memb. Sci.* 2012, 413, 48–61.
17. Liu, Y., Zeng, G., Pan, Y., Lai, Z. Synthesis of highly c-oriented ZIF-69 membranes by secondary growth and their gas permeation properties. *J. Memb. Sci.* 2011, 379, 46–51.
18. Caro, J. Are MOF membranes better in gas separation than those made of zeolites? *Curr. Opin. Chem. Eng.* 2011, 1, 77–83.
19. Tanh Jeazet, H. B., Staudt, C., Janiak, C. Metal–organic frameworks in mixed-matrix membranes for gas separation. *Dalt. Trans.* 2012, 41, 14003.
20. Chung, T.-S., Jiang, L. Y., Li, Y., Kulprathipanja, S. Mixed matrix membranes (MMMs) comprising organic polymers with dispersed inorganic fillers for gas separation. *Prog. Polym. Sci.* 2007, 32, 483–507.
21. Dong, G., Li. H., Chen, V. Challenges and opportunities for mixed-matrix membranes for gas separation. *J. Mater. Chem. A.* 2013, 1, 4610.

22. Baker, R. W. *Membrane Technology and Applications.* John Wiley & Sons, Ltd: New Jersey, 2004.
23. Merkel, T. C., Freeman, B. D., Spontak, R. J., He. Z., Pinnau, I., Meakin, P., Hill, A. J. Ultrapermeable, reverse-selective nanocomposite membranes. *Science.* 2002, 296, 519–522.
24. Jiang, L. Y., Chung, T. S., Cao, C., Huang, Z., Kulprathipanja, S. Fundamental understanding of nano-sized zeolite distribution in the formation of the mixed matrix single- and dual-layer asymmetric hollow fiber membranes. *J. Memb. Sci.* 2005, 252, 89–100.
25. Jiang, L. Y., Chung, T. S., Kulprathipanja, S. An investigation to revitalize the separation performance of hollow fibers with a thin mixed matrix composite skin for gas separation. *J. Memb. Sci.* 2006, 276, 113–125.
26. Mahajan, R., Burns, R., Schaeffer, M., Koros, W. J. Challenges in forming successful mixed matrix membranes with rigid polymeric materials. *J. Appl. Polym. Sci.* 2002, 86, 881–890.
27. Pechar, T. W., Kim, S., Vaughan, B., Marand, E., Tsaptsis, M., Jeong, H. K., Cornelius, C. J J. Fabrication and characterization of polyimide–zeolite L mixed matrix membranes for gas separations. *J. Memb. Sci.* 2007, 277, 195–202.
28. Ismail, A. F., Kusworo, T. D., Mustafa, A. Enhanced gas permeation performance of polyethersulfone mixed matrix hollow fiber membranes using novel Dynasylan Ameo silane agent. *J. Memb. Sci.* 2008, 319, 306–312.
29. Gür, T. M. Permselectivity of zeolite filled polysulfone gas separation membranes. *J. Memb. Sci.* 1994, 93, 283–289.
30. Kim, S., Pechar, T. W., Marand, E. Poly(imide siloxane) and carbon nanotube mixed matrix membranes for gas separation. *Desalination* 2006, 192, 330–339.
31. Funk, C. V., Lloyd, D. R. Zeolite-filled microporous mixed matrix (ZeoTIPS) membranes: Prediction of gas separation performance. *J. Memb. Sci.* 2008, 313, 224–231.
32. Adatoz, E., Avci, A. K., Keskin, S. Opportunities and challenges of MOF-based membranes in gas separations. *Sep. Purif. Technol.* 2015, 152, 207–237.
33. Ahn, J., Chung, W.-J., Pinnau, I., Guiver, M. D. Polysulfone/silica nanoparticle mixed-matrix membranes for gas separation. *J. Memb. Sci.* 2008, 314, 123–133.
34. Anson, M., Marchese, J., Garis, E., Ochoa, N., Pagliero, C. ABS copolymer-activated carbon mixed matrix membranes for CO_2/CH_4 separation. *J. Memb. Sci.* 2004, 243, 19–28.
35. Bastani, D., Esmaeili, N., Asadollahi, M. Polymeric mixed matrix membranes containing zeolites as a filler for gas separation applications: A review. *J. Ind. Eng. Chem.* 2013, 19, 375–393.
36. Bertelle, S., Gupta, T., Roizard, D., Vallières, C., Favre, E. Study of polymer-carbon mixed matrix membranes for CO_2 separation from flue gas. *Desalination* 2006, 199, 401–402.

37. Goh, P. S., Ismail, A. F., Sanip, S. M., Ng, B. C., Aziz, M. Recent advances of inorganic fillers in mixed matrix membrane for gas separation. *Sep. Purif. Technol.* 2011, 81, 243–264.
38. Vu, D. Q., Koros, W. J., Miller, S. J. Mixed matrix membranes using carbon molecular sieves. *J. Memb. Sci.* 2003, 211, 335–348.
39. Zhao, D., Ren, J., Qiu, Y., Li, H., Hua, K., Li, X., Deng, M. Effect of graphene oxide on the behavior of poly(amide-6-b-ethylene oxide)/graphene oxide mixed-matrix membranes in the permeation process. *J. Appl. Polym. Sci.* 2015, 132, 42624.
40. Moore, T. T., Mahajan, R., Vu, D. Q., Koros, W. J. Hybrid membrane materials comprising organic polymers with rigid dispersed phases. *AIChE J.* 2004, 50, 311–321.
41. Tantekin-Ersolmaz, Ş. B., Atalay-Oral, C., Tatlier, M., Erdem-Senatalar, A., Schoeman, B., Sterte, J. Effect of zeolite particle size on the performance of polymer–zeolite mixed matrix membranes. *J. Memb. Sci.* 2000, 175, 285–288.
42. Su, N. C., Smith, Z. P., Freeman, B. D., Urban, J. J. Size-Dependent Permeability Deviations from Maxwell's Model in Hybrid Cross-Linked Poly(ethylene glycol)/Silica Nanoparticle Membranes. 2015, 27, 2421–2429.
43. Ordoñez, M. J. C., Balkus, K. J., Ferraris, J. P., Musselman, I. H. Molecular sieving realized with ZIF-8/Matrimid® mixed-matrix membranes. *J. Memb. Sci.* 2010, 361, 28–37.
44. Koros, W. J., Vu, D. Q., Mahajn, R., Miller, S. J. Mixed matrix membranes with pyrolized carbon sieve particles and methods of making and using the same. (2001).
45. Rezakazemi, M., Ebadi Amooghin, A., Montazer-Rahmati, M. M., Ismail, A. F., Matsuura, T. State-of-the-art membrane based CO_2 separation using mixed matrix membranes (MMMs): An overview on current status and future directions. *Prog. Polym. Sci.* 2014, 39, 817–861.
46. Aroon, M. A., Ismail, A. F., Matsuura, T., Montazer-Rahmati, M. M. Performance studies of mixed matrix membranes for gas separation: A review. *Sep. Purif. Technol.* 2010, 75, 229–242.
47. Li, Y., Guan, H.-M., Chung, T.-S., Kulprathipanja, S. Effects of novel silane modification of zeolite surface on polymer chain rigidification and partial pore blockage in polyethersulfone (PES)–zeolite A mixed matrix membranes. *J. Memb. Sci.* 2006, 275, 17–28.
48. Husain, S., Koros, W. J. Mixed matrix hollow fiber membranes made with modified HSSZ-13 zeolite in polyetherimide polymer matrix for gas separation. *J. Memb. Sci.* 2007, 288, 195–207.
49. Li, Y., Chung, T., Cao, C., Kulprathipanja, S. The effects of polymer chain rigidification, zeolite pore size and pore blockage on polyethersulfone (PES)-zeolite. A mixed matrix membranes. *J. Memb. Sci.* 2005, 260, 45–55.
50. Zhan, X., Lu, J., Tan, T., Li, J. Mixed matrix membranes with HF acid etched ZSM-5 for ethanol/water separation: Preparation and pervaporation performance. *Appl. Surf. Sci.* 2012, 259, 547–556.

51. Cong, H., Radosz, M., Towler, B. F., Shen, Y. Polymer–inorganic nanocomposite membranes for gas separation. *Sep. Purif. Technol.* 2007, 55, 281–291.
52. Şen, D., Kalıpçılar, H., Yilmaz, L. Development of polycarbonate based zeolite 4A filled mixed matrix gas separation membranes. *J. Memb Sci.* 2007, 303, 194–203.
53. Yong, H. H., Park, H. C., Kang, Y. S., Won, J., Kim, W. N. Zeolite-filled polyimide membrane containing 2,4,6-triaminopyrimidine. *J. Memb. Sci.* 2001, 188, 151–163.
54. Duval, J.-M., Kemperman, A.J.B., Folkers, B., Mulder, M. H. V., Desgrandchamps, G., Smplders, C. A. Preparation of zeolite filled glassy polymer membranes. *J. Appl. Polym. Sci.* 1994, 54, 409–418.
55. Mahajan, R., Koros, W. J. Mixed matrix membrane materials with glassy polymers. Part 1. *Polym. Eng. Sci.* 2002, 42, 1432–1441.
56. Shu, Husain, S., Koros, W. J. A General strategy for adhesion enhancement in polymeric composites by formation of nanostructured particle surfaces. *J. Phys. Chem. C* 2007, 111, 652–657.
57. Venna, S. R., Lartey, M., Li, T., Spore, A., Kumar, S., Nulwala, H. B., Luebke, D. R., Rosi, N. L., Albenze, E., Fabrication of MMMs with improved gas separation properties using externally-functionalized MOF particles. *J. Mater. Chem. A* 2015, 3, 5014–5022.
58. Anjum, M. W., Vermoortele, F., Khan, A. L., Bueken, B., De Vos, D. E., Vankelecom, I. F. J. Modulated UiO-66-based mixed-matrix membranes for CO_2 separation. *ACS Appl. Mater. Interfaces.* 2015, 7, 193–201.
59. Smith, S. J. D., Ladewig, B. P., Hill, A. J., Lau, C. H., Hill, M. R. Post-synthetic Ti exchanged UiO-66 metal-organic frameworks that deliver exceptional gas permeability in mixed matrix membranes. *Sci. Rep.* 2015, 5, 7823.
60. Feijani, E. A., Tavasoli, A., Mahdavi, H. Improving gas separation performance of poly(vinylidene fluoride) based mixed matrix membranes containing metal–organic frameworks by chemical modification. *Ind. Eng. Chem. Res.* 2015, 54, 12124–12134.
61. Naseri, M., Mousavi, S. F., Mohammadi, T., Bakhtiari, O. Synthesis and gas transport performance of MIL-101/Matrimid mixed matrix membranes. *J. Ind. Eng. Chem.* 2015, 29, 249–256.
62. Abedini, R., Omidkhah, M., Dorosti, F. Highly permeable poly(4-methyl-1-pentyne)/NH_2 -MIL 53 (Al) mixed matrix membrane for CO_2/CH_4 separation. *RSC Adv.* 2014, 4, 36522–36537.
63. Denny, M. S., Cohen, S. M. In situ modification of metal-organic frameworks in mixed-matrix membranes. *Angew. Chemie Int. Ed.* 2015, 54, 9029–9032.
64. Rodenas, T., van Dalen, M., Serra-Crespo, P., Kapteijn, F., Gascon, J. Mixed matrix membranes based on NH_2-functionalized MIL-type MOFs: Influence of structural and operational parameters on the CO_2/CH_4 separation performance. *Microporous Mesoporous Mater.* 2014, 192, 35–42.
65. Casado-Coterillo, C., Fernandez-Barquin, A., Zornoza, B., Tellez, C., Coronas, J., Irabein, A. Synthesis and characterisation of MOF/ionic liquid/

chitosan mixed matrix membranes for CO_2/N_2 separation. *RSC Adv.* 2015, 5, 102350–102361.
66. Noble, R. D. Perspectives on mixed matrix membranes. *J. Memb. Sci.* 2011, 378, 393–397.
67. Hudiono, Y. C., Carlisle, T. K., Bara, J. E., Zhang, Y., Lin, D. L., Noble, R. D. A three-component mixed-matrix membrane with enhanced CO_2 separation properties based on zeolites and ionic liquid materials. *J. Memb. Sci.* 2010, 350, 117–123.
68. Mohshim, D. F., Mukhtar, H., Man, Z. The effect of incorporating ionic liquid into polyethersulfone-SAPO34 based mixed matrix membrane on CO_2 gas separation performance. *Sep. Purif. Technol.* 2014, 135, 252–258.
69. Nasir, R., Mukhtar, H., Man, Z., Mohshim, D. F. Material advancements in fabrication of mixed-matrix membranes. *Chem. Eng. Technol.* 2013, 36, 717–727.
70. Gorgojo, P., Uriel, S., Téllez, C., Coronas, J. Development of mixed matrix membranes based on zeolite Nu-6(2) for gas separation. *Microporous Mesoporous Mater.* 2008, 115, 85–92.
71. Xu, L. Interfacial engineering of the interphase between carbon fibers and vinyl ester resin. 2003, 64–12, 6285.
72. Broyles, N. S., Verghese, K. N. E., Davis, S. V., Li, H., Davis, R. M., Lesko, J. J., Riffle, J. S. Fatigue performance of carbon fibre/vinyl ester composites: the effect of two dissimilar polymeric sizing agents. *Polymer (Guildf).* 1998, 39, 3417–3424.
73. Rafizah, W. A. W., Ismail, A. F. Effect of carbon molecular sieve sizing with poly(vinyl pyrrolidone) K-15 on carbon molecular sieve–polysulfone mixed matrix membrane. *J. Memb. Sci.* 2008, 307, 53–61.
74. Kessler, J. O., Moody, C. D. Drinking water from sea water by forward osmosis. *Desalination* 1976, 18, 297–306.
75. Elimelech, M. Yale constructs forward osmosis desalination pilot plant. *Membr. Technol.* 2007, 7–8.
76. Achilli, A., Cath, T. Y., Marchand, E. A., Childress, A. E. The forward osmosis membrane bioreactor: A low fouling alternative to MBR processes. *Desalination.* 2009, 239, 10–21.
77. Cornelissen, E. R., Harmsen, D., de Korte, K. F., Ruiken, C. J., Qin, J., Oo, H., Wessels, L. P. Membrane fouling and process performance of forward osmosis membranes on activated sludge. *J. Memb. Sci.* 2008, 319, 158–168.
78. Holloway, R. W., Childress, A. E., Dennett, K. E., Cath, T. Y. Forward osmosis for concentration of anaerobic digester centrate. *Water Res.* 2007, 41, 4005–4014.
79. Basu, S., Cano-Odena, A., Vankelecom, I. F. J. Asymmetric Matrimid®/ [$Cu_3(BTC)_2$] mixed-matrix membranes for gas separations. *J. Memb. Sci.* 2010, 362, 478–487.
80. Junaidi, M. U. M., Leo, C. P., Ahmad, A. L., Ahmad, N. A. Fluorocarbon functionalized SAPO-34 zeolite incorporated in asymmetric mixed matrix

membranes for carbon dioxide separation in wet gases. *Microporous Mesoporous Mater.* 2015, 206, 23–33.
81. Zulhairun, A. K., Ismail, A. F., Matsuura, T., Abdullah, M. S., Mustafa, A. Asymmetric mixed matrix membrane incorporating organically modified clay particle for gas separation. *Chem. Eng. J.* 2014, 241, 495–503.
82. Dai, Y., Johnson, J. R., Karvan, O., Sholl, D. S., Koros, W. J. Ultem®/ZIF-8 mixed matrix hollow fiber membranes for CO_2/N_2 separations. *J. Memb. Sci.* 2012, 401, 76–82.
83. Cohen, M. H., Turnbull, D. Molecular Transport in Liquids and Glasses. *J. Chem. Phys.* 1959, 31, 1164.
84. Winberg, P. Membrane fouling and process performance of forward osmosis membranes on activated sludge. *J. Memb. Sci.* 2005, 319(1–2), 158–168.
85. Ghosal, K., Freeman, B. D. Gas separation using polymer membranes: An overview. *Polym. Adv. Technol.* 1994, 5, 673–697.
86. Kim, J. H., Lee, Y. M. Gas permeation properties of poly(amide-6-b-ethylene oxide)–silica hybrid membranes. *J. Memb. Sci.* 2001, 193, 209–225.
87. Ch.Baerlocher, Lynne McCuster, D. H. O. *Atlas of Zeolite Framework Types.* 6th Edition. Elsevier: Amsterdam, 2007.
88. Caro, J., Noack, M. Zeolite membranes — Recent developments and progress. *Microporous Mesoporous Mater.* 2008, 115, 215–233.
89. Wang, H., Holmberg, B. A., Yan, Y. Homogeneous polymer–zeolite nanocomposite membranes by incorporating dispersible template-removed zeolite nanocrystals. *J. Mater. Chem.* 2002, 12, 3640–3643.
90. Clarizia, G., Algieri, C., Regina, A., Drioli, E. Zeolite-based composite PEEK-WC membranes: Gas transport and surface properties. *Microporous Mesoporous Mater.* 2008, 115, 67–74.
91. Jia, M., Peinemann, K.-V., Behling, R.-D. Molecular sieving effect of the zeolite-filled silicone rubber membranes in gas permeation. *J. Memb. Sci.* 1991, 57, 289–292.
92. Jia, M.-D., Pleinemann, K.-V., Behling, R.-D. Preparation and characterization of thin-film zeolite–PDMS composite membranes. *J. Memb. Sci.* 1992, 73, 119–128.
93. Huang, Z., Li, Y., Wen, R., May Teoh, M., Kulprathipanja, S. Enhanced gas separation properties by using nanostructured PES-Zeolite 4A mixed matrix membranes. *J. Appl. Polym. Sci.* 2006, 101, 3800–3805.
94. Hae-Kwon Jeong, Krych, W., Ramanan, H., Nair, S., Marand, E., Tsaptsis, M. Fabrication of polymer/selective-flake nanocomposite membranes and their use in gas separation. *Chem. Mater.* 2004, 16, 3838–3845.
95. Zornoza, B., Esekhile, O., Koros, W. J., Téllez, C., Coronas, J. Hollow silicalite-1 sphere-polymer mixed matrix membranes for gas separation. *Sep. Purif. Technol.* 2011, 77, 137–145.
96. Khan, A. L., Cano-Odena, A., Gutiérrez, B., Minguillón, C., Vankelecom, I. F. J. Hydrogen separation and purification using polysulfone acrylate–zeolite mixed matrix membranes. *J. Memb. Sci.* 2010, 350, 340–346.
97. Shindo, R., Kishida, M., Sawa, H., Kidesaki, T., Sato, S., Kanehashi, S., Nagai, K. Characterization and gas permeation properties of

polyimide/ZSM-5 zeolite composite membranes containing ionic liquid. *J. Memb. Sci.* 2014, 454, 330–338.

98. Ebadi Amooghin, A., Omidkhah, M., Kargari, A. The effects of aminosilane grafting on NaY zeolite–Matrimid® 5218 mixed matrix membranes for CO_2/CH_4 separation. *J. Memb. Sci.* 2015, 490, 364–379.
99. Zarshenas, K., Raisi, A., Aroujalian, A. Mixed matrix membrane of nano-zeolite NaX/poly (ether-block-amide) for gas separation applications. *J. Memb. Sci.* 2016, 510, 270–283.
100. Peydayesh, M., Asarehpour, S., Mohammadi, T., Bakhtiari, O. Preparation and characterization of SAPO-34 — Matrimid® 5218 mixed matrix membranes for CO_2/CH_4 separation. *Chem. Eng. Res. Des.* 2013, 91, 1335–1342.
101. Adams, R., Carson, C., Ward, J., Tannenbaum, R., Koros, W. Metal organic framework mixed matrix membranes for gas separations. *Microporous Mesoporous Mater.* 2010, 131, 13–20.
102. Xin, Q., Liu, T., Li, Z., Wang, S., Li, Y., Li, Z., Ouyang, J., Jiang, Z., Wu, H. Mixed matrix membranes composed of sulfonated poly(ether ether ketone) and a sulfonated metal–organic framework for gas separation. *J. Memb. Sci.* 2015, 488, 67–78.
103. Sabetghadam, A., Seoane, B., Keskin, D., Duim, N., Rodenas, T., Shahid, S., Sorribas, S., Guillouzer, C. L., Clet, G., Tellez, C., Daturi, M., Coronas, J., Kapteijn, F., Gascon, J. Metal organic framework crystals in mixed-matrix membranes: Impact of the filler morphology on the gas separation performance. *Adv. Funct. Mater.* 2016, 26, 3154–3163.
104. Waqas Anjum, M., Bueken, B., De Vos, D., Vankelecom, I. F. J. MIL-125(Ti) based mixed matrix membranes for CO_2 separation from CH_4 and N_2. *J. Memb. Sci.* 2016, 502, 21–28.
105. Guo, X., Huang, H., Ban, Y., Yang, Q., Xiao, Y., Li, Y., Yang, W., Zhong, C. Mixed matrix membranes incorporated with amine-functionalized titanium-based metal-organic framework for CO_2/CH_4 separation. *J. Memb. Sci.* 2015, 478, 130–139.
106. Hsieh, J. O., Balkus, K. J., Ferraris, J. P., Musselman, I. H. MIL-53 frameworks in mixed-matrix membranes. *Microporous Mesoporous Mater.* 2014, 196, 165–174.
107. Furukawa, H., Cordowa, K., O'Keeffe, M., Yaghi, O. M. The chemistry and applications of metal-organic frameworks. *Science* 2013, 341, 1230444.
108. H. Yehia, T. J. Pisklak, J. P. Ferraris, K. J. Balkus, I. H. M. Methane facilitated transport using copper (II) biphenyl dicarboxylate-triethylenediamine poly(3-acetoxyethylthiophene) mixed matrix membranes. *Polym. Prepr.* 2004, 45, 35–36.
109. Car, A., Stropnik, C., Peinemann, K.-V. Hybrid membrane materials with different metal–organic frameworks (MOFs) for gas separation. *Desalination.* 2006, 200, 424–426.
110. Díaz, K., Garrido, L., López-González, M., Castillo, L. F. del, Riande, E. CO_2 Transport in polysulfone membranes containing zeolitic imidazolate frameworks as determined by permeation and PFG NMR techniques. 2009, 43, 316–325.

111. Serre, C., Milange, F., Thouvenot, C., Nogues, M., Marsolier, G., Louer, D., Ferey, G. Very large breathing effect in the first nanoporous chromium(III)-based solids: MIL-53 or CrIII(OH)•$\{O_2C-C_6H_4-CO_2\}\cdot\{HO_2C-C_6H_4-CO_2H\}_x \cdot H_2O_y$. *J. Am. Chem. Soc.* 2002, 124, 13519–13526.
112. Basu, S., Cano-Odena, A., Vankelecom, I. F. J. MOF-containing mixed-matrix membranes for CO_2/CH_4 and CO_2/N_2 binary gas mixture separations. *Sep. Purif. Technol.* 2011, 81, 31–40.
113. Zornoza, B., Martinez-Joaristi, A., Serra-Crespo, P., Tellez, C., Coronas, J., Gascon, J., Kapteijn, F. Functionalized flexible MOFs as fillers in mixed matrix membranes for highly selective separation of CO_2 from CH_4 at elevated pressures. *Chem. Commun.* 2011, 47, 9522–9524.
114. Higuchi, A., Agatsuma, T., Uemiya, S., Kojima, T., Mizoguchi, K., Pinnau, I., Nagai, K., Freeman, B. D. Preparation and gas permeation of immobilized fullerene membranes. *J. Appl. Polym. Sci.* 2000, 77, 529–537.
115. Kim, S., Chen, L., Johnson, J. K., Marand, E. Polysulfone and functionalized carbon nanotube mixed matrix membranes for gas separation: Theory and experiment. *J. Memb. Sci.* 2007, 294, 147–158.
116. Shen, J., Zhang, M., Liu, G., Guan, K., Jin, W. Size effects of graphene oxide on mixed matrix membranes for CO_2 separation. *AIChE J.* 2016, 62, 2843–2852.
117. Zhao, D., Ren, J., Li, H., Li, X., Deng, M. Gas separation properties of poly(amide-6-b-ethylene oxide)/amino modified multi-walled carbon nanotubes mixed matrix membranes. *J. Memb. Sci.* 2014, 467, 41–47.
118. McAllister, M. J., Li, J., Adamson, D. H., Schniepp, H. C., Asdala, A. A., Liu, J., Herrera,-Alonso, M., Milius, D. L., Car, R., Prudhomme, R. K., Aksay, I. Single sheet functionalized graphene by oxidation and thermal expansion of graphite. *Chem. Mater.* 2007, 19, 4396–4404.
119. Bunch, J. S., Verbridge, S. S., Alden, J. S., Can der Zande, A. M.,Parpia, J. M., Craighead, H. G., McEuen, P. L. Impermeable atomic membranes from graphene sheets. 2008, 8, 2458–2462.
120. Yang, Y.-H., Bolling, L., Priolo, M. A., Grunlan, J. C. Super gas barrier and selectivity of graphene oxide-polymer multilayer thin films. *Adv. Mater.* 2013, 25, 503–508.
121. Dawson, R., Cooper, A. I., Adams, D. J. Nanoporous organic polymer networks. *Prog. Polym. Sci.* 2012, 37, 530–563.
122. Kaur, P., Hupp, J. T., Nguyen, S. T. Porous organic polymers in catalysis: Opportunities and Challenges. 2011, 1, 819–835.
123. Budd, P. M., Msayib, K. J., Tattershll, C. E., Ghanem, B. S., Reynolds, K. J., McKeown, N. B., Fritsch, D. Gas separation membranes from polymers of intrinsic microporosity. *J. Memb. Sci.* 2005, 251, 263–269.
124. Lau, C. H., Nguyen, P. T., Hill, M. R., Thornton, A. W., Konstas, K., Doherty, C. M., Mulder, R. J., Bourgeois, L., Liu, A. C., Sprouster, D. J., Sullivan, J. P., Bastow, T. J., Hill, A. J., Gin, D. L., Noble, R. D. Ending aging in super glassy polymer membranes. *Angew. Chemie Int. Ed.* 2014, 53, 5322–5326.

125. Yong, W. F., Li, F. Y., Chung, T. S., Tong, Y. W. Highly permeable chemically modified PIM-1/Matrimid membranes for green hydrogen purification. *J. Mater. Chem. A* 2013, 1, 13914–13925.
126. Kang, Z., Peng, Y., Qian, Y., Yuan, D., Addicoat, M. A., Heine, T., Hu, Z., Tee, L., Guo, Z., Zhao, D. Mixed matrix membranes (MMMs) comprising exfoliated 2D covalent organic frameworks (COFs) for efficient CO_2 separation. *Chem. Mater.* 2016, 28, 1277–1285.
127. Rabbani, M. G., El-Kaderi, H. M. Synthesis and characterization of porous benzimidazole-linked polymers and their performance in small gas storage and selective uptake. (2012).
128. A.K. Sekizkardes, V.A. Kusuma, G. Dahe, E.A. Roth, L.J. Hill, A. Marti, M. Macala, S.R. Venna, D. H. Carbon dioxide separation from flue gas by mixed matrix membranes with dual phase microporous polymeric constituents. *Chem. Commun.* Submitted (2016).
129. Moghadam, F., Omidkhah, M. R., Vasheghani-Farahani, E., Pedram, M. Z., Dorosti, F. The effect of TiO2 nanoparticles on gas transport properties of Matrimid5218-based mixed matrix membranes. *Sep. Purif. Technol.* 2011, 77, 128–136.
130. Hosseini, S. S., Li, Y., Chung, T.-S., Liu, Y. Enhanced gas separation performance of nanocomposite membranes using MgO nanoparticles. *J. Memb. Sci.* 2007, 302, 207–217.
131. Merkel, T. C., He. Z., Piannau, I. Effect of nanoparticles on gas sorption and transport in poly(1-trimethylsilyl-1-propyne). *Macromolecules.* 2003, 36, 6844–6855.
132. Vu, D. Q., Koros, W. J., Miller, S. J. Mixed matrix membranes using carbon molecular sieves: II. Modeling permeation behavior. *J. Memb. Sci.* 2003, 211, 335–348.
133. Zhao, Z., Ma, X., Li, Z., Lin, Y. S. Synthesis, characterization and gas transport properties of MOF-5 membranes. *J. Memb. Sci.* 2011, 382, 82–90.
134. Perez, E. V., Balkus, K. J., Ferraris, J. P., Musselman, I. H. Mixed-matrix membranes containing MOF-5 for gas separations. *J. Memb. Sci.* 2009, 328, 165–173.
135. Seoane, B., Coronas, J., Gascon, I., Benavides, M. E. Metal–organic framework based mixed matrix membranes: a solution for highly efficient CO_2 capture? *Chem. Soc. Rev.* 2015, 44, 2421–2454.
136. Zornoza, B., Tellez, C., Coronas, J., Gascon, J., Kapteijn, F. Metal organic framework based mixed matrix membranes: An increasingly important field of research with a large application potential. *Microporous Mesoporous Mater.* 2013, 166, 67–78.
137. Xin, Q., Ouyang, J., Liu, T., Li, Z., Liu, Y., Wamg, S., Wu, H., Jiang, Z., Cao, X. Enhanced interfacial interaction and CO_2 separation performance of mixed matrix membrane by incorporating polyethylenimine-decorated metal–organic frameworks. *ACS Appl. Mater. Interfaces.* 2015, 7, 1065–1077.

138. Pal, R. Permeation models for mixed matrix membranes. *J. Colloid Interface Sci.* 2008, 317, 191–198.
139. Watanabe, T., Sholl, D. S. Accelerating applications of metal-organic frameworks for gas adsorption and separation by computational screening of materials. *Langmuir.* 2012, 28, 14114–14128.
140. Bouma, R. H. B., Checchetti, A., Chidichimo, G., Drioli, E. Permeation through a heterogeneous membrane: the effect of the dispersed phase. *J. Memb. Sci.* 1997, 128, 141–149.
141. Maxwell, J. C. *Treatise on electricity and Magnetism.* Oxford University Press: London, 1873.
142. Petropoulos, J. H. A comparative study of approaches applied to the permeability of binary composite polymeric materials. *J. Polym. Sci. Polym. Phys. Ed.* 1985, 23, 1309–1324.
143. Bruggeman, D. A. G. Berechnung verschiedener physikalischer Konstanten von heterogenen Substanzen. I. Dielektrizitätskonstanten und Leitfähigkeiten der Mischkörper aus isotropen Substanzen. *Ann. Phys.* 1935, 416, 636–664.
144. Nielsen, L. E. Thermal conductivity of particulate-filled polymers. *J. Appl. Polym. Sci.* 1973, 17, 3819–3820.
145. Lewis, T. B., Nielsen, L. E. Dynamic mechanical properties of particulate-filled composites. *J. Appl. Polym. Sci.* 1970, 14, 1449–1471.
146. Pal, R. Permeation models for mixed matrix membranes. *J. Colloid Interface Sci.* 2008, 317, 191–198.
147. Felske, J. D. Effective thermal conductivity of composite spheres in a continuous medium with contact resistance. *Int. J. Heat Mass Transf.* 2004, 47, 3453–3461.
148. Shimekit, B., Mukhtar, H., Murugesan, T. Prediction of the relative permeability of gases in mixed matrix membranes. *J. Memb. Sci.* 2011, 373, 152–159.
149. Theodorou, D. N., Suter, U. W. Detailed molecular structure of a vinyl polymer glass. *Macromolecules.* 2002, 18, 1467–1478.
150. Wang, Q., Yan, Q., Nealey, P. F., Pablo, J. J. D. Monte Carlo simulations of diblock copolymer thin films confined between chemically heterogeneous hard surfaces. (2000).
151. Feng, J., Ruckenstein, E. The Morphology of symmetric triblock copolymer melts confined in a slit: A Monte Carlo simulation. *Macromol. Theory Simulations* 2002, 11, 630.
152. Yu, B., Chen, T., Jin, Q., Ding, D., Li, B., Shi, A-C. Confinement-induced novel morphologies of block copolymers. *Phys. Rev. Lett.* 2006, 96, 138306.
153. Feng, J., Ruckenstein, E. Morphologies of AB diblock copolymer melts confined in nanocylindrical tubes. *Macromolecules.* 2006, 39, 4899–4906.
154. Feng, J., Ruckenstein, E. Morphology transitions of AB diblock copolymer melts confined in nanocylindrical tubes. *J. Chem. Phys.* 2006, 125, 164911.
155. Feng, J., Cavicchi, K. A., Heinz, H. Control over self-assembly of diblock copolymers on hexagonal and square templates for high area density circuit boards. *ACS Nano.* 2011, 5, 9413–9420.

156. Termonia, Y. Monte-Carlo modeling of dense polymer melts near nanoparticles. *Polymer (Guildf)*. 2009, 50, 1062–1066.
157. Gao, Y., Liu, J., Zhang, L., Cao, D. Existence of a glassy layer in the polymer-nanosheet interface: Evidence from molecular dynamics. *Macromol. Theory Simulations.* 2014, 23, 36–48.
158. Jie, F., Surendar Venna, D. H. Interactions at the interface of polymer matrix-filler particle composites. *Polymer*. Submitted.

Chapter 2

Ceramic-Supported Organic Composite Membranes for Gas Separation

Gongping Liu and Wanqin Jin*

State Key Laboratory of Materials-Oriented Chemical Engineering, Nanjing Tech University, 5 Xinmofan Road, Nanjing 210009, P. R. China
**wqjin@njtech.edu.cn*

Organic–inorganic composite membranes have received increasing attention for molecular separation. In this chapter, recent progresses on ceramic-supported organic composite membranes for gas separation are briefly reviewed. The organic separation layer of the composite membranes includes traditional polymeric membrane materials such as polydimethylsiloxane (PDMS), poly(ethylene glycol) (PEG), and poly(amide-b-ethylene oxide) (PEBA), as well as emerging two-dimensional graphene-based membrane materials. Fabrication method, physicochemical property, and gas separation performance of the membranes are discussed. Compared with the state-of-the-art, the developed ceramic-supported organic composite membranes exhibited excellent gas separation performance for carbon dioxide and hydrogen separation, suggesting great potential for practical application.

Although the gas permeation performance for various membranes mainly depends on the properties of the membrane materials, the types of membrane structure also have an influence on the membrane performance which should not be neglected. In industrial application, composite membranes that consist of a porous support layer with a thin dense skin layer on top are predominantly used because of their higher gas permeability and mechanical strength compared to symmetric dense membranes. The stability of the composite membranes is determined by not only the separation layer but also the interface between the separation and support layers. In the cases of high temperature or high pressure, polymeric supports cannot constrain the shear stress at the interface; thus the separation layer may delaminate from the support with elapse of the operation time. However, the rigid ceramic supports can confine the polymer that penetrates into the pores, which improves the stability of the composite membrane with the confinement effect. In addition, ceramic supports can provide sufficient mechanical stiffness to support a thin selective polymeric layer even at high pressure. The low transport resistance of ceramic support could enhance the gas permeance as well. Because of the above-mentioned advantages of ceramic supports, numerous studies have been focused on the preparation of organic–inorganic composite membranes, which combine the advantages of both organic and inorganic membranes. Our group has developed a series of ceramic-supported polymeric composite membranes.[1] The separation materials include PDMS,[2–4] poly(vinyl alcohol) (PVA),[5] and polyelectrolytes.[6] The applications of these composite membranes contain the bio fuels recovery,[7] gasoline desulfuration,[8] pervaporation coupled process,[7] and dehydration of alcohols and esters.[9,10] Those ceramic-supported polymeric composite membranes exhibit remarkable high performance, especially high permeate flux. This is attributed to the thin polymer separation layer and ceramic support with low transport resistance. This chapter summarized recent progresses in ceramic-supported organic composite membranes for gas separation. The separation layer includes polymers such as PDMS, PEG and PEBA, and also two-dimensional materials such as graphene oxide.

1. Ceramic-supported Polymeric Composite Membranes

Our group developed ceramic supported PDMS and poly(ethylene glycol) diacrylate (PEGDA) composite membranes for CO_2 separation from light gases.[11] For ceramic-supported PDMS composite membrane, tubular asymmetric ZrO_2–Al_2O_3 composite membranes were used as supports. PDMS polymer was dissolved in *n*-heptane; then cross-linking agent TEOS and catalyst dibutyltin dilaurate were added into the polymer solution. The solution was stirred at 30°C. Subsequently, the cross-linked PDMS solution was coated onto the outer surface of ceramic supports by a dip-coating method. The membranes were dried for 24 h under room temperature and then introduced into an oven to remove any residual solvent and to complete cross-linking at 120°C for 12 h. The preparation conditions including polymer concentration and amount of the cross-linking agent are systematically optimized: 15 wt.% PDMS concentration, 20 wt.%

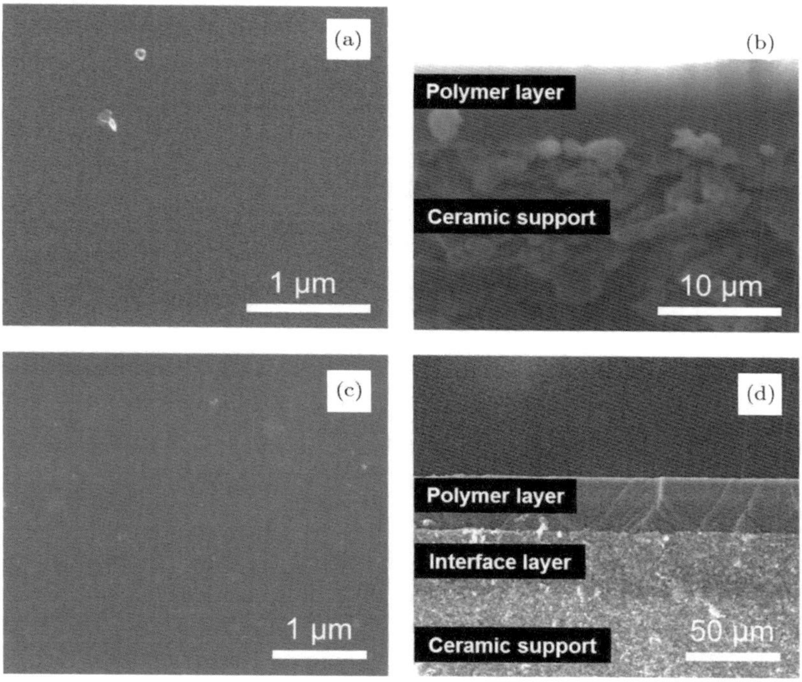

Fig. 1. SEM images of the ceramic-supported (a) surface and (b) cross-section of PDMS composite membrane; and (c) surface and (d) cross-section of PEGDA composite membrane.

Source: Reproduced with permission from Ref. 11. Copyright (2013) Elsevier.

cross-linking agent amount. Figures 1(a) and 1(b) show the SEM micrographs of the ceramic-supported PDMS composite membrane. It can be seen that the active PDMS layer is uniformly coated on the surface of ceramic supports. The surface of the composite membrane is dense and defect-free, and the membrane structures (ceramic layer, PDMS layer) can be observed clearly. The PDMS layer is well adhered to the porous ceramic support layer.

Alumina disks were used as supports for PEGDA composite membranes. Alumina powders were uniaxially pressed at 200 MPa to prepare green disks. The green disks were then sintered at 1200°C for 2 h to form the porous supports. 0.1% (by mass) initiator (HCPK) was added into PEGDA in an amber glass jar to prepare pre-polymerization solution. After stirring, the solution was mixed with a predetermined amount of water to form the target composition. The resulting polymer solution was coated onto one surface of the Al_2O_3 support by dip-coating method for 30 s, and then the resulted polymer layer on the surface was polymerized by exposure to UV light for 90 s. Various PEGDA concentrations in pre-polymerization solution were used for the preparation of cross-linked PEGDA composite membranes. As shown in Figs. 1(c) and 1(d) — when PEGDA mass concentration is 100% — the polymer separation layer is successfully coated onto the support. It is obvious that the membrane surface is dense and crack-free. From the cross-section image, it is found that there is clearly an interface layer formed by the penetration of polymer solution into the ceramic support. This interface layer reinforced the adhesion strength between the coating layer and ceramic support.

Based on the comparison of PDMS and PEGDA membranes using other polymeric supports or freestanding membranes reported in literatures, the ceramic-supported composite membrane can be considered to have higher gas permeance, while the separation factor is also relatively good. Compared with ceramic-supported PDMS composite membrane, PEGDA composite membrane presented smaller gas permeance but much higher selectivity for CO_2/N_2. The ceramic-supported composite membrane can be expected as a candidate for CO_2 separation from light gases.

Cheng *et al.*[12] fabricated ceramic hollow fiber-supported poly(amide-b-ethylene oxide)/polyethylene glycol dimethylether (PEBA/PEGDME) composite membrane for CO_2 separation from biohythane. About 4 g of Pebax MH 1657 pellets was dissolved in 76 g of the solvent mixture consisting of 70/30 w/w ethanol/water. The polymer solution was stirred under reflux at 80°C for 2 h until it was completely dissolved. After cooling the solution to room temperature, 4 g of PEGDME was added, and the solution was stirred for 1 h at room temperature. Finally, the obtained homogeneous solution was filtered through a stainless steel filter with a pore size of 32 μm. The composite membranes were fabricated via a two-time dip coating method. After pre-wetting by de-ionized water for about 10 s, the ceramic hollow fiber was immersed into the PEBA/PEGDME coating solution for 20 s. The viscosity of the coating solution was 13.2 mPa s. The hollow fiber was then dried at 80°C for 2 h. The hollow fiber was then immersed again into the coating solution for another 20 s. The viscosity of the coating solution increased to 21.8 mPa s because the solution slowly crystallized. Finally, the composite hollow fiber was dried at 313 K for 15 h and at 293 K for 5 days before testing the gas separation performance.

As shown in the SEM image (Fig. 2(b)), numerous particulate matters with diameters ranging from 1 to 5 μm are observed on the surface of the PEBA/PEGDME selective layer. This result is due to the polymer Pebax® MH 1657 containing 40 wt.% crystalline polyamide (polyamide 6), which contributes to the existence of large particle structures. Meanwhile, the agglomeration of polymers in the coating solution probably happens during the time interval between the first and second coating of the ceramic hollow fiber caused by the intrinsic van der Waals force. Figure 2(b) reveals that 25 μm-thick PEBA/PEGDME selective layer closely adheres on the support with no gap in the cross-section. Given that the support was pre-wetted by de-ionized water before coating the selective layer, no obvious interface layer is formed by the penetration of the polymer solution into the ceramic support.

Fig. 2. (a) Surface and (b) cross-section of ceramic hollow fiber-supported PEBA/PEGDME composite membrane.

Source: Reproduced with permission from Ref. 12. Copyright (2015) RSC.

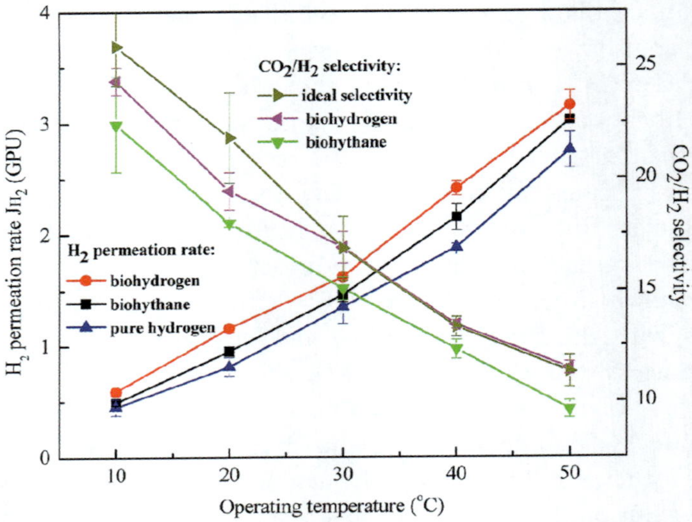

Fig. 3. H_2 permeation rate and CO_2/H_2 selectivity in the porous ceramic hollow fiber-supported PEBA/PEGDME composite membrane (feed pressures of pure or mixed gases were constant at 0.12 MPa) biohythane: 40% CO_2, 48% CH_4, 12% H_2; biohydrogen: 40% CO_2, 60% H_2; ideal selectivity: $J_{\text{pure}\,CO_2}/J_{\text{pure}\,H_2}$.

Source: Reproduced with permission from Ref. 12. Copyright (2015) RSC.

Figure 3 shows the variation of H_2 permeation rates under different feed systems. It shows: $J_{\text{pure H2}} < J_{\text{biohythane}} < J_{\text{biohydrogen}}$. The H_2 permeation rates in biohydrogen are generally 40% higher than that in pure hydrogen, which is due to the CO_2-induced plasticization on the polymer membranes. Furthermore, the H_2 permeation rates significantly increased with the operating temperature, and a 6-fold increase is observed for pure hydrogen. This is owing to mobility enhancement of gas molecules and the increase of fractional free volume of the polymer membrane. The CO_2/H_2 selectivity decreases with increasing operating temperature due to the decreased selectivity in solubility and diffusivity. In biohydrogen, the plasticization and competitive sorption improve the CO_2 and H_2 permeance, while the improvement in H_2 permeability is more significant at low temperature. Thus, the CO_2/H_2 selectivity in biohydrogen is lower than the ideal CO_2/H_2 selectivity. Compared to the other PEBA/PEGs membranes, the ceramic hollow fiber-supported PEBA/PEGDME composite membrane showed higher CO_2/H_2 selectivity.

Air separation has attracted increasing attention owing to its potential industrial application, including combusting enhancement, fuel cell processes, emission control, and medical application. As we know, cryogenic distillation and pressure swing adsorption are the traditional methods to obtain oxygen-enriched air, which however are highly energy-intensive. The membrane gas separation technology is attractive and alternative to conventional technologies for oxygen and nitrogen separation mainly because of its simplicity in operation, low cost, small size, and low energy consumption. Our group developed a new kind of hollow fiber ceramic-supported PDMS composite membrane[13] (Figs. 4(a) and 4(b)). Besides low transport resistance, many other advantages are found for hollow fiber membranes when compared with other configurations, such as high packing density, cost-effectiveness, and a self-support structure. The prepared hollow fiber PDMS–ceramic composite membranes with high flux and selectivity showed great potential in pervaporation process. The hollow fiber ceramic-supported PDMS composite membranes were also applied in separation of oxygen from air.[14]

The influence of PDMS layer thickness, pressure difference, temperature, and feed concentration on the oxygen-enriching performance were systematically investigated. The results in Fig. 4(c) indicated that the membrane (with the optimal PDMS layer thickness of $25\,\mu m$) achieved oxygen permeance of 104 GPU ($1\ GPU = 10^{-6}\ cm^3\ (STP)/cm^2{\cdot}s{\cdot}cmHg$) with O_2/N_2 ideal selectivity of 2.0 at room temperature and 0.1 MPa difference in pressure. It showed low temperature was good for gas separation and the influence of pressure difference is small. Besides, over 10 days of continuous operation, this enrichment performance was stable, in which the oxygen concentration was enriched from air and maintained at about 30% by single-stage separation through the membrane. The result suggests that the hollow fiber ceramic-supported PDMS composite membranes can be expected to be promising candidates for oxygen enrichment process.

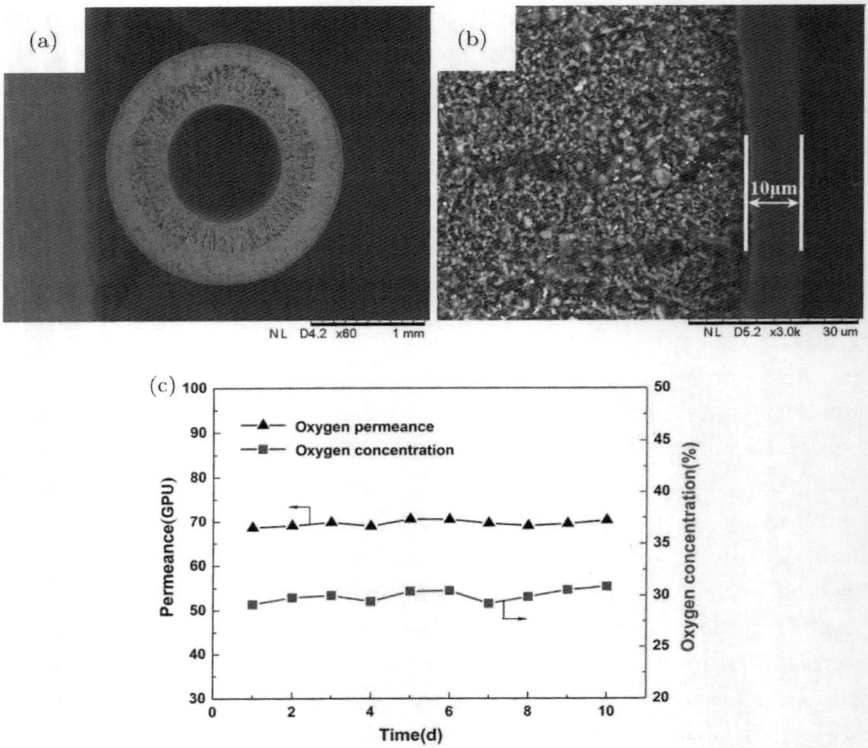

Fig. 4. (a)–(b) SEM cross-section of ceramic hollow fiber-supported PDMS composite membrane; (c) long-term stability of the PDMS membrane at room temperature for air separation.

Source: Reproduced with permission from Ref. 14. Copyright (2016) John Wiley & Sons.

2. Ceramic-supported GO Composite Membranes

The rise of the typical two-dimensional material graphene and its derivatives has opened a brand new era for developing the next generation of membranes with extraordinary separation properties.[15] Ultrathin single-atom thickness and nearly frictionless surface enable graphene nanosheets to form membranes which minimize transport resistance and maximize permeance.[16] The structural nanopores with homogenous dispersion and the intrinsic interlayer spacing between the graphene layers endow the graphene-based membranes with molecular sieving channels for fast–selective gas permeation.[17] The pioneering work of Geim *et al.*[18] found that graphene oxide (GO) membranes are impermeable to gases and liquids but solely allow unimpeded evaporation of water. Afterwards, the graphene-based membranes have been reported for water purification[19–23] and solvent dehydration.[24–26] In addition, graphene-based laminates can also exhibit gas separation characteristics if their stacking structures are carefully controlled.[27,28] GO composite membranes exhibit similar structure of polymeric composite membranes that consist of an active layer on top of porous substrate. Thus, the concept of "organic–inorganic composite membrane" was applied for developing graphene-based membranes.[16,29]

2.1. *Vacuum-filtration Method*

Free standing GO papers were fabricated by the flow-directed (i.e., filtration) assembly of GO sheets, resulting in tightly packed interlocking sheets that subtly undulate along the paper surface. The layered structure with sub-nanosized inter-layer spacing and the excellent macroscopic flexibility and stiffness promise the practical utilization of such GO laminates for molecular separation. The slowly flowing water in the confined galleries, together with the electrostatic and van der Waals attractive forces between the very large aspect ratio compliant sheets of GO, is largely responsible for their sequential deposition into the laminar structure.[16]

Li *et al.*[27] used single-layered GO flake to prepare ultrathin GO membranes on the anodic aluminum oxide (AAO) support via vacuum filtration. Centrifugation and dilution of GO dispersions were found to be important for preparing high-quality GO membranes. Figure 5(a) shows a ~9-nm thick GO membrane with a permeation area of ~4 cm^2 on AAO support. X-ray diffraction shows the characteristic peak of GO at 2θ of 11.1°, and GO flakes are ~500 nm in size and single-layered, as confirmed by atomic force microscopy (AFM) (Fig. 5(b)), which also shows the height profile of a GO flake

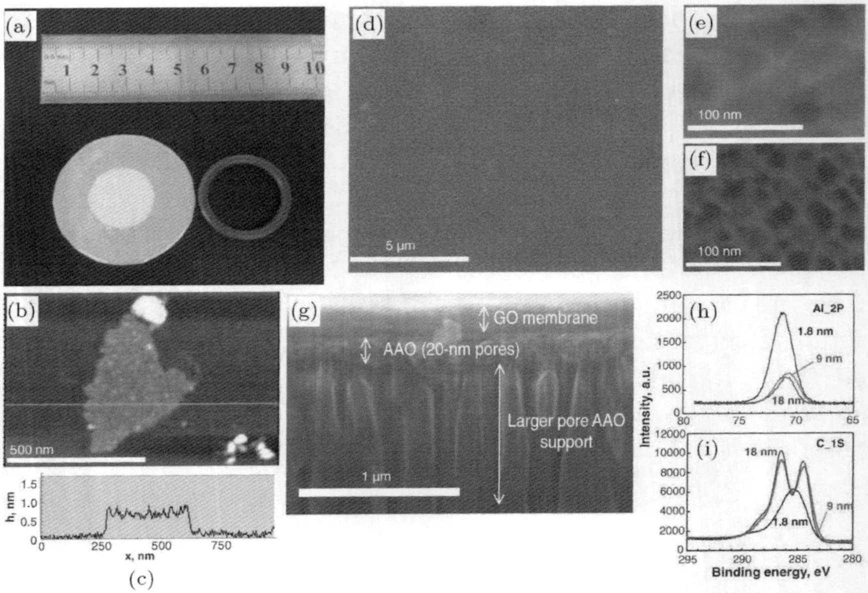

Fig. 5. GO membranes supported on porous AAO. (a) Digital picture of an ultrathin GO membrane on AAO (~9 nm). The white circular area is the permeation area (~4 cm^2) with supported GO; the yellow Kapton tape is used for GO protection and sealing by an O-ring during permeation measurements. (b) AFM image of a GO flake on freshly cleaved mica. (c) The height profile across the green line in (b) is shown here. h, height; x, position. (d) Field-emission scanning electron microscopy (FE-SEM) image of the surface of a GO membrane (~18 nm thick) on porous AAO. (e) FE-SEM image of the GO membrane surface (~18 nm thick) with higher magnification. (f) FE-SEM image of the AAO surface without the GO membrane. (g) FE-SEM image of the cross-sectional view of a thick GO membrane (~180 nm). (h) Al 2P and (i) C 1S XPS spectra of ultrathin GO membranes (~1.8, 9, and 18 nm thick) supported on porous AAO. a.u., arbitrary units.

Source: Reproduced with permissions from Ref. 27. Copyright (2013) the American Association for the Advancement of Science.

(Fig. 5(c)). Figures 5(d) and 5(e) show the surface of an 18 nm-thick GO membrane on AAO in which a very thin GO coating can be seen compared with the AAO support (Fig. 5(f)). A relatively thick GO membrane (~180 nm (Fig. 5(g))) was deposited to correlate the GO deposition with the membrane thickness. GO dispersion for this 180 nm-membrane preparation was then diluted 100, 20, and 10 times to obtain the above 1.8, 9 and 18 nm thick GO membranes, respectively, assuming no GO loss during filtration and constant membrane density. X-ray photoelectron spectroscopy (XPS) was used to detect surface elements for these ultrathin GO membranes on AAO (Figs. 5(h) and 5(i)). For a 1.8-nm thick membrane, a substantial amount of aluminum can be seen, because the mean free path of excited electrons is longer than the surface GO membrane thickness. However, for thicker membranes (9 and 18 nm), much smaller amounts of the underlying aluminum can be seen, because GO thickness is larger than the excited electron mean free path. This finding is consistent with surface carbon detection by XPS as well (Fig. 5(i)).

Permeation tests with different light gas molecules were conducted to probe pore sizes of the ultrathin GO–AAO composite membrane. Hydrogen (kinetic diameter: 0.289 nm) permeated ~300 times faster than did CO_2 (0.33 nm) through an ~18-nm thick GO membrane at 20°C. Their kinetic diameter difference is only 0.04 nm, suggesting that the average size of pores for permeation in the GO membrane may be between 0.289 and 0.33 nm. Generally, gas permeance is inversely proportional to the membrane thickness for conventional membranes. It is interesting to find that H_2 and He permeances decrease exponentially as membrane thickness increases from 1.8 to 180 nm. The authors speculated that the major transport pathway for these molecules is selective structural defects within GO flakes, instead of spacing between GO flakes. They reduced GO membrane's thickness from 1.8 to 20 nm and conducted pressure-driven water permeation. It was found that water permeance decreased approximately three orders of magnitude; for example, water permeance through a 3-nm GO membrane was 1370 $L/(m^2{\cdot}h{\cdot}bar)$, whereas it was 0.5 $L/(m^2{\cdot}h{\cdot}bar)$ through a reduced GO membrane. It suggests that interlayer spacing has been eliminated or considerably narrowed by reduction. They then measured single-gas permeation through 18 nm reduced GO membranes but found no obvious gas permeance change, which suggests that interlayer spacing is not the major transport pathway and permeation of molecules is attributed to the selective structural defects within GO flakes.

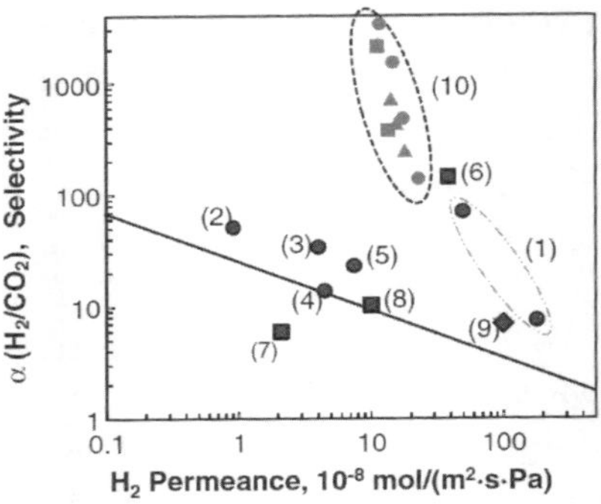

No.	Symbol	Membrane material	Temperature range, °C
1	●	Silica	100-300
2	●	Silicon carbide	
3	●	MFI Zeolite (modified)	
4	●	ZIF-7	
5	●	Zeolite composite	
6	■	MFI zeolite (modified)	>300
7	■	DDR zeolite	
8	■	Silica-titania	
9	◆	MOF	<100
10	■	GO (1.8-nm)	20-100
	●	GO (9-nm)	
	▲	GO (18-nm)	

Fig. 6. Comparison of ultrathin GO membranes with polymeric membranes and inorganic microporous membranes for H_2–CO_2 mixture separation: selectivity versus H_2 permeance. The black line denotes the 2008 upper bound of the polymeric membrane for H_2/CO_2, assuming membrane thickness is 0.1 μm. Blue points (1–9) represent the reported microporous inorganic membranes, and red points (10) indicate ultrathin GO membranes. The table at right explains points 1–10. ZIF, zeolitic imidazolate framework; MOF, metal-organic framework.

Source: Reproduced with permissions from Ref. 27. Copyright (2013) the American Association for the Advancement of Science.

These ultrathin GO–AAO composite membranes showed mixture separation selectivities as high as 3400 and 900 for H_2–CO_2 and H_2–N_2 mixtures, respectively, through selective structural defects on GO. Figure 6 shows a comparison of ultrathin GO membranes with polymeric membranes and inorganic membranes for H_2–CO_2 mixture separation. Typically, for separation using polymeric membranes, permeance decreases as separation selectivity increases. An upper bound can typically be used to compare the separation performance of a new membrane with that of previous membranes. Ultrathin GO membranes are far above the upper bound for polymeric membranes (black line) and show superior separation performance compared with representative inorganic membranes.

2.2. *Spray-evaporation Method*

In order to fabricate laminar GO membranes, vacuum filtration may be most frequently adopted in membrane preparation for the reason that the vacuum filter or substrates can act naturally as a support to combine with the GO separation layer. However, it still has to face the challenge that the method is only dependent on the property of GO suspension, and it is hard to acquire precise structure controlling during the fabrication process. Therefore, a facile method utilizing spray evaporation-induced self-assembly was developed in our group to fabricate substrate-based GO membrane. As shown in the schematic of the proposed GO deposition process in Fig. 7, the sprayed ultra small droplets can afford coating uniformity and high evaporation efficiency. Furthermore, the spray-evaporation self-assembly process takes merely several minutes, which is more productive and effective compared with filtration methods taking hours or much longer time to obtain GO membrane with similar thickness.[30] Such GO membrane with optimized nanostructure is expected to exhibit separation property for gas molecules. Typically, certain amount of GO nanosheets were added into the solvent with the concentration of

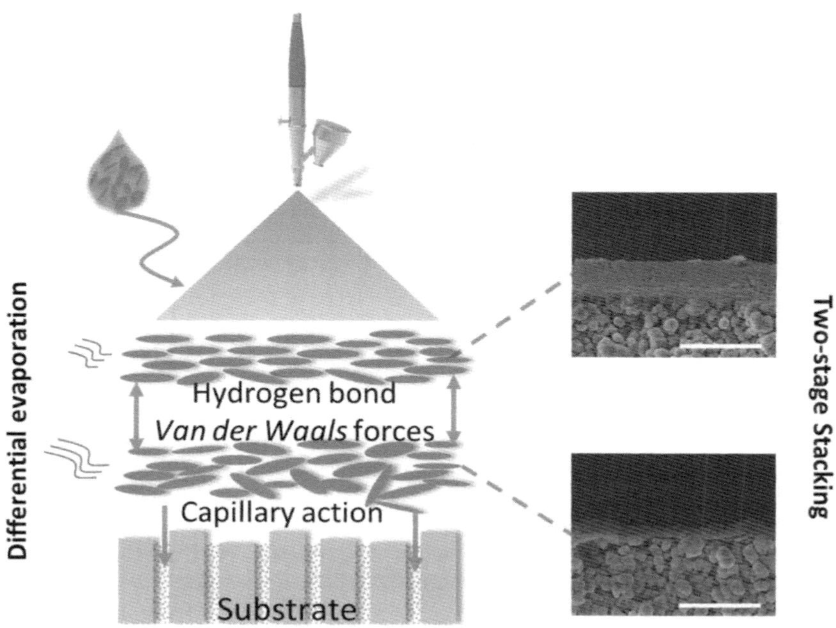

Fig. 7. Schematic process of GO stacking on alumina substrate.

1 mg/g, then followed by ultrasonication to disperse. The lateral size of GO nanosheets utilized to form the membrane is around 1 μm. To accelerate the evaporation process at a relatively low temperature, GO nanosheets were dispersed in ethanol/water binary solvents with the ethanol mass fractions of 0.1, 0.3, 0.5, 0.7, 0.9, and 1. The GO membranes derived from these solvents are thus denoted as GO-0.1, GO-0.3, GO-0.5, GO-0.7, GO-0.9, and GO-1, respectively. Alumina substrate with pore size of 200 nm (smaller than the size of GO nanosheets to prevent the pore filling) was heated to the setting temperature by the heating plate. And GO suspension was sprayed vertically onto the substrate at various times. It should be noted that the next spraying was not carried out until the previous sprayed suspension was totally dried.

By finely controlling the microstructure, the GO-based membrane can also be applied for gas separation. The spray-evaporation induced GO membrane with optimized stacking structure shows selective H_2 transport behaviors. Changing the evaporation rate during membrane fabrication process can cause different gas permeation properties. As shown in Fig. 8(a), for the non-uniform membranes of GO-0.1 and GO-0.3 (which indicate that the mass fraction of ethanol in solvent are 10 and 30 wt.%), the existence of defects leads to relatively higher H_2 permeance and lower selectivity. GO-0.5 and GO-0.7 membranes are more appropriate for gas separation due to the mild stacking rate and the relatively good dispersibility to obtain a uniform membrane as well as laminar structure. When the evaporation process is excessively shortened and GO dispersibility gets much worse in solvent with high ethanol mass fraction, the H_2/CO_2 selectivities of GO-0.9 and GO-1 membranes are close to that of blank alumina substrate possibly due to the fatal cracks or defects caused by extraordinarily rapid evaporation. The transport mechanism in the spray-evaporation induced GO membrane was summarized and schematically diagrammed in Fig. 8(b). For a not well-stacked GO layer, larger defects make the transport pathway their domain. But for laminar GO layer, there are two main ways for gases to travel through stacked GO nanosheets. One is through the tortuous pathway formed by parallel-stacked GO nanosheets,

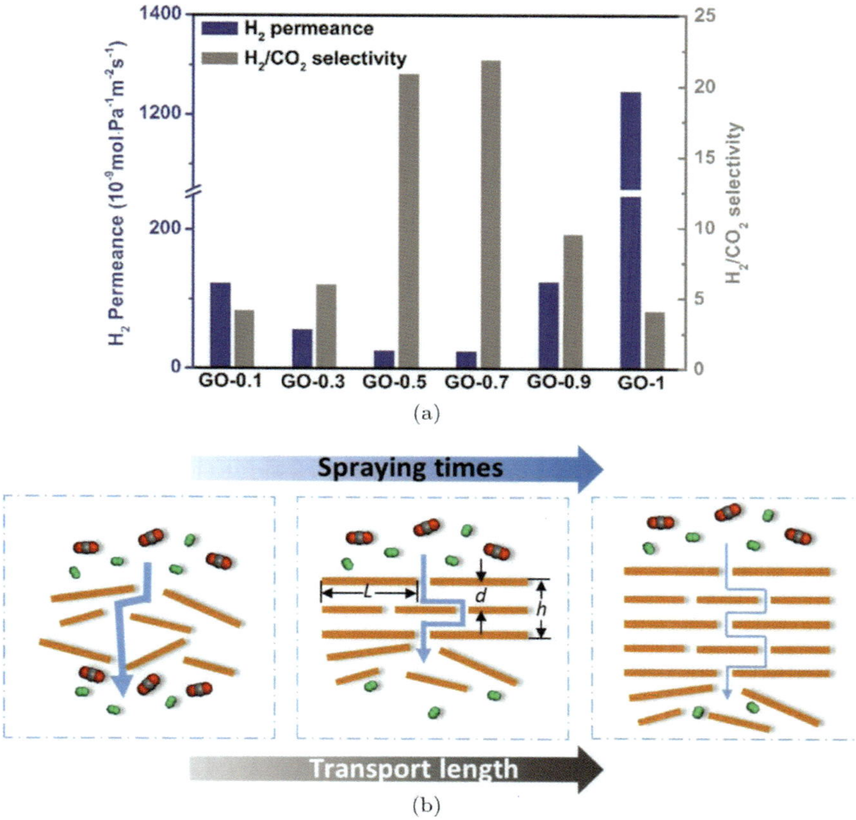

Fig. 8. (a) Gas permeation of GO membranes derived by varied evaporation rate; (b) schematic process of gas molecules permeating through the deposited GO layer.

and the other is through the small defects or pinholes in GO nanosheets. An augmentation in spraying times results in much more complicated transport pathways and the consequent higher transfer resistance.

At a mild evaporation rate and proper spraying times, laminar-structured GO membranes can be obtained which own relatively high H_2 selectivities over CO_2. The detailed gas permeation test of GO-0.5 membrane with 20 times of spraying was carried out with H_2/CO_2

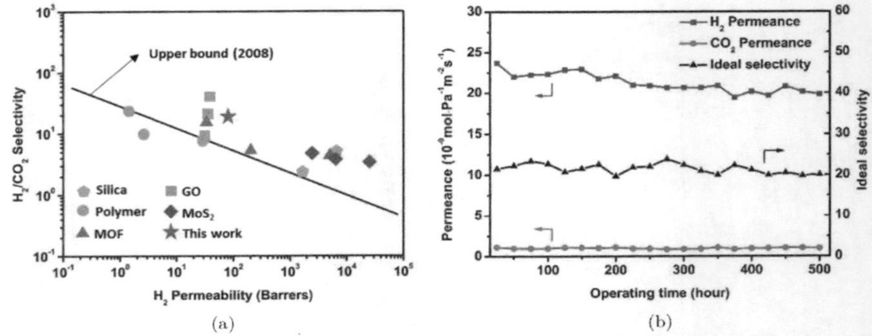

Fig. 9. (a) Pure gas permeation of GO-0.5 membrane with the thickness of 1 μm. (b) Long-term gas permeance investigation.

selectivity of 20.9 (3.3 times higher than that of blank alumina substrate) and H_2 permeance of 2.7×10^{-8} mol$\cdot$Pa$^{-1}\cdot$m$^{-2}\cdot$s^{-1} (Fig. 9(a)). Moreover, the desirable long-term gas permeation stability of this GO membrane was confirmed as shown in Fig. 9(b), indicating that the GO membrane was stable during a 500-h continuously pressurized operation. This also proves that such GO membrane is rather stable in dry state solving the problem of wet-state storage.

2.3. *External Force-driven Assembly Method*

Generally, molecular transport through the assembled graphene oxide laminate occurs in in-plane slit-like pores and then plane-to-plane inter-galleries. However, it still has to confront the challenge of regulating the slit-like pores and inter-galleries within sub-nanometer size that are necessary for precise sieving of monovalent ions (water desalination) and small gases (gas separation). In our group, a facile external force driven assembly approach was developed to realize sub-nanometer two-dimensional channels readily assembled by GO building blocks.[31] As shown in Fig. 10, the designed external forces are applied both outside and inside the GO laminate. They collaboratively overcome the intrinsic repulsive electrostatic interactions within GO, with the aim of eliminating non-selective stacking defects and stimulating highly ordered assembly of GO nanosheets.

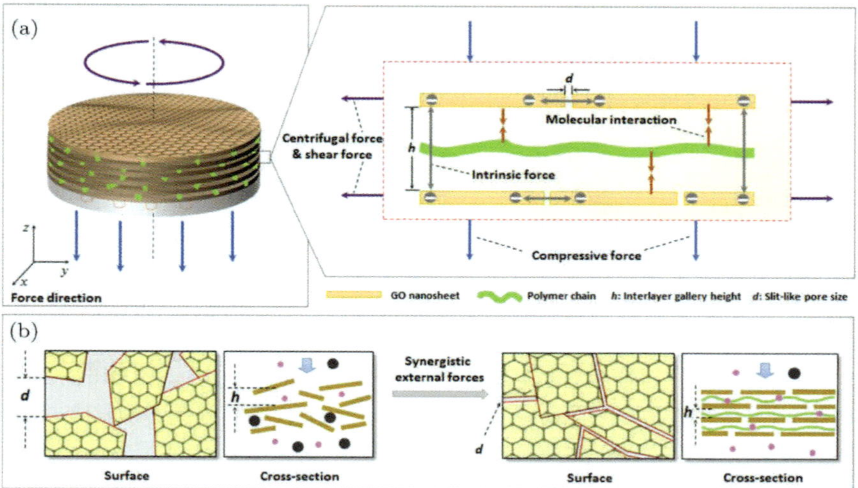

Fig. 10. Design and construction of GO channels. (a) External force-driven assembly approach for fabricating GO channels. It involves three-dimensional external forces in x, y and z axes. Enlarged schematic shows force analysis for one two-dimensional channel unit consisting of GO nanosheets and polymer chain. Three main types of forces are included: intrinsic force, "outer" external forces (compressive force, centrifugal force, and shear force) which are applied outside the two-dimensional channel unit, and "inner" external force (GO−polymer molecular interactions) which are applied inside the GO channel unit. (b) Hypothetical evolution of surface and cross-section of GO-assembled two-dimensional channels from intrinsic force-induced disordered structure (left) to highly ordered laminar structures (right) driven by introduced synergistic external forces.

Source: Reproduced with permission from Ref. 31. Copyright (2016) American Chemical Society.

The carboxyl groups on the edges of the graphene layer, on the one side, prevent GO aggregation caused by attractive van der Waals' force and hydrogen bonds between GO nanosheets. Thus, GO nanosheets were easily dispersed in water solution with GO concentration of 0.1–2 mg/mL. On the other side, the stronger repulsive forces would hinder close packing of GO nanosheets during self-assembly process.[30] That is supposed to give rise to local out-of-order accumulation and several large voids in the GO laminate. A schematic of this disorderly laminar structure driven by intrinsic forces of GO is given in Fig. 5(b). Here, the introduced external forces

(EFs) were classified into two types: "outer" EFs and "inner" EFs. The "outer" EFs refer to compressive, centrifugal, and shear forces, which are applied outside GO laminate; the "inner" EFs refer to molecular interactions that are applied inside the laminate. As shown in Fig. 10(a), the concept of three-dimensional "outer" EFs was demonstrated by employing vacuum-spin technique for GO assembly process, in which the vacuum suction provided EF along z axis while the centrifugal rotation supplied EFs along x and y axes. The negative effect of repulsive interactions in x, y, and z directions can be effectively suppressed by the introduced in-plane (along x and$y$ axes) centrifugal force and air shear force, and plane-to-plane (along z axis) compressive force. Consequently, the stacking of GO nanosheets begins to be ordered both in in-plane and plane-to-plane directions. The quasi-ordered laminar structure offers a prerequisite for controlling the slit-like pore sizes (d) and interlayer height (h) for selective transporting molecules. In order to further weaken the negative intrinsic repulsive electrostatic forces, meanwhile, "inner" EFs were added into the assembly by means of building up molecular interactions via step-by-step deposition of GO nanosheets and polyethylenimine (PEI) polymer molecules. Then, electrostatic interaction, covalent binding, and hydrogen bonding can be expected between GO and PEI. The synergic function of "outer" EFs and "inner" EFs would realize the highly ordered laminar structure (Fig. 10(b)).

Figure 11(a) shows typical digital photos of blank ceramic substrate and the as-prepared EFDA-GO membrane. By step-by-step deposition of GO nanosheets driven by EFs, the white disk with a diameter of 3 cm was entirely covered by a continuous brown GO layer. As displayed in Fig. 11(b), our home made ceramic substrate has a porous surface with average pore size of ~100 nm. The GO layer was uniformly distributed on the substrate surface, and no in-plane defect was observed by SEM. From the cross-sectional view (Fig. 11(c)), we can find a porous support layer corresponding to the ceramic substrate, providing sufficient mechanical strength while owning negligible transport resistance for gases. And a 1-μm thick well-stacked GO laminate layer is firmly adhered onto the porous substrate. Only driven by the "outer" EFs, GO nanosheets tend

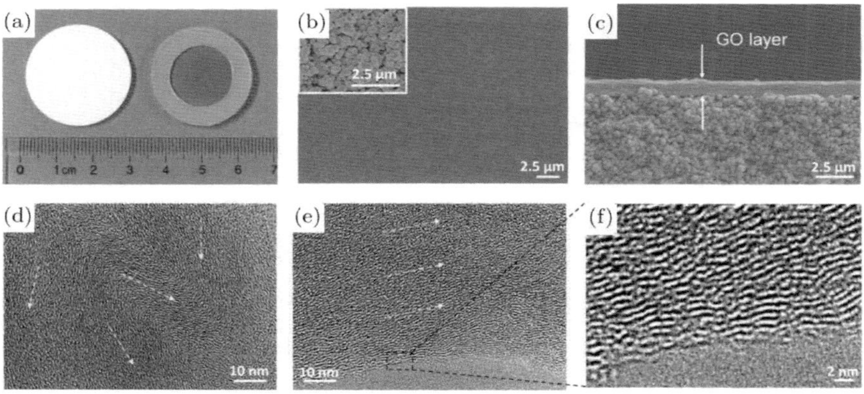

Fig. 11. (a) Digital photos of blank Al_2O_3 substrate and EFDA–GO membrane. The rubber washer was used for GO protection during the permeation test. (b) Surface and (c) cross-section SEM images of EFDA–GO membrane with blank Al_2O_3 substrate inserted; TEM images of (d) GO membrane prepared by "outer" EFs, (e) and (f) EFDA–GO membrane (PEI solution concentration of 0.25 wt.%). The yellow dashed arrows are eye-guiding lines indicating the orientations of GO.

Source: Reproduced with permission from Ref. 31. Copyright (2016) American Chemical Society.

to assemble in several different orientations, indicating a relatively random packing manner (Fig. 11(d)). This random packing could lead to tiny defects and disordered channel arrangement in the in-plane and plane-to-plane directions. On the contrary, after tuning by additional "inner" EFs (GO-polymer molecular interactions), GO nanosheets were assembled in almost the same direction to form a highly ordered laminar structure with sub-nanometer spaces among them (Figs. 11(e) and 11(f)). It implies formation of plenty of well-organized two-dimensional channels in the GO layer.

GO membrane prepared by external force driven assembly (EFDA–GO membrane) with highly ordered structure was applied for gas separation. Gas permeation measurement was first carried out by used hydrogen (H_2) as the small gas molecule, with kinetic diameter of 0.29 nm, carbon dioxide (CO_2) as the large gas molecule with kinetic diameter of 0.33 nm (only 0.04 nm larger than that of H_2). As shown in Fig. 12(a), the as-prepared EFDA–GO membranes

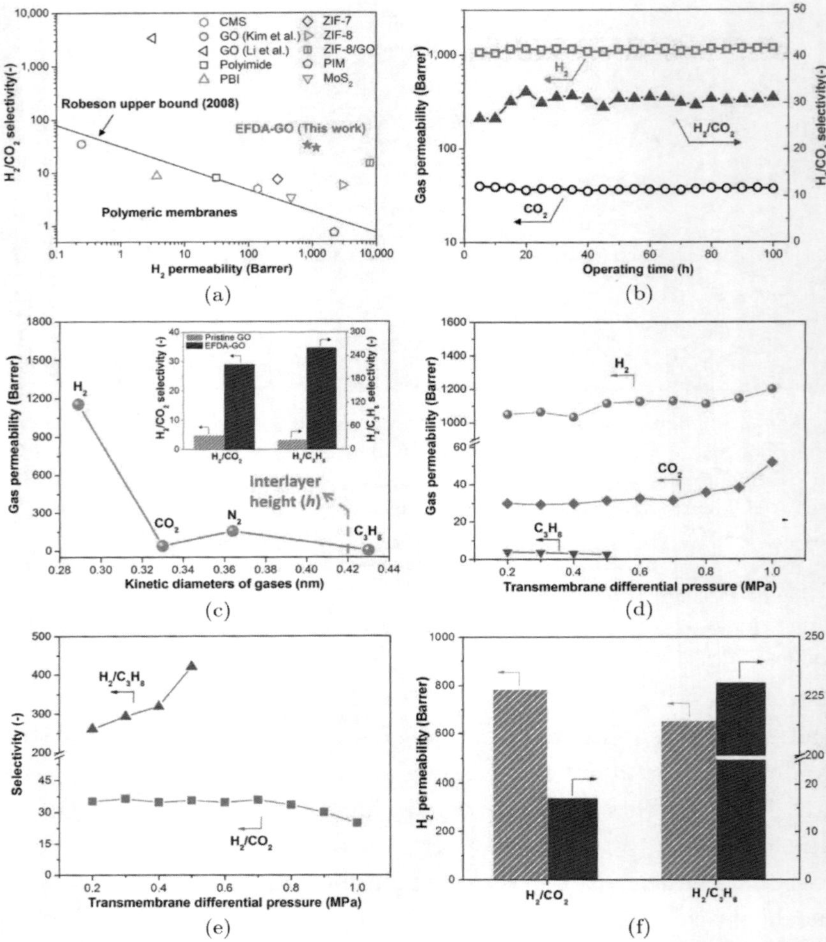

Fig. 12. (a) H_2–CO_2 separation performance of EFDA–GO membranes compared with state-of-the-art gas separation membranes. (b) H_2–CO_2 long-term operation test for EFDA–GO membrane. (c) Gas permeation behaviors of EFDA–GO membrane and the comparison of H_2–CO_2 and H_2–C_3H_8 selectivities with those of pristine GO membrane. The gas permeation tests were measured at 0.2 MPa and 25°C. (d) Gas permeability and (e) ideal selectivities for H_2–CO_2 and H_2–C_3H_8 under different transmembrane pressures. (f) Mixed gas separation tests of EFDA–GO membrane. Mixed gases (50:50 H_2–CO_2, 50:50 H_2–C_3H_8, vol.%) were used as the feed gases with total flux of 30 mL/min, while N_2 was selected as the sweep gas with flux of 30 mL/min.

Source: Reproduced with permission from Ref. 31. Copyright (2016) American Chemical Society.

exhibit extraordinary molecular-sieving properties for H_2–CO_2 with H_2 permeability of 840–1200 Barrer [1 Barrer = 10^{-10} cm^3 (STP) cm cm^{-2} s^{-1} $cmHg^{-1}$] and H_2–CO_2 selectivity of 29–33. This performance clearly surpasses the permeability–selectivity upper bound for polymeric membranes.[32] In particular, the EFDA–GO membranes showed 2–3 orders of magnitude higher H_2 permeability than that of commercialized polybenzimidazole (PBI) and polyimide (PI) membranes, accompanied with 3-fold increase of selectivity.[33] And our membranes are also competitive compared with emerging advanced membranes including polymers of intrinsic microporosity (PIMs),[34] carbon molecular sieves (CMSs),[35] MOFs,[36–38] and laminar GO and MoS_2 membranes.[27,28,39] The gas channels at sub-nanometer scale make our GO membrane with ultra fast and highly selective permeation of H_2 promising for practical applications which need effective high throughput.[40] They show great potential for industrial application especially for H_2 purification (methanol reforming process) and CO_2 capture. Furthermore, the membrane showed a steady excellent performance during 100 h continuous permeation test (Fig. 12(b)). After the long-term operation, no defects were found in the membrane. The result demonstrates that the GO composite membranes with highly interlocked structure are able to deal with flow of high volumetric gases and streams. The gas-sieving property of the membrane was further probed by permeation of larger gas molecules: N_2 (0.364 nm) and C_3H_8 (0.43 nm). In particular, C_3H_8 permeability is significantly low ($<$4 Barrer) and H_2–C_3H_8 selectivity is as high as 260. It is worth noting that the H_2–CO_2 and H_2–C_3H_8 selectivities of EFDA–GO membrane are 6 and 11 times higher than those of the pristine GO membrane, respectively (inset in Fig. 12(c)).

The effect of operation pressure on the gas separation performance was also studied. It can be found from Figs. 12(d)–12(f) that our EFDA–GO membrane exhibited stable H_2–CO_2 separation performance with the feed pressure from 0.2 to 0.7 MPa. Nevertheless, when the feed pressure went larger than 0.7 MPa, CO_2 permeability began to increase while the H_2–CO_2 selectivity decreased gradually. It is known that GO shows high affinity with CO_2 molecules due to the abundant oxygen-containing groups. The permeability increase and selectivity decrease under high pressure might be caused by plasticization which is usually found in gas separation membranes when penetrant molecules can form strong interaction with membrane materials. For C_3H_8, it can be found that the gas permeability decreased with pressure, and the selectivity of H_2–C_3H_8 increased significantly. This can be owing to the condensation of C_3H_8 molecules at high pressure. Mixed gas separations of H_2–CO_2 and H_2–C_3H_8 were conducted as well. As shown in Fig. 12(f), the as-prepared EFDA–GO membrane still showed good molecular sieving performance, despite a reasonably declined selectivity and permeability compared with the pure gas permeation results, which is probably due to the competitive adsorption of gas mixtures.

Moreover, it can be demonstrated that gas sieving behaviors through the EFDA–GO membrane could be readily modulated by controlling the parameters of external force driven approach. The effects of GO concentration and lateral size, polymer concentration, and deposition numbers on gas permeability and selectivity are shown in Fig. 13. It was found that either high GO concentration or large GO nanosheets benefits high GO stacking density that suppresses non-selective defects while it would increase the gas permeation resistance. A suitable amount of polymer could enhance a firm and ordered laminar structure, thus improving gas-sieving selectivity. Additionally, the thickness of GO laminate can be tailored by varying the number of depositions. Ten depositions are sufficient for fabricating a thin and dense GO membrane.

2.4. *Ceramic Hollow Fiber-supported GO Composite Membrane*

As discussed above, the ceramic disk-supported GO composite membranes exhibited excellent gas separation performance. In order to push the GO composite membranes into practical application, our group employed an industrially preferred membrane configuration — hollow fiber to fabricate GO composite membrane. Our group developed a convenient and rapid vacuum suction method to prepare GO membranes on the ceramic hollow fiber, which exhibited excellent pervaporation separation performance.[41] The special configuration makes GO membranes very easy to be scaled up. In addition, GO

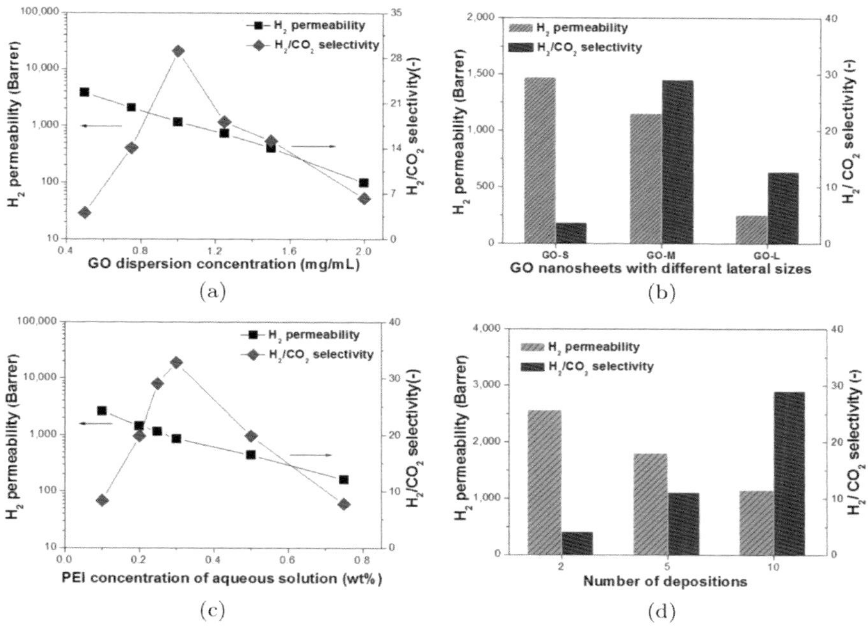

Fig. 13. Effect of (a) GO dispersion concentration, (b) GO lateral size, (c) PEI concentration of aqueous solution and, (d) number of depositions on the gas permeation properties of EFDA–GO membranes.

Source: Reproduced with permission from Ref. 31. Copyright (2016) American Chemical Society.

nanosheets stacked to form a cylindrical shell around the ceramic hollow fiber, keeping it more stable than flat GO membranes. The ceramic hollow fiber-supported GO composite gas separation membranes were prepared by the vacuum-suction method. One side of the ceramic hollow fiber was sealed, and the other side was connected to a vacuum pump. The whole hollow fiber was immersed in the GO aqueous solution. With driving pressure, GO flakes were stacked on the surface in order.

By controlling the preparation time, GO membranes with different thicknesses were achieved. Figure 14 (top) shows that the thickness of the GO membranes increases with increase in operation time. However, there is a nonlinearity between the membrane thickness and operation time. The reason is that the resistance was reinforced

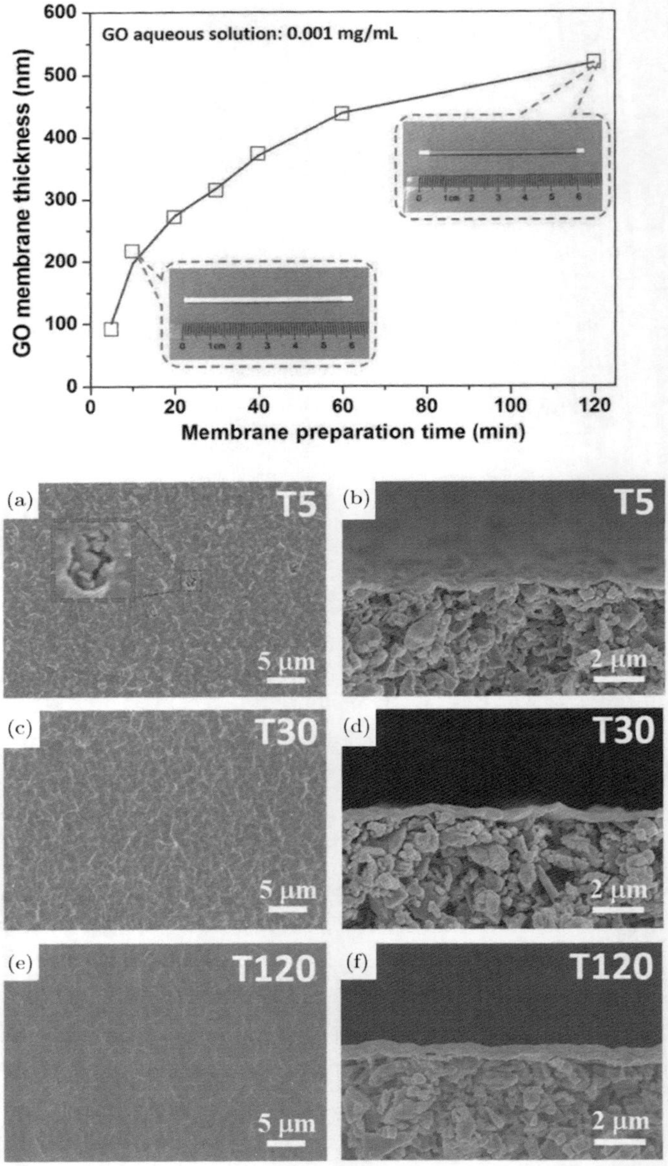

Fig. 14. Top: the thickness of the ceramic hollow fiber-supported GO membrane as a function of the membrane-preparation time. Bottom: SEM images of the GO membrane prepared at operation times of (a) and (d) 5 min, (b) and (e) 30 min, and (c) and (f) 120 min; (a)–(c) membrane surface; (e) and (f) membrane cross-section.

by increasing the membrane thickness, which inhibited more GO sheets to stack on the surface. The inset in Fig. 14 (top) presents two typical GO membranes prepared at 10 and 120 min, respectively. Obviously, the membrane colour becomes darker when the thickness increases. Additionally, both of them show a continuous and uniform layer, indicating that the vacuum suction method is very effective to fabricate ceramic hollow fiber-supported GO composite membranes. Figure 14 (bottom) shows the microstructures of three typical GO membranes which were prepared at 5, 30, and 120 min, respectively (the corresponding membranes are marked as T5, T30, and T120, respectively). When the time is too short, there are some pinholes on the surface of the GO membrane (Fig. 14(a)). The inset of Fig. 14(a) gives a clearer and enlarged image. From Figs. 14(b) and 14(c), continuous and complete membranes can be observed. And the surface becomes smoother with the growth of thickness. Figs. 14(d)–14(f) exhibit the corresponding cross-section images of T5, T30, and T120 membranes, respectively. All of them attach well with the ceramic hollow fiber, which may be attributed to the hydrogen bond between the oxygen-containing functional groups of the GO membrane and the hydroxy group on the surface of the ceramic hollow fiber.

Gas separation measurements of single H_2 and CH_4 were utilized to examine the membrane's quality. As shown in Fig. 15(a), the gas permeances of H_2 and CH_4 decline together, when the membrane thickness increases. But the H_2–CH_4 selectivity of T30 GO membrane gets a peak, and meanwhile the membrane still has a good H_2 permeance. Considering permeance and selectivity, a 30-min operation time was selected as the most optimized condition to prepare the ceramic hollow fiber-supported GO membranes.

Single gas permeations of T30 GO membrane, including H_2, CO_2, O_2, N_2, and CH_4, were tested in detail. As shown in Fig. 15(b), the permeances of these small gas molecules decrease in the order $H_2 > CH_4 > N_2 > O_2 > CO_2$, with increase in the molecular weight. The corresponding ideal selectivity of H_2–CO_2, H_2–O_2, H_2–N_2 and H_2–CH_4 are 14.98, 7.48, 7.21, and 6.37, respectively. In contrast to other gases, CO_2 shows a sharp decrease in the permeance and the highest hydrogen selectivity, which can be attributed to the chemical

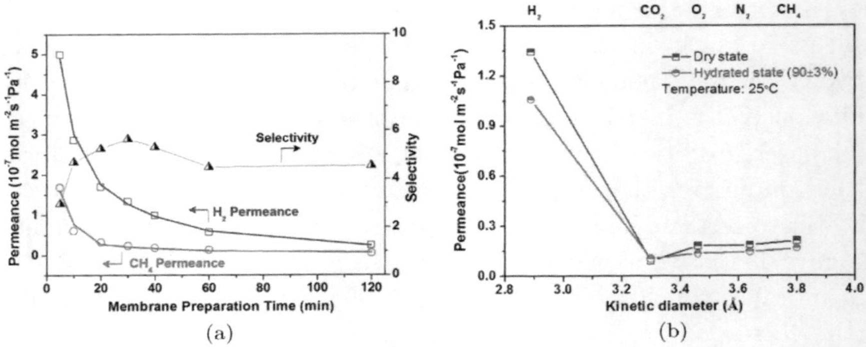

Fig. 15. (a) H_2 and CH_4 permeance and corresponding H_2–CH_4 selectivity of GO membrane as a function of the membrane preparation time; (b) single gas permeance (H_2, CO_2, O_2, N_2, and CH_4) with dry feed or hydrated feed as a function of the molecular kinetic diameter.

nature of GO material. As we know, there are numerous carboxylic acid groups distributed at the edge of the GO flakes. A strong interplay between these polar groups and C–O bonds in the non-polar CO_2 molecules would happen. For CO_2 transfer, CO_2 as a Lewis acid or a Lewis base participates in hydrogen bonding, which inhibits it from getting transferred within the stacked GO structure.[28] A similar phenomenon was also found in the porous metal-organic framework (MOF) ZIF-78 membrane.[42] The two polar functional groups, $-NO_2$, in ZIF-78 structure made it exhibit the highest affinity for CO_2, which blocked the diffusion of CO_2 molecules through the ZIF-78 channels.

The influence of humidity on the gas transfer was further investigated. Most gas permeances (except CO_2) decrease when humid steam is added in the feed, because the water molecules in the GO channels limits the gas transfer. However, a slight increase in CO_2 is observed. This result further confirms that CO_2 molecules have special interaction with the carboxylic acid groups in GO. Because of the growth of CO_2 permeance, the corresponding H_2–CO_2 ideal selectivity declines obviously under the wet condition, indicating that wet gas has disadvantages for hydrogen recovery from H_2–CO_2 mixture.

Single-gas measurements only can give the ideal selectivity, because the gas molecular transfer in this process is relatively

Table 1. Single gases and binary mixtures' separation performances of ceramic hollow fiber-supported-GO composite membrane in dry state.

	Permeance ($\times 10^{-7}$ mol·m^{-2}s^{-1}Pa^{-1})	H_2 ideal selectivity	H_2 separation factor
Single gas			
H_2	1.34	—	
CO_2	0.09	14.98	
O_2	0.18	7.48	
N_2	0.19	7.21	
CH_4	0.21	6.37	
Mixed gas			
H_2 (50 vol.% in CO_2)	1.20		10.22
CO_2 (50 vol.% in H_2)	0.11		
H_2 (50 vol.% in O_2)	1.04		4.69
O_2 (50 vol.% in H_2)	0.20		
H_2 (50 vol.% in N_2)	1.03		4.43
N_2 (50 vol.% in H_2)	0.21		
H_2 (50 vol.% in CH_4)	1.24		4.67
CH_4 (50 vol.% in H_2)	0.24		

independent. However, interplays between different gases are generally not negligible, which may result in a prominent deviation of the mixture-separation factor from the ideal selectivity. Table 1 lists all the single gases and binary mixtures' separation performances. Like single-gas test, the ceramic hollow fiber-supported GO membrane shows a same regular for mixtures. But the corresponding permeance and separation factor decrease a bit because of the competitive adsorption and diffusion between different gas molecules. This phenomenon was often observed in zeolite membranes. Generally speaking, although the H_2–CO_2 separation factor drops, it still reveals a useful separation ability (10.2) for practical hydrogen-recovery application.

2.5. *Ceramic-supported GO Mixed Matrix Membranes*

Generally, the preparation of GO membranes has been mostly based on aqueous ambience. GO nanosheets are dispersed in water and then processed using filtration, dip coating, or spin coating; after evaporating the water, the GO membrane is finally formed. However, there are still many challenges these GO membranes have to confront. On one hand, free standing GO membranes without substrates or brittle GO membranes with no reliable modules are difficult for practical application. On the other hand, the assembly of GO into membranes driven by external forces (e.g., pressure, centrifugal force) generally face the difficulty of precise and reproducible tuning of the membrane-processing parameters. The repulsive electrostatic interactions caused by negatively charged carboxyl groups are easily prone to generate inevitable non-selective defects in the GO membrane.

Therefore, our group developed a mixed matrix membranes (MMMs)-based methodology for fabricating GO-based membranes that involves GO assembly in polymeric ambience (Fig. 16(a)).[43] The

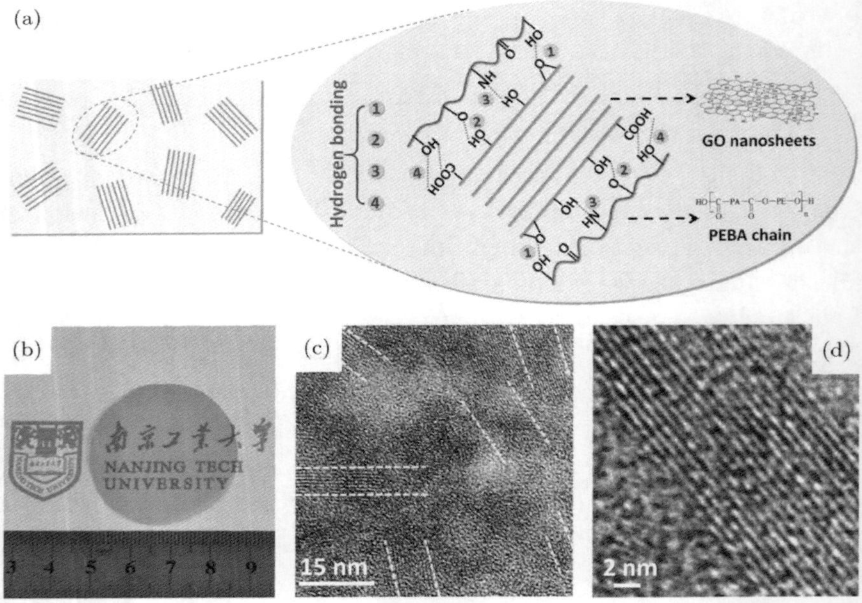

Fig. 16. (a) Schematic of the assembly of GO nanosheets in polymeric ambience based on hydrogen bonds formed between different groups on GO and the PEBA chain, (b) digital photographs of the membrane with 0.1 wt.% GO, (c) overview, and (d) amplified TEM image of the cross-section of GO-1 membrane.

Source: Reproduced with permission from Ref. 43. Copyright (2015) John Wiley & Sons.

key concept of this methodology is the construction and utilization of favorable molecular interactions between GO and the polymer to stimulate the stacking of GO nanosheets into well-defined GO laminates, in which the interlayer spaces could provide molecular-sieving channels. GO nanosheets contain many hydroxyl, carbonyl, and carboxyl groups that are all highly polarized, with oxygen atoms being the negative center. This chemical characteristic is beneficial for the generation of hydrogen bonding, especially in the case of a high O/C ratio on GO. Thus PEBA, containing –N–H–, H–N–C=O, and O–C=O groups, and an ethanol–water-mixed solvent were selected to disperse GO nanosheets in the polymeric ambience. Membranes with GO concentrations of 0.05, 0.075, and 0.1 wt.% were fabricated and denoted as GO-05, GO-075, and GO-1, respectively.

The morphology of the GO–PEBA membranes was confirmed by TEM observation of the membrane cross-section, shown in Fig. 16(c). The laminar GO nanosheets are surrounded by the amorphous PEBA polymeric domains, and their thickness ranges from 6 to 15 nm. The regular laminar structures are clearly seen in the amplified image shown in Fig. 16(d), indicating that the *d*-spacing (d) of the GO laminate given by TEM is ca. 0.7 nm, which is smaller than that of the reported GO gas separation membrane derived from aqueous ambience ($d = 0.79$ nm).[27] This is mainly attributed to the confinement effects of the polymeric chains and hydrogen bonding between GO and PEBA. The reduced size of the GO *d*-spacing would lead to enhanced molecular sieving properties for gas separation. Considering that the *d*-spacing also includes the graphene thickness (0.35nm), the interlayer spacing can be calculated to approximately 0.35 nm, which can be expected to be empty space for molecular transport. The size of the space is in the range of the molecular kinetic diameters of industrial gases such as CH_4 (0.38 nm), N_2 (0.36 nm), CO_2 (0.33 nm), and H_2 (0.29 nm). Thus, the membrane with GO laminates would provide selective transport channels for gas separation.

The fast selective gas channels of the GO laminates were studied by measuring the gas permeation properties through the as-prepared membranes. As shown in Fig. 17, the membranes exhibit fastest transport for CO_2 molecules and very low permeation of other gases,

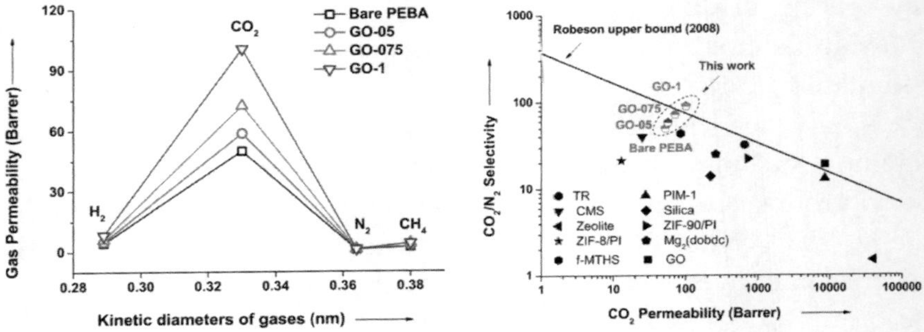

(a) (b)

Fig. 17. Gas permeation through the membranes with different GO concentrations: (a) permeabilities of H_2, CO_2, N_2, and CH_4; (b) CO_2–N_2 separation performance of PEBA, GO-05, GO-075, and GO-1 membranes as compared with representative membranes reported in the literature, gas permeation tests were performed at 0.3 MPa and 25°C.

Source: Reproduced with permission from Ref. 43. Copyright (2015) John Wiley & Sons.

with a gas permeability order of $CO_2 > H_2 > CH_4 > N_2$. Moreover, by increasing the number of GO nanosheets, both the CO_2 permeability and selectivity of the membrane are significantly enhanced, breaking the permeability–selectivity trade-off relation in polymeric membranes. The gas transport behavior means that fast selective CO_2 channels were provided by the regular laminar GO structures in these membranes. The CO_2 permeability reaches 100 Barrer, and CO_2–N_2 selectivity is 91. As illustrated in Fig. 17(b), the separation performance of GO-1 for CO_2–N_2 transcends the upper bound for the reported polymeric membranes, including thermally rearranged (TR) polymers, polymers of intrinsic microporosity (PIM), and inorganic membranes such as carbon, silica, and zeolite, as well as MMMs. Moreover, a continuous permeation of CO_2 and N_2 was made to last for 6000 min; it was then voluntarily stopped, and no defect was found in the GO-1, indicating that these membranes are very stable for practical application. The unprecedented preferential CO_2 permeation could be due to the unique characteristics of the GO laminates.

Generally, molecular permeation through gas separation membranes is determined by two crucial properties: solubility and diffusivity. Strong specific affinity and molecular sieving effects, in addition to straight diffusion pathways, are beneficial for making highly permeable and selective membranes. The overall size distribution of cavities in the as-prepared membranes was analyzed by using the positron annihilation lifetime spectroscopy technique. In general, a longer *o*-positronium (*o*-Ps) lifetime suggests larger cavity sizes. As displayed in Fig. 18(a), the lifetime distribution is narrowed in GO-1, accompanied by a reduction in the *o*-Ps lifetime, and an improvement of the *o*-Ps intensity, I_3, which could exactly enhance the molecular sieving ability and cavity numbers for diffusion. The cavity numbers and/or size of the bare PEBA membrane are thought to decrease owing to the abundant hydrogen bonds between GO and PEBA. This can be inferred from the gradual improvement of the glass transition temperature of PEBA with increasing GO amount. Thus, the GO laminates are considered to create a large number of cavities that are expected to be more numerous than the ones corresponding to the increased *o*-Ps intensity (table inserted in Fig. 18(a)).

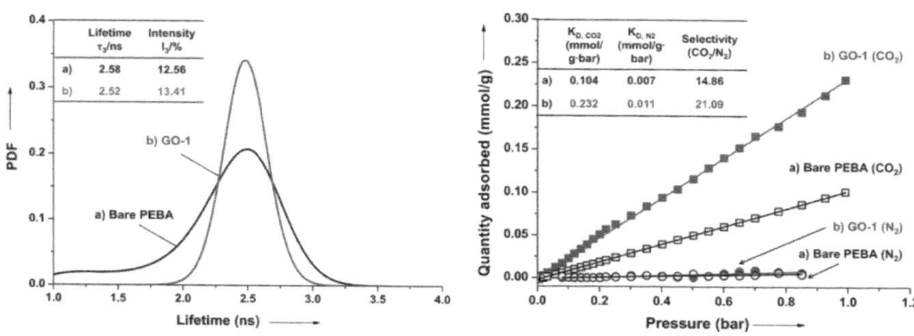

(a) (b)

Fig. 18. (a) Positron annihilation lifetime spectra of bare PEBA and GO-1 membrane; (b) pure-component CO_2 and N_2 adsorptions on PEBA and GO-1 membrane, both of which employ single-site Langmuir models; K_D is Henry's constant.

Source: Reproduced with permission from Ref. 43. Copyright (2015) John Wiley & Sons.

Moreover, TEM observation shows that the interlayer spacing of the GO laminates is ca. 0.35 nm, which allows a cutoff in permeation for molecules with kinetic diameters greater than this value. The channels exhibit a molecule-sieving effect, and thus gas could diffuse selectively through the membrane.

In addition, the specific adsorption of CO_2 in the GO laminates further accelerates the preferential CO_2 transport through the membrane. The gas adsorption behaviors on GO-1 was measured, and a typical result of CO_2–N_2 adsorption is shown in Fig. 18(b). Because of the favorable interactions between the polar groups such as –COOH, –OH on GO, and the polar individual C–O bonds on CO_2, the GO laminates in the membrane contribute to 2.2 times higher CO_2 adsorption than what could be obtained with the bare PEBA membrane. Meanwhile, the CO_2–N_2 adsorption selectivity also increased by 42%, which was obtained through the table in Fig. 18(b). Although H_2 molecules are supposed to diffuse much faster than CO_2 due to their smaller kinetic diameter, the permeability of CO_2 is 12 times higher than that of H_2. This is owing to the high adsorption selectivity of CO_2–H_2 resulting from the GO laminates. Therefore, it's speculated that by exhibiting preferential CO_2 adsorption and diffusion, the laminar GO nanosheets in the as-prepared membrane create fast selective transport channels for CO_2 molecules.

The fabrication feasibility and structural stability are two significant criteria for pushing GO-based membranes into practical application. GO polymeric dispersion was shown to be a superior approach in that membranes can be easily coated on various substrates of different configurations and materials, resulting in outstanding processability. The preparation of two popular industrial composite membranes with polymeric flat and inorganic hollow fiber substrates was demonstrated. FESEM images of the membrane cross-sections (Fig. 19) show that uniform active layers are firmly adhered onto the porous substrates, with an average thickness of ca. 5 μm. The separation performance of the hollow fiber-membrane with the GO concentration of 0.1 wt.% is comparable to that of the flat one (GO-1).

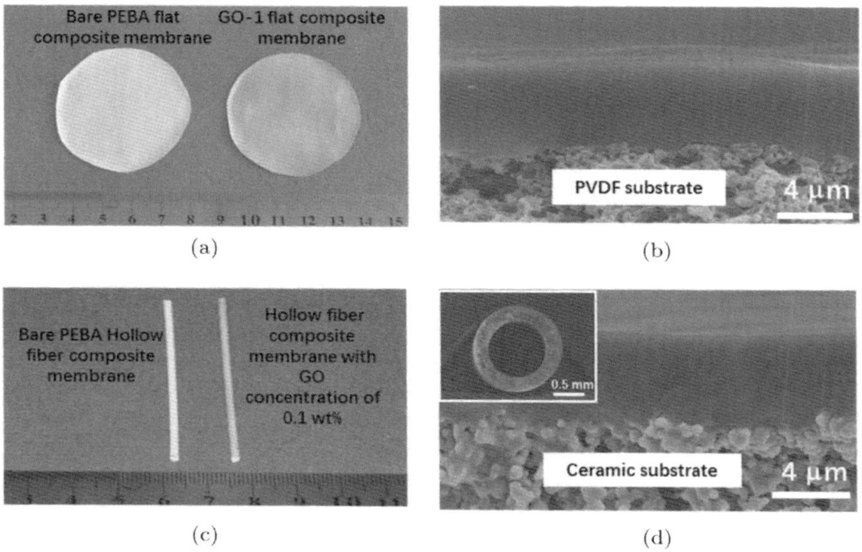

Fig. 19. Digital photographs of (a) bare PEBA (left, white) and GO-1 (right, light yellow) flat composite membranes using PVDF substrate; (c) bare PEBA (left, white) and (right, light yellow) hollow-fiber composite membrane with GO concentration of 0.1 wt.% using ceramic substrate. FESEM cross-section images of (b) GO-1 flat composite membrane and (d) hollow fiber composite membrane with GO concentration of 0.1 wt.% (inset: overview of the hollow fiber).

Source: Reproduced with permission from Ref. 43. Copyright (2015) John Wiley & Sons.

3. Conclusion and Outlook

This chapter summarizes recent advances in design and preparation of ceramic-supported organic composite membranes for gas separation. It is demonstrated that the integration of organic separation layer and ceramic support layer can achieve favorable nanostructures. Ceramic-supported polymeric composite membranes exhibited high permeance and good selectivity for carbon dioxide and air separation. The introduction of 2D graphene oxide highly improved the gas permeability and selectivity, showing great potential for hydrogen purification and carbon dioxide capture.

Current challenges for ceramic-supported organic composite membranes include fabricating ultra thin separation layer, optimizing organic–inorganic interfacial morphology, transport mechanism, long-term stability, etc. Novel membrane materials and/or nanostructures are required to extend the application spectrum of these membranes into other molecular separation process such as water desalination and ion exchange processes. The multi discipline (chemistry, materials science, physical and chemical engineering) study is necessary to deepen the understanding of ceramic-supported organic composite membranes. Meanwhile, novel characterization techniques and molecular simulations will provide insights into the nanostructures and transport mechanism of this emerging membrane.

Acknowledgments

This work was financially supported by the National Basic Research Program of China (No. 2009CB623406), the National Natural Science Foundation of China (Nos. 21476107, 21490585, 21406107), the Innovative Research Team Program of the Ministry of Education of China (No. IRT13070), the Natural Science Foundation of Jiangsu Province (No. BK20140930), Top-notch Academic Programs Project (TAPP) of Jiangsu' higher education institutions.

References

1. Liu, G., Wei, W., Jin, W., Xu, N. Polymer/ceramic composite membranes and their application in pervaporation process. *Chinese J. Chem. Eng.* 2012, 20(1), 62–70.
2. Xiangli, F., Chen, Y., Jin, W., Xu, N. Polydimethylsiloxane (PDMS)/ceramic composite membrane with high flux for pervaporation of ethanol–water mixtures. *Ind. Eng. Chem. Res.* 2007, 46(7), 2224–2230.
3. Xiangli, F., Wei, W., Chen, Y., Jin, W., Xu, N. Optimization of preparation conditions for polydimethylsiloxane (PDMS)/ceramic composite pervaporation membranes using response surface methodology. *J. Memb. Sci.* 2008, 311(1–2), 23–33.
4. Wei, W., Xia, S., Liu, G., Dong, X., Jin, W., Xu, N. Effects of polydimethylsiloxane (PDMS) molecular weight on performance of PDMS/ceramic composite membranes. *J. Memb. Sci.* 2011, 375(1–2), 334–344.
5. Zhu, Y., Xia, S., Liu, G., Jin, W. Preparation of ceramic-supported poly(vinyl alcohol)–chitosan composite membranes and their applications in pervaporation dehydration of organic/water mixtures. *J. Memb. Sci.* 2010, 349(1–2), 341–348.

6. Chen, Y., Xiangli, F., Jin, W., Xu, N. Organic–inorganic composite pervaporation membranes prepared by self-assembly of polyelectrolyte multilayers on macroporous ceramic supports. *J. Memb. Sci.* 2007, 302(1–2), 78–86.
7. Liu, G., Wei, W., Wu, H., Dong, X., Jiang, M., Jin, W. Pervaporation performance of PDMS/ceramic composite membrane in acetone butanol ethanol (ABE) fermentation–PV coupled process. *J. Memb. Sci.* 2011, 373(1–2), 121–129.
8. Xu, R., Liu, G., Dong, X., Wanqin, Jin. Pervaporation separation of n-octane/thiophene mixtures using polydimethylsiloxane/ceramic composite membranes. *Desalination.* 2010, 258(1–3), 106–111.
9. Xia, S., Dong, X., Zhu, Y., Wei, W., Xiangli, F., Jin, W. Dehydration of ethyl acetate–water mixtures using PVA/ceramic composite pervaporation membrane. *Sep. Purif. Technol.* 2011, 77(1), 53–59.
10. Xia, S., Wei, W., Liu, G., Dong, X., Jin, W. Pervaporation properties of polyvinyl alcohol/ceramic composite membrane for separation of ethyl acetate/ethanol/water ternary mixtures. *Korean J. Chem. Eng.* 2012, 29(2), 228–234.
11. Liu, S., Liu, G., Wei, W., Xiangli, F., Jin, W. Ceramic supported PDMS and PEGDA composite membranes for CO2 separation. *Chinese J. Chem. Eng.* 2013, 21(4), 348–356.
12. Cheng, J., Hu, L., Ji, C., Zhou, J., Cen, K. Porous ceramic hollow fiber-supported Pebax/PEGDME composite membrane for CO2 separation from biohythane. *RSC Advances.* 2015, 5(74), 60453–60459.
13. Dong, Z., Liu, G., Liu, S., Liu, Z., Jin, W. High performance ceramic hollow fiber supported PDMS composite pervaporation membrane for bio-butanol recovery. *J. Memb. Sci.* 2014, 450, 38–47.
14. Pian, C., Shen, J., Liu, G., Liu, Z., Jin, W. Ceramic hollow fiber-supported PDMS composite membranes for oxygen enrichment from air. *Asia-Pac. J. Chem. Eng.* 2016, 11(3), 460–466.
15. Novoselov, K. S., Geim, A. K., Morozov, S. V., Jiang, D., Zhang, Y., Dubonos, S. V., Grigorieva, I. V., Firsov, A. A. Electric field effect in atomically thin carbon films. *Science.* 2004, 306(5696), 666–669.
16. Liu, G., Jin, W., Xu, N. Graphene-based membranes. *Chem. Soc. Rev.* 2015.
17. Mi, B. Graphene oxide membranes for ionic and molecular sieving. *Science.* 2014, 343(6172), 740–742.
18. Nair, R. R., Wu, H. A., Jayaram, P. N., Grigorieva, I. V., Geim, A. K. Unimpeded permeation of water through helium-leak–tight graphene-based membranes. *Science.* 2012, 335(6067), 442–444.
19. Han, Y., Xu, Z., Gao, C. Ultrathin graphene nanofiltration membrane for water purification. *Adv. Funct. Mater.* 2013, 23(29), 3693–3700.
20. Huang, H., Song, Z., Wei, N., Shi, L., Mao, Y., Ying, Y., Sun, L., Xu, Z., Peng, X. Ultrafast viscous water flow through nanostrand-channelled graphene oxide membranes. *Nat. Commun.* 2013, 4.
21. Joshi, R. K., Carbone, P., Wang, F. C., Kravets, V. G., Su, Y., Grigorieva, I. V., Wu, H. A., Geim, A. K., Nair, R. R. Precise and ultrafast molecular sieving through graphene oxide membranes. *Science.* 2014, 343(6172), 752–754.

22. Goh, K., Jiang, W., Karahan, H. E., Zhai, S., Wei, L., Yu, D., Fane, A. G., Wang, R., Chen, Y. All-carbon nanoarchitectures as high-performance separation membranes with superior stability. *Adv. Funct. Mater.* 2015, 25(47), 7348–7359.
23. Hu, M., Mi, B. Enabling graphene oxide nanosheets as water separation membranes. *Environ. Sci. Technol.* 2013, 47(8), 3715–3723.
24. Huang, K., Liu, G., Lou, Y., Dong, Z., Shen, J., Jin, W. A graphene oxide membrane with highly selective molecular separation of aqueous organic solution. *Angew. Chem. Int. Edit.* 2014, 53(27), 6929–6932.
25. Hung, W.-S., Tsou, C.-H., De Guzman, M., An, Q.-F., Liu, Y.-L., Zhang, Y.-M., Hu, C.-C., Lee, K.-R., Lai, J.-Y. Cross-linking with diamine monomers to prepare composite graphene oxide-framework membranes with varying d-spacing. *Chem. Mater.* 2014, 26(9), 2983–2990.
26. Tang, Y. P., Paul, D. R., Chung, T. S., Free-standing graphene oxide thin films assembled by a pressurized ultrafiltration method for dehydration of ethanol. *J. Memb. Sci.* 2014, 458(0), 199–208.
27. Li, H., Song, Z., Zhang, X., Huang, Y., Li, S., Mao, Y., Ploehn, H. J., Bao, Y., Yu, M. Ultrathin, molecular-sieving graphene oxide membranes for selective hydrogen separation. *Science.* 2013, 342(6154), 95–98.
28. Kim, H. W., Yoon, H. W., Yoon, S.-M., Yoo, B. M., Ahn, B. K., Cho, Y. H., Shin, H. J., Yang, H., Paik, U., Kwon, S., Choi, J.-Y., Park, H. B. Selective gas transport through few-layered graphene and graphene oxide membranes. *Science.* 2013, 342(6154), 91–95.
29. Liu, G., Jin, W., Xu, N. Two-dimensional-material membranes: a new family of high-performance separation membranes. *Angew. Chem. Int. Edit.* **2016**.
30. Chen, C., Yang, Q.-H., Yang, Y., Lv, W., Wen, Y., Hou, P.-X., Wang, M., Cheng, H.-M. Self-assembled free-standing graphite oxide membrane. *Adv. Mater.* 2009, 21(29), 3007–3011.
31. Shen, J., Liu, G., Huang, K., Chu, Z., Jin, W., Xu, N. Subnanometer two-dimensional graphene oxide channels for ultrafast gas sieving. *ACS Nano.* 2016, 10(3), 3398–3409.
32. Robeson, L. M. The upper bound revisited. *J. Memb. Sci.* 2008, 320(1–2), 390–400.
33. Yang, T., Xiao, Y., Chung, T.-S. Poly-/metal-benzimidazole nano-composite membranes for hydrogen purification. *Energ. Environ. Sci.* 2011, 4(10), 4171–4180.
34. Carta, M., Malpass-Evans, R., Croad, M., Rogan, Y., Jansen, J. C., Bernardo, P., Bazzarelli, F., McKeown, N. B. An efficient polymer molecular sieve for membrane gas separations. *Science.* 2013, 339(6117), 303–307.
35. Shiflett, M. B., Foley, H. C. Ultrasonic deposition of high-selectivity nanoporous carbon membranes. *Science.* 1999, 285(5435), 1902–1905.
36. Zhang, X., Liu, Y., Li, S., Kong, L., Liu, H., Li, Y., Han, W., Yeung, K. L., Zhu, W., Yang, W., Qiu, J. New membrane architecture with high performance: ZIF-8 membrane supported on vertically aligned ZnO nanorods for gas permeation and separation. *Chem. Mater.* 2014, 26(5), 1975–1981.

37. Li, Y.-S., Liang, F.-Y., Bux, H., Feldhoff, A., Yang, W.-S., Caro, J. Molecular sieve membrane: Supported metal–organic framework with high hydrogen selectivity. *Angew. Chem.* 2010, 122(3), 558–561.
38. Huang, A., Liu, Q., Wang, N., Zhu, Y., Caro, J. Bicontinuous zeolitic imidazolate framework ZIF-8@GO membrane with enhanced hydrogen selectivity. *J. Am. Chem. Soc.* 2014, 136(42), 14686–14689.
39. Wang, D., Wang, Z., Wang, L., Hu, L., Jin, J. Ultrathin membranes of single-layered MoS2 nanosheets for high-permeance hydrogen separation. *Nanoscale.* 2015, 7(42), 17649–17652.
40. Gin, D. L., Noble, R. D. Designing the next generation of chemical separation membranes. *Science.* 2011, 332(6030), 674–676.
41. Huang, K., Liu, G., Lou, Y., Dong, Z., Shen, J., Jin, W. A graphene oxide membrane with highly selective molecular separation of aqueous organic solution. *Angew. Chem. Int. Ed. Engl.* 2014, 53(27), 6929–6932.
42. Dong, X., Huang, K., Liu, S., Ren, R., Jin, W., Lin, Y. S. Synthesis of zeolitic imidazolate framework-78 molecular-sieve membrane: defect formation and elimination. *J. Mater. Chem.* 2012, 22(36), 19222.
43. Shen, J., Liu, G., Huang, K., Jin, W., Lee, K.-R., Xu, N. Membranes with fast and selective gas-transport channels of laminar graphene oxide for efficient CO_2 capture. *Angew. Chem. Int. Edit.* 2015, 54(2), 578–582.

Chapter 3

Molecular Modeling of MOF Membranes for Gas Separations

Ilknur Erucar* and Seda Keskin†,‡

*Department of Natural and Mathematical Sciences, Faculty of Engineering, Ozyegin University, Cekmekoy, 34794, Istanbul, Turkey

†Department of Chemical and Biological Engineering, Koc University, Rumelifeneri Yolu, Sariyer, 34450, Istanbul, Turkey

‡skeskin@ku.edu.tr

Metal organic frameworks (MOFs) have received significant attention as membrane materials in the last decade. MOFs typically have high porosities, tunable pore sizes, large surface areas, reasonable thermal and mechanical stabilities. These properties make them ideal membrane materials for gas separations. Considering the very large number of available MOF structures, it is not practically possible to test membrane-based gas-separation performance of MOFs using purely experimental means. Molecular modeling methods are highly useful to complement experimental studies by providing insights into the gas-separation mechanisms of MOFs. Molecular simulation studies have been recently used to screen large numbers of MOFs for gas separation applications to identify the most promising MOF membranes. In this way, experimental efforts, time, and resources can be directed to the materials with the highest separation performance. The aim of this chapter is to review the recent advances in molecular modeling of MOF membranes for gas-separations and address the current opportunities and challenges in using molecular simulations for MOF membranes.

1. Introduction

Developing energy-efficient gas separation systems has gained significant interest due to the economic and environmental concerns. Membrane-based gas separations are widely used in many industrial processes such as air separation (N_2–O_2), hydrogen recovery (H_2–N_2, H_2–CO, H_2–CO_2, H_2–CH_4), natural-gas purification (CO_2–CH_4) and flue gas separation (CO_2/N_2) due to their lower energy requirements and lower operating costs compared to conventional gas-separation methods such as cryogenic distillation, condensation, adsorption, and absorption.[1] It was recently discussed that membrane-based gas separation market has grown into a $150 million/year business, and it is still growing.[2] Permeability and selectivity are the two important properties for a membrane-based gas separation process. Permeability is defined as the rate of fluid flow per unit area of the membrane per unit driving force such as concentration, pressure, or electrical potential. A membrane with high gas permeability requires less surface area hence smaller capital cost. Selectivity is simply defined as the ratio of permeabilities of the two competing gas components. A selective membrane provides high purity and thus requires less complex units. Therefore, both high permeability and high selectivity are desired for an efficient and economic membrane-based gas separation process.[3]

Membranes can be broadly divided into three groups: polymeric membranes, inorganic membranes, and the combination of the two, mixed matrix membranes.(MMMs)[1] Polymeric membranes are widely used in industrial applications due to their easy synthesis, low cost, and easy scalability. Polymeric membranes can be classified into three classes as porous, non-porous (dense), and asymmetric membranes.[4] Porous polymeric membranes are rigid and highly voided structures generally prepared by solution casting, sintering, and phase-separation methods. These membranes provide high gas permeability but low selectivity because of the indistinct voids which limits their size-selective property.[5] Non-porous or dense polymer membranes exhibit high selectivity but low permeability, and the gas transport in these polymers is governed by the solution-diffusion mechanism.[6] The gas molecules first dissolve in the polymeric membranes and then diffuse through the membrane due to the concentration gradient. Asymmetric polymeric membranes consist of different layers including selective thin and porous layers.[4] Cellulose acetates, aromatic polycarbonates, and polyimides have been widely used in industrial membrane applications as asymmetric membranes.[7]

Polymeric membranes are highly dominant in the current membrane market, and they are used for many different chemical separations such as acid-gas removal, nitrogen enrichment, ammonia purge gas recovery, refinery gas purification, oxo-chemical synthesis, and dehydration.[8] However, they have two main disadvantages: permeability–selectivity trade-off and low thermal–chemical stability under harsh chemical conditions.[9] Robeson[10] collected a large amount of experimental data for polymeric membranes and showed the inverse relationship between their selectivity and permeability. Polymers that exhibit high gas selectivity generally have low gas permeability and polymers that have high permeability, are not selective enough. This trade-off resulted in description of the upper bounds for different gas separations. There is an ongoing research for the design and development of new membranes that can exceed these upper bounds to achieve high-efficiency gas separations.

Inorganic membranes such as carbons, zeolites, and silica membranes offer high gas permeability due to their molecular-sieve properties and exhibit better thermal and mechanical stabilities at high temperatures compared to polymeric membranes.[11] They can be used as membrane reactors because they are chemically inert and have good wear resistance. Silica-based inorganic membranes have been widely used to separate H_2 from CO_2 or CH_4 whereas zeolites are used to preferentially separate CO_2 from N_2 and CH_4.[12] Carbon molecular-sieve membranes are also used in small gas-separation applications because of their narrow pore sizes (3–5 Å).[13] Although inorganic membranes possess good separation characteristics, they are not able to compete with the polymeric membranes for large-scale commercial use since they are significantly expensive than the polymeric membranes. It is also challenging to fabricate large, defect-free inorganic membranes because inorganic materials are generally brittle and difficult to process. Another problem of inorganic membranes is the lack of tunable pore sizes which hinders their usage in separation of large molecules such as air/hydrocarbons.[13]

MMMs have been developed to combine advantages of polymers and inorganic materials. MMMs are heterogeneous membranes in which organic or inorganic materials are incorporated into polymer matrices as filler particles. The main idea behind MMMs is to use the commercial polymer membrane technology and enhance the separation performance of low-cost polymer-membranes by using highly permeable and selective filler particles. Porous inorganic materials such as carbon molecular sieves,[14] zeolites,[15] carbon nanotubes[16] and metal organic frameworks (MOFs)[16,17] have been used to synthesize MMMs. MMMs can overcome the permeability–selectivity trade-off of polymeric membranes and the brittleness problem of inorganic membranes. MMMs can be successfully implemented at large scale since their fabrication requires only minor adaptation of commercial polymer–membrane technology. Physical properties of the inorganic fillers and morphology of the polymer–filler interface affect the separation performances of MMMs.[18] Over the past 16 years, many different MMMs have been fabricated and tested for gas separations, but there are still problems related with poor polymer–filler compatibility and filler agglomerations at the interface which hinder the separation performance of MMMs.[18,19] Current challenges in fabrication of MMMs such as interface void formation, particle sedimentation, pore blockage, and chain rigidification have been summarized in the literature.[18,19b]

In this chapter, we will focus on a new group of membrane materials, MOFs. MOFs are a relatively new group of nanoporous materials which are composed of metal ions or clusters connected by organic linkers to create highly porous networks. The advent of MOFs was triggered in 1999 by Yaghi *et al.*[20] who synthesized a crystalline 3D framework called MOF-5. Figure 1 illustrates the topology of MOF-5, one of the most widely studied MOFs in the literature. MOF-5 has zinc oxide clusters as the metal corners, and it is a 2D cubic network connecting the large cages at the center of each unit cell. The surface area of MOF-5 ($3631\,m^2/g$) was reported to be much higher than the zeolite with the highest surface area, zeolite Y ($904\ m^2/g$),[21] and activated carbons ($526–2386\ m^2/g$).[22] The "building block approach" was developed to synthesize large families of isoreticular MOFs having the same topology with different pore sizes and surface areas.[23] This approach has received widespread attention in recent years since it provided the opportunity to synthesize numerous materials, over 50 000, with different chemical compositions and functionalities.[24]

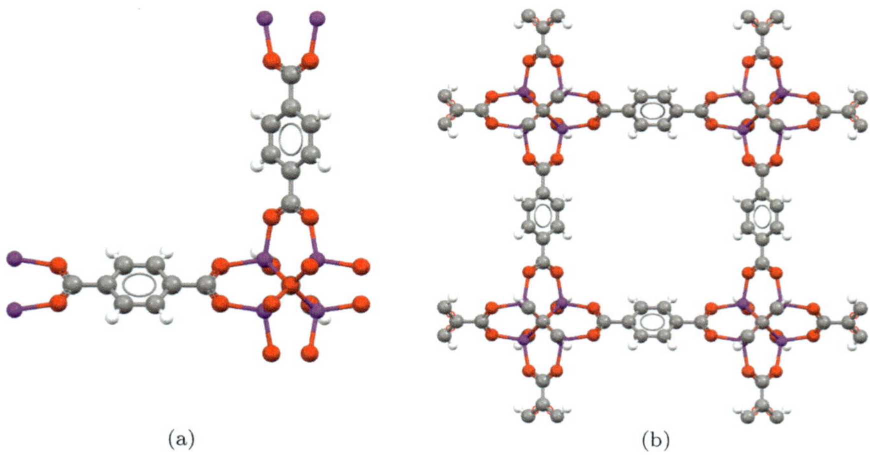

Fig. 1. The structure of cubic MOF-5 (IRMOF-1): (a) ZnO_4 tetrahedra joined by benzene dicarboxylate linkers (O: red, C: gray, H: white, and Zn: purple). (b) The primitive cubic net structure of MOF-5 in (1,0,0) direction.

Theoretically unlimited numbers of MOFs can be synthesized by changing the combination of organic linkers and metal ions.[25] Crystal structures of thousands of synthesized MOFs are deposited in a large database, Cambridge Crystallographic Data Centre (CCDC).[24] Several MOF families have been reported such as zeolitic imidazolate frameworks (ZIFs), porous coordination networks (PCNs), covalent organic frameworks (COFs). MOFs have gained considerable attention as promising nanoporous materials due to their exceptional physical and chemical properties such as very large surface areas (500–6500 m^2/g), high pore volumes (1–4 cm^3/g), tunable pore sizes (1–98 Å) and reasonable thermal and mechanical stabilities.[26] These outstanding properties of MOFs have been tested in many applications including gas storage,[27] gas separation,[28] drug storage[29] and delivery,[30] catalysis,[31] and sensing.[32] Among these different applications, gas separation has received significant interest due to tunable pore sizes of MOFs. MOFs have been used to fabricate defect-free, thin-film membranes and tested for gas separation applications in addition to being used as filler particles in polymers to make MOF-based MMMs.

The first thin-film MOF membrane was fabricated using MOF-5 on a porous α-alumina substrate.[33] The H_2 permeability of this membrane was high due to the fast diffusion of gas molecules through the large pores (8Å × 15Å) of MOF-5. Many other experimental studies also fabricated and reported gas permeance, permeability normalized by the thickness of the membrane, of MOF-5 membranes.[34] CuBTC membrane was also reported to exhibit high H_2 permeability and H_2 selectivity in the separation of H_2–N_2, H_2–CO_2, and H_2–CH_4 mixtures.[35] Following this study, several other CuBTC membranes were reported in the literature.[36] Other than MOF-5 and CuBTC, ZIF-8 and ZIF-90 were widely used to make thin-film MOF membranes. Although there are thousands of different MOF structures that have been synthesized, only a few different types of thin-film MOF membranes have been fabricated to date, and most of these membranes were fabricated using the same MOF materials such as MOF-5, CuBTC, ZIF-8. In a recent review, we discussed that only 29 different types of thin-film MOF membranes exist in the literature, and we provided experimental gas permeability and selectivity measurements of these membranes together with their operating conditions.[37] The small number of fabricated MOF membranes can be attributed to the difficulty of making defect-free, continuous MOF films. Thin-film MOF membranes are fabricated using expensive supports such as alumina and titania, and the synthesis time is long, almost 2 days.[38] The poor intergrowth at the membrane-support interface, moisture-limited stability, and reproducibility, which is the fabrication of the same MOF membrane with the same separation performance, are the main challenges in this area.[39] A wide variety of synthesis methods such as direct growth, layer-by-layer growth, secondary growth and other novel fabrication methods were used in fabrication of thin-film MOF membranes and these methods were summarized in several reviews.[38–40] Considering the diffuculties in fabricating defect-free, thin-film MOF membranes, it is not possible to assess the potential of MOF membranes using purely experimental means. Computational methods play a critical role to examine a large number of MOF membranes in a short time and to identify the most promising MOF materials for targeted gas separations.

An alternative strategy to overcome the challenges of making thin-film MOF membranes is the fabrication of MOF-based MMMs. MOFs are embedded into polymer membranes to combine the advantage of low cost and easy processability of polymers with the advantages of high gas permability and selectivity of MOFs.[41] Using MOFs as fillers in MMMs is much more economic than making thin-film MOF membranes because MOFs are not synthesized in large amounts, generally not more than 1 g. Synthesis of MOFs generally requires long reaction times (12–48 h) and expensive solvents.[42] Only six MOFs, ZIF-8 (Basolite Z1200), MIL(Materials of Institut Lavoisier)-53(Al) (Basolite A100), Fe-BTC (BasoliteF300), MOF-177 (Basolite Z377), HKUST-1 (Basolite C300), and MIL-88a (Basolite A520) are currently synthesized in large amounts (maximum 500 g) by BASF and sold by Sigma Aldrich.[43] Therefore, using MOFs as fillers in polymer membranes can overcome the current limitations of making thin-film MOF membranes.

We searched the literature for publications featuring the terms "MOF", "gas separation" and "membrane". Figure 2 shows the quick

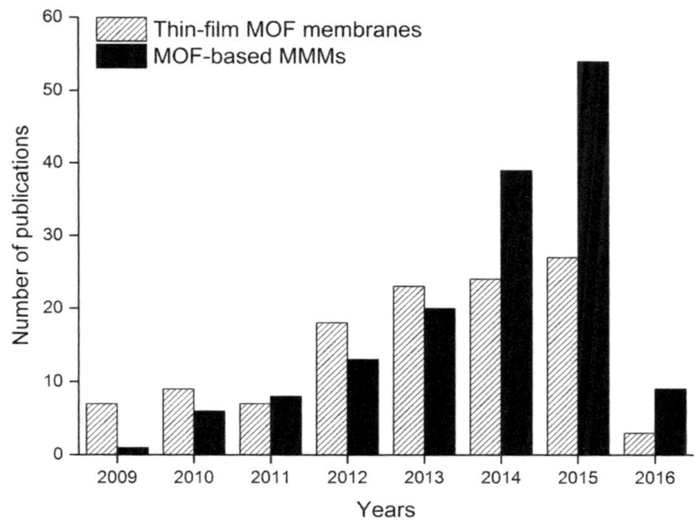

Fig. 2. Number of publications featuring the terms "metal organic framework", "gas separation" in their topics and "membrane" in titles, Accessed: 2016–04–12.

growth of number of publications on thin-film MOF membranes and MOF-based MMMs. Since fabrication of MOF-based MMMs is relatively easier than that of thin-film MOF membranes, the number of studies on MOF-based MMMs is higher than the one on pure MOF membranes. Majority of the publications on thin-film MOF membranes points out the same MOF, ZIF-8 whereas the publications on MOF-based MMMs cover more different types of MOF membranes such as UiOs (University of Oslo), MILs, and ZIFs. The first MOF-based MMM was synthesized using Cu(II)BPDC-TED (copper (II) biphenyl dicarboxylate diethylenediamine) in poly3-acetoxyethylthiophene.[44] In this study, CH_4 permeability was enhanced, but CO_2 permeability and CO_2–CH_4 selectivity decreased with the addition of MOF fillers.[44] Later, MOF-5 fillers were used to fabricate MOF-5/Matrimid MMMs.[45] Results showed that CO_2, O_2, N_2, and CH_4 permeabilities of MMMs at different MOF-5 loadings increased compared to pure Matrimid membrane due to the faster diffusion of gas molecules but selectivities remained almost constant.[45] Many MOF-based MMMs exhibited high separation performance and exceeded the Robeson's upper bound. For example, MIL-68(Al)/Matrimid MMM was reported to be highly selective (80) for an equimolar CO_2–CH_4 mixture and it overperformed the previously reported MOF-based MMMs by exceeding the Robeson's upper bound.[46] There are still critical issues which should be addressed to achieve high separation performance of MOF-based MMMs such as the compatibility between MOF and polymer interface to reduce the formation of particle aggregation or interfacial defects.[19a,47] Therefore, appropriate selection of MOF–polymer pairs is important to achieve high gas separation performance. Considering the very large numbers of available MOFs and polymers, many different MOF–polymer MMMs can be synthesized. Even when only a single polymer is used, thousands of MOF materials are available to be used as filler particles in that polymer. Computational methods that can accurately help to match MOF–polymer pairs for high efficiency MMMs are very useful. In this way, experimental efforts, time, and resources can be directed to the most promising MOF–polymer pairs. Among computational approaches, molecular simulations play a very useful role to provide molecular insights about the gas-separation properties of MOFs. These simulations can accurately predict the gas separation performances of both pure MOF membranes and MOF-based MMMs.

In this chapter, we aim to review the molecular-simulation methods that have been used to examine the potential of MOF membranes in gas separations. Section 2 discusses molecular simulation methods, Grand Canonical Monte Carlo (GCMC) and equilibrium molecular dynamics (EMDs), which have been used to model thin-film MOF membranes. Gas-permeation models used to describe MOF-based MMMs are also reviewed in Sec. 2. The accuracy of these molecular modeling methods is discussed by comparing predictions of computational methodologies with the experimental measurements for both thin-film membranes and MOF-based MMMs in Sec. 3. Section 3 discusses the effects of functionalization, structural flexibility, electrostatic interactions between gas molecules and MOF atoms, and type of force fields used in molecular simulations on the separation performance of MOF membranes. Several computational approaches to examine separation properties in thin-film membranes and MOF-based MMMs were also reviewed. Finally, opportunities and challenges in using computational approaches to assess gas-separation potential in MOF membranes are addressed in Sec. 4.

2. Computational Methods

In this section, computational methods used to estimate gas permeability and gas selectivity of thin-film MOF membranes and MOF-based MMMs are discussed. In order to estimate gas permeability and selectivity of membranes, data for gas adsorption and diffusion in the pores of MOFs are required. Molecular-simulation methods have been widely used to assess single-component and mixture gas adsorption and diffusion in nanoporous materials such as MOFs. We introduced two molecular-simulations methods, GCMC and EMD, and showed how the gas adsorption and diffusion data of MOFs obtained from these simulations can be used to predict the membrane properties.

2.1. Molecular Simulations of Thin-film MOF Membranes

Selectivity is the most widely studied parameter to assess gas-separation performances of membranes. Gas selectivity of a thin-film MOF membrane depends on the ratio of gas permeabilities of each gas species through the membrane. Gas permeabilities can be estimated using both adsorption and diffusion data of gas molecules in the pores of MOFs. GCMC simulations are performed to obtain adsorption data whereas EMD simulations are required to provide the data for diffusion of gas molecules. Results obtained from GCMC and EMD simulations are then used to predict overall separation performances of MOF membranes.

GCMC simulations are commonly used to determine the average number of adsorbed gas molecules in MOFs.[48] Here, MOF is the adsorbent and gas molecules are the adsorbates. In GCMC simulations, the chemical potential (μ), temperature (T), and volume (V) are kept constant and the number of particles is fluctuated based on a series of different moves. These moves are translation, rotation, and displacement of a molecule. For mixture simulations, an exchange move of particles is also performed to accelerate the equilibrium. The energy of each configuration of the system is computed using force fields that account for the intermolecular interactions between gas molecules and MOFs. In most GCMC simulations, it is reasonable to assume rigid framework, therefore, the intramolecular interactions including bond stretching, angle bending, and bond rotation can be neglected. However, checking the structural flexibility of MOFs using a sophisticated algorithm developed by Sarkisov *et al.*[49] can be an ideal methodology prior to neglecting the bonded terms to calculate the potential energy of the system.

The intermolecular interactions between two particles generally consist of short-range van der Waals forces and long-range coulombic interactions. Lennard–Jones (LJ) 12–6 potential is widely used to consider non-bonded interactions.[50] The functional form of the LJ potential given in Eq. (1) is based on two parameters, σ, the size parameter and ε, the energy parameter.

$$U_{LJ} = 4\varepsilon_{ij}\left[\left(\frac{\sigma_{ij}}{r_{ij}}\right)^{12} - \left(\frac{\sigma_{ij}}{r_{ij}}\right)^{6}\right]. \quad (1)$$

In Eq. (1), U_{LJ} is the intermolecular potential between two particles (i and j), r is the distance of separation from the center of one particle to the center of the other particle, ε_{ij} is the well depth and represents how strongly the two particles attract each other and σ_{ij} defines a molecular length scale related to the particle diameter and controls the scale of interaction.

Parameters of the LJ function are generally taken from the generic force fields which describe the same types of atoms in all MOFs with the same parameter.[51] The universal force field (UFF)[52] and Dreiding[53] force field are commonly used as transferable force fields in molecular simulations of MOFs, and the validity of the force fields is shown by many studies.[54] For van der Waals terms, atom-based summation method is used, and the interactions between all pair of atoms which are further beyond the cut-off radius are neglected. Periodic boundary conditions are used in molecular simulations to eliminate the surface effects. Lorentz–Berthelot mixing rules are employed to calculate cross-interactions between two dissimilar LJ sites.[55] Electrostatic potential energy (U_{ij}) of quadrupolar gas molecules such as CO_2 and N_2 are estimated using the Coulomb potential given in Eq. (2) where ε_0, q_A, and q_B represent the electric constant, partial atomic charges of A and B, respectively.

$$U_{ij} = \frac{1}{4\pi\varepsilon_0}\frac{q_i q_j}{r_{ij}}. \quad (2)$$

The partial charges of atoms are assigned using different methods from the literature such as (connectivity-based atomic contribution CBAC) method,[56] (extended charge equilibration method EQeq) and high-level quantum chemical calculation methods.[57] High-level quantum chemical calculations consider the electronic structure of atoms, and calculations of the multielectron wavefunctions are computationally costly. The CBAC and EQeq methods are used in molecular simulations especially for large-scale material screening purposes because these approximate methods provide a quick source for partial charges, and the computational cost is lower compared to that of quantum chemical calculations. The CBAC method assumes that atoms with the same connectivity have the same charges in different MOFs, and the accuracy of this method has been shown by calculating the CO_2 adsorption isotherms of different MOFs in the literature.[56] The EQeq method uses the ionization energies and electron affinities for every atom in the periodic table to calculate the partial charges of MOF atoms.[57]

Gas molecules are represented using different models in molecular simulations. For example, H_2 and CH_4 molecules are modeled as single spheres with united atom model. H_2 is generally represented using Buch potential[58] whereas TraPPE force field[59] is used for CH_4. CO_2 is represented as a rigid three-site molecule with partial point charges located at the center of each site.[60] Similarly, N_2 molecule is modeled as a three-site model with two sites located at two N atoms and the third one located at its center of mass (COM) with partial point charges.[61]

Outcome of GCMC simulations is the adsorption amount of gases inside MOFs' pores. Once the GCMC simulations are performed for single-component gases, ideal adsorption selectivity (IAS) can be calculated. IAS is defined as the ratio of single-component adsorbed amount of different gases, c_i^{pure} and c_j^{pure} in MOFs:

$$\text{IAS}_{(i/j)} = \frac{c_i^{\text{pure}}}{c_j^{\text{pure}}}. \tag{3}$$

Results obtained from mixture GCMC simulations are used to predict mixture adsorption selectivity (MAS) using adsorbed amounts of gas components i and j in their binary mixtures:

$$\text{MAS}_{(i/j)} = (c_i^{\text{mixure}}/c_j^{\text{mixure}})/(y_i/y_j). \tag{4}$$

In Eqs. (3) and (4), the subscripts i and j represent the strongly adsorbed component and the weakly adsorbed component, respectively. If the adsorption selectivities are greater than 1, the MOF is selective for component i over j.

EMD simulations are generally used to understand diffusion behavior of gas molecules in MOFs.[48] In EMD simulations, intermolecular and intramolecular interactions are used to compute the forces on the molecules. Then, Newton's equations of motion are solved numerically, and positions and velocities of gas molecules are computed until the equilibrium is reached.[62] The self-diffusivities and corrected diffusivities of gas species in MOFs' pores are computed from the results of EMD trajectories. The self-diffusivity ($D_{i,\mathrm{self}}$) is described as the motion of individual tagged particles. In an isotropic 3D medium, self-diffusivity is related to the mean-squared displacement (MSD) of the tagged particles by the Einstein relation,[63]

$$D_{i,\mathrm{self}} = \lim_{t\to\infty} \frac{1}{dNt} \left\langle \sum_{i=1}^{N} [\vec{r}_i(t) - \vec{r}_i(0)]^2 \right\rangle, \tag{5}$$

where $\vec{r}(t)$ is the position of the tagged particle at time t, N is the number of particles, d is the number of spatial dimensions for diffusion ($d = 2$, 4, or 6 for one, two, or three dimensions, respectively) and the angular brackets represent the ensemble average.

It is more convenient to define the transport diffusivity (D_t) for the macroscopic diffusion of gas molecules in MOFs. Transport diffusivity is calculated by using the corrected diffusivity (D_0) and the thermodynamic correction factor:[64]

$$D_t(c_i^{\mathrm{pure}}) = D_0(c_i^{\mathrm{pure}}) \cdot \frac{\partial \ln f_i^{\mathrm{pure}}}{\partial \ln c_i^{\mathrm{pure}}}. \tag{6}$$

The thermodynamic correction factor ($\partial \ln f_i^{\mathrm{pure}}/\partial \ln c_i^{\mathrm{pure}}$) depends on the adsorbed concentration of gas (c) and fugacity (f), and it is fully defined once the single-component gas adsorption isotherm is known from the GCMC simulations. The corrected diffusivity (D_0) includes information on the collective motion of multiple adsorbed molecules that is relevant to net mass transport and defined as follows:

$$D_{0,i} = \lim_{t\to\infty} \frac{1}{dNt} \left\langle \left(\sum_{l=1}^{N_i} [r_{il}(t) - r_{il}(0)] \right)^2 \right\rangle. \tag{7}$$

In Eq. (7), $r_{i1}(t)$ is the 3D position vector of molecule 1 of species i at time t. Well-developed approaches exist for calculating the corrected diffusion coefficient from MD simulations.[65] The self-diffusivity, transport diffusivity, and corrected diffusivity are concentration dependent and only equal in the limit of zero concentration.[66] In some extreme cases, the self and corrected diffusivities may vary by orders of magnitude.[67]

Results obtained from EMD simulations can be used to calculate both single-component and mixture diffusion selectivities. Single-component diffusion is driven by the concentration gradient of the particular component, and the ratio of single-component, self-diffusivities gives the ideal diffusion selectivity (IDS):

$$\text{IDS}_{(i/j)} = \frac{D_{i,\text{self}}^{\text{pure}}}{D_{j,\text{self}}^{\text{pure}}}. \tag{8}$$

In mixture diffusion, diffusion is governed by both the concentration of the particular component and the other component in the mixture.[62] An equilibrated configuration obtained from a Monte Carlo run is the starting point of an EMD simulation, and longer equilibration time is required for mixture components. Therefore, performing mixture EMD simulations, is more challenging than that of single-component simulations. It is important to note that prior to EMD simulations, GCMC simulations are required to determine the equilibrium gas amount inside the MOF at the specified pressure and temperature. In order to predict mixture diffusion selectivity (MDS), the ratio of self-diffusivities of each gas species in their binary mixture is used:

$$\text{MDS}_{(i/j)} = \frac{D_{i,\text{self}}^{\text{mixture}}}{D_{j,\text{self}}^{\text{mixture}}}. \tag{9}$$

In order to accelerate modeling of MOF membranes, membrane selectivity is approximated by the multiplication of adsorption and diffusion selectivities obtained from the GCMC and EMD simulations, respectively. Ideal membrane selectivity (IMS) is defined as follows:[68]

$$\text{IMS}_{(i/j)} = \text{IAS}_{(i/j)} \cdot \text{IDS}_{(i/j)}. \tag{10}$$

Similarly, membrane's mixture selectivity (MMS) can be calculated as follows:

$$\mathrm{MMS}_{(i/j)} = \mathrm{MAS}_{(i/j)} \cdot \mathrm{MDS}_{(i/j)}. \tag{11}$$

Equations (10) and (11) assume that self-diffusivities of gases are estimated at the feed side of the membrane, and the permeate side is kept under vacuum. Details of this approach can be found in the literature.[68] The validity of this approach to predict the separation performances of MOF membranes was shown by comparing the results with the experiments.[68,69] The CH_4–H_2 selectivity of MOF-5 membrane was accurately estimated using this approach, but the predictions were reported to be less accurate for CuBTC membrane.[68] Recently, validity of this approach was assessed for CH_4–H_2 separation performance of NU-125, NU-140, MOF-5, and ZIF-8 membranes, and results were found to be in good agreement with the experimental measurements.[69] At that point, it is important to note that experiments generally report ideal selectivity rather than mixture selectivity of membranes because of the challenges in measuring transport rates of gas mixtures through MOF membranes. In order to compare predictions of computational methods with the experimental measurements, ideal selectivity calculation as shown in Eq. (10) is commonly used. However, gas molecules exist in the form of mixtures in industrial processes. In order to examine the difference between ideal and mixture selectivities, both were calculated for 15 different ZIF membranes for separation of CO_2–CH_4, CO_2–N_2, CO_2–H_2, and CH_4–H_2.[70] Results showed that ideal selectivities calculated from single-component gas adsorption and diffusion data can be used to make initial predictions about the gas separation performances of materials if mixture data is not available. However, if one of the gas components is strongly favored over another in adsorption such as CO_2 in CO_2–H_2, multi-component effects between different species become important.[70] For example, strongly adsorbing gas species can reduce the diffusion rate of the other species in the mixture. These multicomponent effects play a significant role in determining separation performance of membrane materials. As a result, ideal and mixture selectivities can be different especially at high gas loadings.[71] Therefore, performing GCMC and EMD simulations on the gas mixtures rather than single-component gases is required to assess the separation performance of MOFs under realistic conditions.

Permeability is as important as the selectivity to assess the gas-separation performances of MOF membranes. Single-component gas permeabilities through an MOF membrane are calculated using the shell model[72] which assumes that diffusivities of the adsorbed components are constant throughout the membrane at the average concentration of the feed and permeate sites. Steady state flux (J) of a gas across an MOF membrane is calculated based on Fick's law,[73]

$$J_i = -D_t(c_i^{\text{ave}}) \cdot \nabla c_i, \tag{12}$$

where ∇c_i is the concentration gradient of the adsorbed gas based on the difference between the feed and permeate pressures of the membrane. The shell model[72] assumes that the Fickian diffusivities are estimated at the average adsorbate loadings:

$$c_i^{\text{ave}} \cong (c_i^{\text{permeate}} + c_i^{\text{feed}})/2. \tag{13}$$

Steady state fluxes are then converted to single-component gas permeabilities (P) using the pressure drop (Δp) and membrane thickness (L):[74]

$$P_i^{\text{pure}} = J/\Delta p/L. \tag{14}$$

Mixture gas permeabilities are calculated as follows:

$$P_i^{\text{mixture}} = \frac{\phi \cdot D_{i,\text{self}}^{\text{mixture}} \cdot c_i^{\text{mixture}}}{f_i^{\text{mixture}}}, \tag{15}$$

where P_i is the permeability of species i, ϕ is the porosity of the MOF, $D_{i,\text{self}}$ is the self-diffusivity of species i in the mixture, c_i is the concentration of species i at the feed side of the membrane and f_i is the bulk phase fugacity of species i.[75] Gas permeabilities are generally reported in the unit of Barrers (10^{-10} cm^3 (STP)cm/(cm$^2 \cdot$ s$\cdot$ cm Hg)) in the literature. Once gas permeabilities are calculated, single-component and mixture selectivities can be also found using the ratio of single and mixture gas permeabilities, respectively, and the results are same with Eqs. (10) and (11). Predictions for single-component and mixture gas permeation through MOF membranes were compared with the experimental gas permeance data to validate the accuracy of the computational approach as we will discuss in the next section.[76] Results showed that predictions agree remarkably well with the experimental measurements of mixture gas permeance through several MOF membranes such as ZIF-69, ZIF-78, ZIF-90, and ZIF-95[76] and single-component gas permeance through IRMOF-1, Ni-MOF-74, and MIL-53(Al) membranes.[76a]

2.2. *Computational Modeling of MOF-based MMMs*

Separation performances of MMMs are commonly examined by using "two-phase" description which includes a continuous polymer phase and a dispersed filler phase. Therefore, modeling of MOF-based MMMs requires the knowledge of the gas permeabilities in the polymer phase and dispersed phase (MOFs). Gas permeabilities of many polymeric membranes have been already reported in the literature,[10] however, similar type of data is not available for MOF membranes. Therefore, predicting gas permeability of MOFs is a prerequisite for predicting permeability of MOF-based MMMs. Once the gas permeabilities of both MOF and polymer are known, theoretical permeation models can be used to predict the gas permeability of MOF-based MMMs. Selectivity of the MOF-based MMM can be then calculated as the ratio of the gas permeabilities of two different species.

Several theoretical permeation models are used to predict gas permeabilities of MMMs. These permeation models can be classified into two groups: models considering ideal morphology such as Maxwell,[77] Bruggeman,[78] Lewis-Nielson,[79] Pal[80] and models considering non-ideal morphology such as modified Maxwell,[81] Felske,[82] and modified Felske.[80] Ideal morphology concept assumes that there are no defects, void formation, or polymer rigidification around the particles at the polymer–filler interface. Non-ideal approach, on the other hand, considers all these deficiencies using additional parameters such as packing factor and the matrix rigidification factor. We compared predictions of different permeation models with the available experimental measurements for pure gas and mixed-gas permeation of MOF-based MMMs in order to determine the best predicting permeation model.[83] Results showed that the Maxwell model is the best predicting permeation model among the ones considering ideal morphology and modified Felske model is the best predicting model among the ones considering non-ideal morphology. Therefore, we will mainly discuss the Maxwell, modified Maxwell, Felske, and modified Felske models throughout this section. Detailed descriptions of the other permeation models are available in several review papers in the literature.[19b,84]

The Maxwell model[77] is widely used to estimate gas permeability of MMMs. Maxwell observed the close analogy between dielectrics and transport through membranes, and assuming dilute dispersion of ellipsoids, Maxwell–Wagner–Sillars equation can be written for MOF-filled polymer membranes as follows:[85]

$$P_{\mathrm{MMM}} = P_{\mathrm{POLYMER}} \cdot \left[\frac{n \cdot P_{\mathrm{MOF}} + (1-n) \cdot P_{\mathrm{POLYMER}} + (1-n) \cdot (P_{\mathrm{MOF}} - P_{\mathrm{POLYMER}}) \cdot \phi}{n \cdot P_{\mathrm{MOF}} + (1-n) \cdot P_{\mathrm{POLYMER}} - n \cdot (P_{\mathrm{MOF}} - P_{\mathrm{POLYMER}}) \cdot \phi}\right], \tag{16}$$

where ϕ is the volume fraction of the MOF particles, n is the shape factor of the fillers, P_{MOF} is the MOF's permeability, P_{POLYMER} is the polymer's permeability, and P_{MMM} is the permeability of the MOF–polymer MMM. The Maxwell model assumes that $n = 1/3$ for random dispersion of the spherical filler particles, and the equation reduces to the following where λ is defined as the permeability ratio, $P_{\mathrm{MOF}}/P_{\mathrm{POLYMER}}$:

$$P_{\mathrm{MMM}} = P_{\mathrm{POLYMER}} \cdot \left[\frac{2 \cdot (1-\phi) + (1+2\phi) \cdot \lambda}{(2+\phi) + (1-\phi) \cdot \lambda}\right]. \tag{17}$$

The Maxwell model is valid for low filler volume fractions ($0 < \phi < 0.2$) since it assumes that a particle's flux pattern is not affected by the presence of nearby filler particles. This model does not consider packing limit of particles, the effect of particle size distribution, particle shape, and aggregation of particles. The first computational methodology to estimate gas permeability through an MOF-based MMM was reported in 2010.[86] In that study, Matrimid was used as the polymer, and permeability of gas molecules through Matrimid was taken from the experimental data. The filler was IRMOF-1, and gas permeabilities of IRMOF-1 were predicted using molecular simulations as described in Sec. 2.1. These permeability values were then inserted into the Maxwell and Bruggeman models to predict P_{MMM}, and predictions were in good agreement with the experimental data for IRMOF-1/Matrimid MMM.[86]

The modified Maxwell model[81] was developed to include interfacial voids, rigidified polymer layer, and particle pore blockage. The model is based on the two-phase description: the polymer matrix is one phase, and the combined insert and interphase is the other phase ("pseudo insert phase"). The modified Maxwell model can be written as:[81]

$$P_{\mathrm{MMM}} = P_{\mathrm{POLYMER}} \cdot \left[\frac{P_{\mathrm{eff}} + 2\cdot P_{\mathrm{POLYMER}} - 2\cdot(\phi_d + \phi_I)\cdot(P_{\mathrm{POLYMER}} - P_{\mathrm{eff}})}{P_{\mathrm{eff}} + 2\cdot P_{\mathrm{POLYMER}} + (\phi_d + \phi_I)\cdot(P_{\mathrm{POLYMER}} - P_{\mathrm{eff}})}\right], \tag{18}$$

$$P_{\mathrm{eff}} = P_I \cdot \left[\frac{P_{\mathrm{MOF}} + 2\cdot P_I - 2\phi_{\mathrm{s}}\cdot(P_I - P_{\mathrm{MOF}})}{P_{\mathrm{MOF}} + 2\cdot P_I + \phi_{\mathrm{s}}\cdot(P_I - P_{\mathrm{MOF}})}\right], \tag{19}$$

$$\phi_s = \frac{\phi_d}{\phi_d + \phi_I} = \frac{r_d^3}{(r_d + l_I)^3}, \tag{20}$$

where, P_{eff} is the effective permeability of the pseud insert phase, P_I is the permeability of the interphase, ϕ_s is the volume fraction of the dispersed phase within the pseudo-insert phase, ϕ_d is the overall volume fraction in the membrane of the insert, ϕ_I is the overall volume fraction in the membrane of the interphase, r_d is the insert radius, and I_1 is the interphase thickness. The modified Maxwell model is valid for low to moderate values of ϕ like the Maxwell model.

The Felske model[82] assumes the dispersed particles as a core and the surrounding interfacial layer (rigidified interfacial layer or voids or particle pore blockage) as a shell. The model can be written as follows:

$$P_{\mathrm{MMM}} = P_{\mathrm{POLYMER}} \cdot \left[\frac{2 \cdot (1-\phi) + (1+2\phi) \cdot (\beta/\gamma)}{(2+\phi) + (1-\phi) \cdot (\beta/\gamma)}\right], \tag{21}$$

$$\begin{aligned} \beta &= (2+\delta^3) \cdot \lambda - 2 \cdot (1-\delta^3) \cdot \lambda_{I/\mathrm{POLYMER}} \\ \gamma &= (1+2\delta^3) - (1-\delta^3) \cdot \lambda_{\mathrm{MOF}/I}. \end{aligned} \tag{22}$$

In these equations, $\lambda_{I/\mathrm{POLYMER}} = P_{\mathrm{INTERFACIAL\ SHELL}}/P_{\mathrm{POLYMER}}$, $\lambda_{\mathrm{MOF}/I} = P_{\mathrm{MOF}}/P_{\mathrm{INTERFACIAL\ SHELL}}$, $\lambda = P_{\mathrm{MOF}}/P_{\mathrm{POLYMER}}$, and δ is the ratio of outer radius of interfacial shell to the core radius. When $\delta = 1$, this model, reduces to the Maxwell model, and it has the similar disadvantages.

The Modified Felske model[80] was developed by Pal to consider the morphology and packing factor of the filler particles,

$$\begin{aligned} P_{\mathrm{MMM}} &= P_{\mathrm{POLYMER}} \cdot \left[\frac{1 + 2 \cdot ((\beta-\gamma)/(\beta+2\gamma)) \cdot \phi}{1 - ((\beta-\gamma)/(\beta+2\gamma)) \cdot \phi \cdot \varphi}\right], \\ \varphi &= 1 + \left[\frac{(1-\phi_m)}{(\phi_m)^2}\right] \cdot \phi, \end{aligned} \tag{23}$$

where ϕ_m is the maximum packing volume fraction of filler particles, assumed to be 0.64 for random close packing of uniform spheres.[19b] This model reduces to the original Felske model when the maximum packing volume fraction of the core–shell particles is equal to 1, and it turns to the Lewis–Nielson model when $\delta = 1$. If both parameters are equal to 1, the model reduces to the Maxwell model. The accuracy of the permeation models will be discussed in the next section by comparing model predictions with the experimental measurements for MOF-based MMMs.

3. Assessment of MOF Membranes using Computational Approaches

In this section, predictions of the computational methods were compared with the experimental data for the gas separation performances of pure, thin-film MOF membranes and MOF-based MMMs. Effects of molecular simulation parameters such as force fields on the predicted separation performances of MOF membranes, intramolecular interactions in MOFs, and electrostatic interactions between MOFs and gas molecules were discussed. Computational methods that have been used to screen large numbers of MOFs which have not yet been fabricated as membranes to identify the most promising materials were also reviewed.

3.1. *Thin-film MOF Membranes*

Computational methods that we described in Sec. 2 are used to estimate gas permeabilities and selectivities of thin-film MOF membranes, and results were compared with the experimental measurements. Figure 3 demonstrates comparison of molecular simulation results[87] with the experiments[88] for single-component (CH_4, CO_2,

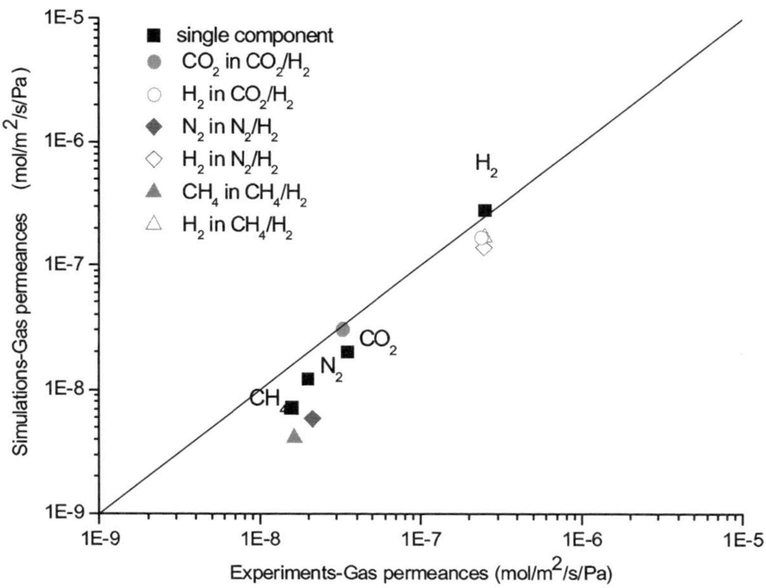

Fig. 3. Comparison of the experimental data[88] with the predictions of the molecular simulations for permeation of single gases and mixed gases through the ZIF-90 membrane at 200°C and 1 bar. Reprinted from Ref. 87 with permission from the American Chemical Society.

H_2, and N_2) and equimolar gas mixture (H_2–CH_4, H_2–CO_2, H_2–N_2) permeances through ZIF-90 membrane. If the simulation results are in good agreement with the experimental measurements, data points should be close to the diagonal line. Figure 3 shows that there is a good agreement between simulations and experiments both for single gases, H_2, CO_2, and CO_2–H_2 mixture permeances. Single component and mixture permeances of larger molecules such as CH_4 and N_2 were slightly underestimated by simulations due to neglecting flexibility of the framework as we will discuss later.[87] The accuracy of computational methods was also shown for several other MOF membranes such as IRMOF-1, Ni-MOF-74, MIL-53(Al), ZIF-69, ZIF-78, ZIF-95, and ZIF-8 by comparing the simulation results with the experimentally measured gas permeance data.[76a] These results demonstrated that computational methods can provide quick and accurate information about the gas-separation properties of MOF membranes.

Gas-separation performances of several MOF membranes such as MgMOF-74, ZnMOF-74, and MOF-177 were compared with well-known zeolite membranes such as NaX, NaY, and DDR for CO_2–CH_4, CO_2–N_2, CO_2–H_2 and CH_4–H_2 mixtures.[89] Results showed that MgMOF-74 and ZnMOF-74 are promising membranes due to their high gas selectivity and permeability. Krishna and van Baten[90] compared gas-separation performances of MgMOF-74 and ZnMOF-74 for CO_2–CH_4, CO_2–N_2, CO_2–H_2, and CH_4–H_2 mixtures considering a range of feed pressures. MgMOF-74 and ZnMOF-74 exhibited almost the same separation performance for CH_4–H_2 mixture because they have similar adsorption strengths for CH_4. For CO_2–CH_4, CO_2–N_2, and CO_2–H_2 mixtures, selectivities of MgMOF-74 were found to be higher than those for ZnMOF-74, which was attributed to higher CO_2 binding energy of MgMOF-74. MgMOF-74 was identified as the best membrane candidate for CO_2 separations in that study.

Computational approaches have been used to assess gas separation performance of large numbers of MOF membranes that have not been fabricated yet. Separation performances of different ZIF membranes were predicted by molecular simulations.[76b] ZIFs are considered as a subclass of MOFs which consist of metals such as Zn or

Co and imidazole ligands with a tetrahedral coordination geometry, and they show high thermal stability and chemical robustness which make them strong candidates as membranes.[91] Selectivities and permeabilities of ZIF membranes were calculated for CH_4–H_2, CO_2–CH_4 and CO_2–H_2 mixtures and compared with well-known zeolite and MOF membranes in Fig. 4.[76b] In this figure, permeation selectivity is the membrane's mixture selectivity. Figure 4(a) shows that ZIF-65 and ZIF-90 membranes are H_2 selective membranes like traditional zeolite membranes, LTA and CHA, due to their molecular sieving properties. ZIF-2 and ZIF-79 outperformed zeolite and other MOF membranes for CH_4–H_2 separation due to their higher CH_4 selectivities and permeabilities.

Figure 4(b) shows the CO_2–CH_4 permeation selectivity and CO_2 permeability of ZIFs, zeolites, and MOFs. Robeson's upper bound[10] was shown to compare the separation performance of these materials with those of polymeric membranes. Materials exceeding the upper bound are considered as promising membranes for this separation. ZIF-90 was reported to exceed the upper bound due to its high CO_2 permeation selectivity which was attributed to the fact that both adsorption and diffusion selectivity favor CO_2 over CH_4. Figure 4(c) shows that ZIF-65 and ZIF-90 membranes show high H_2 selectivity over CO_2 due to high diffusion selectivities of these materials toward H_2. As can be seen from Fig. 4, MgMOF-74 and ZnMOF-74 membranes outperformed zeolite membranes, CHA, DDR, MFI, MOF membranes, CuBTC, MOF-177, and other ZIF membranes.

Similar computational studies were done for 20 different porous coordination network (PCN) membranes for separation of CO_2–H_2, CO_2–N_2, CO_2–CH_4 and CH_4–H_2 mixtures.[92] These PCN membranes had not been fabricated yet using computational approaches their separation properties were predicted. Several PCNs (PCN-10, PCN-11 and PCN-131) were identified to exceed the upper bound established for CO_2–CH_4 and CO_2–N_2 separations because both adsorption and diffusion favor CO_2 in the mixture in these PCNs.[92] It was concluded that a promising membrane must have both high diffusion selectivity and high adsorption selectivity for the same gas species

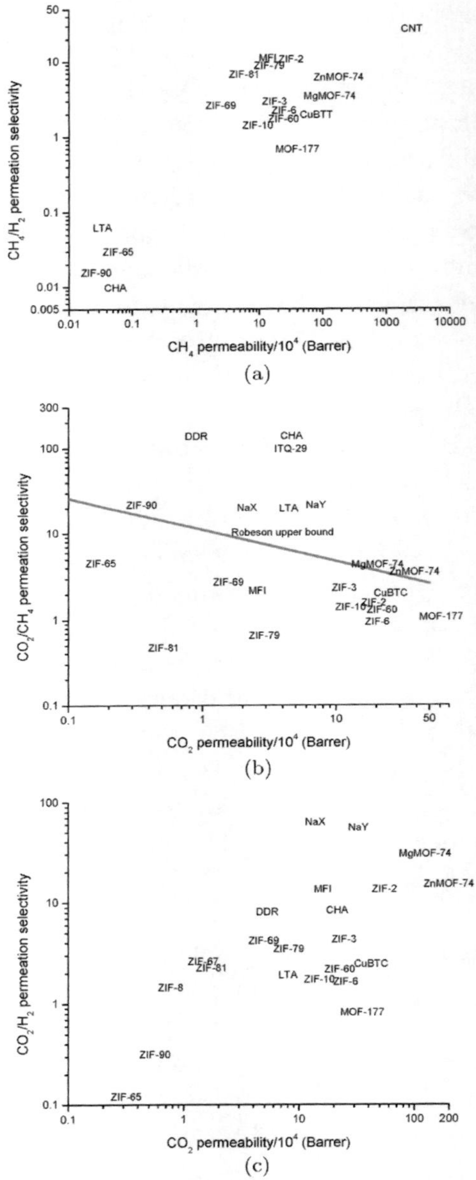

Fig. 4. Permeation-based separation performance of ZIFs for (a) CH_4–H_2, (b) CO_2–CH_4, and (c) CO_2–H_2 mixtures. The bulk gas mixtures are CH_4–H_2: 10–90, CO_2–CH_4:10–90, and CO_2–H_2:1/99 for ZIFs at 298 K and (a) 50–50, (b) 50–50, and (c) 15–85 for zeolites at 300 K. Reprinted from the Ref. 76b with permission from the American Chemical Society.

in the mixture since the membrane selectivity is estimated as the product of the adsorption and diffusion selectivities.

Qiao *et al.*[93] computationally screened a large number of hypothetical MOFs and 29 real MOFs such as IRMOFs, MILs, M-MOF-74, ZIFs, University of Michigan crystalline materials (UMCMs), UiOs, CuBTC for adsorption-based CO_2 separation. Gas permeabilities and membrane selectivities were calculated for 20 different amine functionalized M-MOF-74 membranes which had the most remarkable CO_2 adsorption selectivity. Among M-MOF-74 series, unmodified and ethylene diamine-functionalized Mg-MOF-74 membranes were also experimentally fabricated, and simulation results were compared with the experimental results as shown in Fig. 5. Predictions of the unmodified Mg-MOF-74 membrane were in good agreement with the experimental results. Amine functionalization of M-MOF-74 membranes enhanced CO_2 separation. For example, the CO_2–H_2 selectivity of unmodified Mg-MOF-74 membrane increased with the amine functionalization. M-MOF-74 series which were modified by

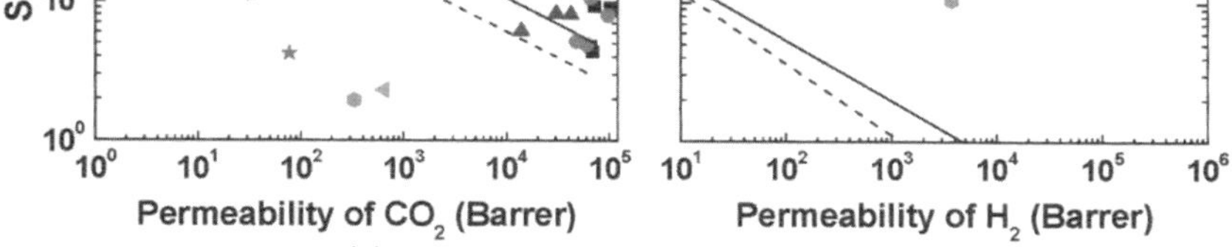

(a) (b)

Fig. 5. Comparison of permeability and selectivity of different $(NH_2CH_2CH_2NH)_n$-M-MOF-74 at 298 K and 1 bar for (a) for CO_2–CH_4 and (b) CO_2–H_2. The numbers of amine functional groups are represented as squares: 1; circles: 3; triangles pointing up: 6; triangles pointing down: 9; diamond: 18. Permeation selectivity and permeability data of Mg-MOF-74 series are shown as: unmodified Mg-MOF-74: triangle pointing left; experimental unmodified Mg-MOF-74: hexagon and the experimental ethylene-diamine functionalized Mg-MOF-74: star. Reprinted from the Ref. 93 with permission from of The Royal Society of Chemistry.

nine amine functional groups exceeded the CO_2–CH_4 and CO_2–H_2 upper bounds. Results of that study suggest that computational methods are highly useful to direct experimental studies for functionalization of MOF membranes.

Computational studies that we reviewed so far assumed a rigid framework in molecular simulations of MOFs which means crystallographic positions of MOF atoms are fixed during the molecular simulations. This assumption is reasonable since it saves a tremendous amount of computer time. However, recent studies showed that considering the structural flexibility of MOFs is important because it may affect adsorption and diffusion gas molecules.[94] For example, adsorption isotherms and diffusion coefficients of CO_2 and CH_4 molecules in ZIF-8 were computed using both rigid and flexible force fields, and results showed that structure flexibility has much more influence on the diffusion of gases especially if the pore size of the structure is close to the kinetic diameter of the gas molecule.[94] In order to examine the effect of flexibility on the separation performance of MOF membranes, predictions from molecular simulations were compared with the experimentally fabricated bio-MOF-1, bio-MOF-13, and bio-MOF-14 membranes.[95] These bio-MOFs were synthesized using adenine as a biomolecular ligand and zinc salts as metal centers.[95] The presence of Lewis-basic sites, including an amino group and pyrimidine nitrogens leads to the strong confinement of CO_2 molecules.[95] and makes them appealing for CO_2 separations. Figure 6 shows that results of flexible EMD simulations of bio-MOF-1 agree much better with the experimental data compared to the results of rigid simulations. Predictions for bio-MOF-13 and bib-MOF-14 membranes were poor because these MOFs have two- and three-chain configurations, respectively, and it was reported to be challenging to identify which configuration was dominant in the fabricated membranes.

The type of the force field used in molecular simulations may strongly affect the results of simulations. Effects of force field parameters on the predicted adsorption-based and membrane-based gas separation performances of ZIF-68 and ZIF-69 were examined for CH_4–H_2, CO_2–CH_4, CO_2–H_2, and CO_2–N_2, mixtures using three

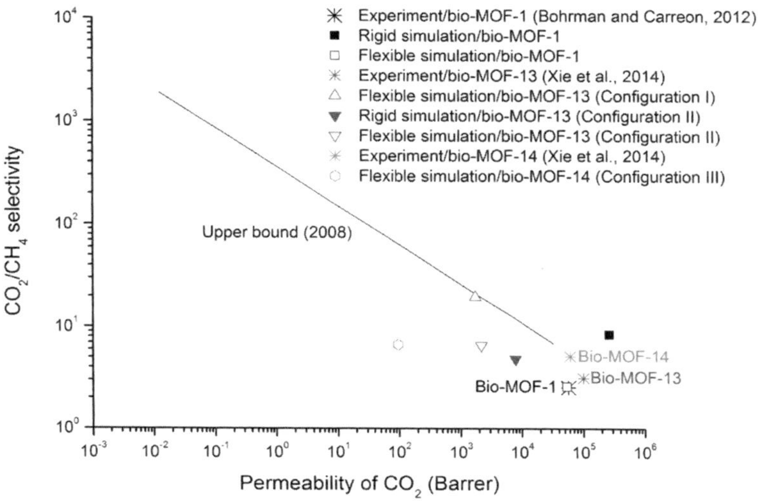

Fig. 6. Comparison of experiments[95a,95b] and molecular simulations for bio-MOF-1, bio-MOF-13, and bio-MOF-14 membranes in CO_2–CH_4 (50–50) separation. Feed pressure is 2.4 bar, and temperature is 298 K for bio-MOF-1, 295 K for bio-MOF-13 and bio-MOF-14. Simulation results for rigid bio-MOF-13 (configuration I) and rigid bio-MOF-14 (configuration III) were not shown since gas diffusion was not measurable in these materials. Reprinted from the Ref. 95c with permission from the Elsevier.

different force fields including Dreiding, UFF, and a modified force field developed by Liu and Smit for these ZIFs to match the experimental gas uptakes.[96] Results showed that predicted membrane selectivities were not affected by the choice of force field. Generic force fields, Dreiding, and UFF predicted higher adsorption selectivity values compared to the modified force field.[96a] On the other hand, generic force fields predicted lower diffusion selectivities compared to the modified force field since then strongly adsorbed component diffuses much slowly than the weakly adsorbed component. As a result, adsorption selectivity and diffusion selectivity compensated each other, and predicted membrane selectivities obtained from three force fields became similar. This result suggests that generic force fields can be safely used to screen MOF membranes.

Computational studies we reviewed so far used EMD simulations to obtain gas diffusivity data through MOF membranes. Several different computational approaches are used in the literature to examine separation properties of MOF membranes. For example, Haldoupis *et al.*[97] introduced a steric model to calculate diffusion coefficients of gas molecules in MOFs. In this model, the size of the largest cavity and the smallest opening in the pore network are assumed as the critical parameters for adsorption and diffusion of gas molecules in MOFs. This model was tested for more than 500 MOFs to predict the Henry's constant for adsorption and diffusion coefficients of gas molecules. Using the same geometric approach, Watanabe and Sholl[98] screened 1163 MOFs to identify the possible promising candidates for CO_2–N_2 separation. This geometric approach is highly useful for quick screening of large numbers of materials. It is important to note that geometric approach is valid in the limit of low pressures at low gas loadings when the mixture effects are neglected. Mixture GCMC and EMD simulations are required to obtain adsorption and diffusion data of gases under realistic operating conditions, generally at high pressures. Secondly, geometric approach is valid for only nonpolar and spherical adsorbates. It is necessary to include electrostatic interactions between polar adsorbates and MOFs to make accurate predictions. Sitprasert *et al.*[99] compared predictions of membrane selectivities calculated using the geometric approach and predictions calculated using the GCMC-flexible EMD simulations with the experimental results of ZIF-8 membrane. Results obtained from GCMC-flexible EMD simulations were found to be in better agreement with the experiments compared to those obtained from the geometric approach. From these results, it can be concluded that geometric approach is a very useful and quick method for initial screening of MOFs to identify the promising MOFs for separation of interest. However, conventional GCMC and EMD simulations are required to obtain more precise results about the MOF membranes.

More sophisticated computational methods to mimic transport behaviors of gas molecules through thin-film MOF membranes have been recently introduced in the literature.[100] For example, CO_2–N_2 separation performance of 2D COF membranes was assessed using MD simulations to understand separation mechanisms of few-layered porous membranes.[100a] In MD simulations, 2D COF membrane was constructed using rectangular nanosheets. To mimic the real separation process, molecules that pass through the permeate side were removed at a specified time, and a certain number of molecules were added to the feed side. Results showed that two- and four-layered-COF membranes are highly permeable for CO_2, and they can exceed the Robeson's upper bound.[100a] Since water molecules are present in the flue gas, the effect of humidity on gas-separation performance of MOF membranes was also discussed in that study.[100a] Figure 7 shows the effect of water content on the CO_2–N_2 selectivity and CO_2 permeance. Separation performance of four-layered COF membrane did not change in the presence of water.[100a] That computational study has been useful to model few-layered ultrathin membranes prior to experimental efforts and suggests that MD simulations can provide the information about the diffusion process of gas molecules through few-layered ultrathin membranes.

Similarly, Cabrales-Navarro *et al.*[100b] simulated MOF membranes performing MD simulations for CO_2–N_2 and CO_2–CH_4 separations. They monitored the dynamic behavior of CO_2–N_2 and CO_2–CH_4 mixtures through IRMOF-1 and CuBTC membranes. Figure 8 shows the distribution of gas molecules through permeate, membrane, and feed phases. The number of molecules passing through the permeate side as a function of time was investigated, and

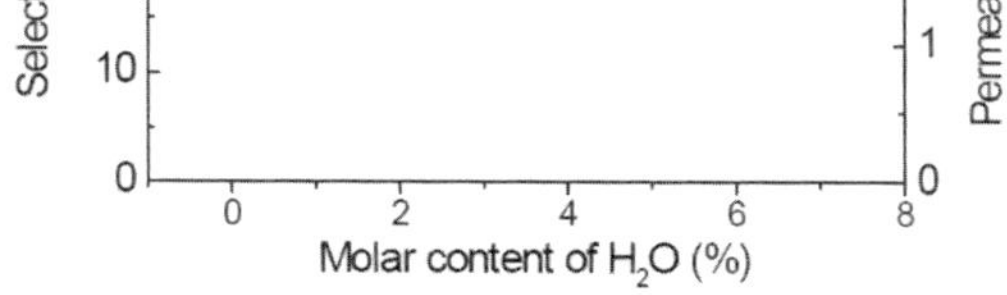

Fig. 7. CO_2–N_2 selectivity (red, top) and the CO_2 permeance (black, bottom) of the four-layered COF membrane as a function of water content in the feed CO_2–N_2–H_2O gas mixture. Reprinted from the Ref. 100a with permission of The Royal Society of Chemistry.

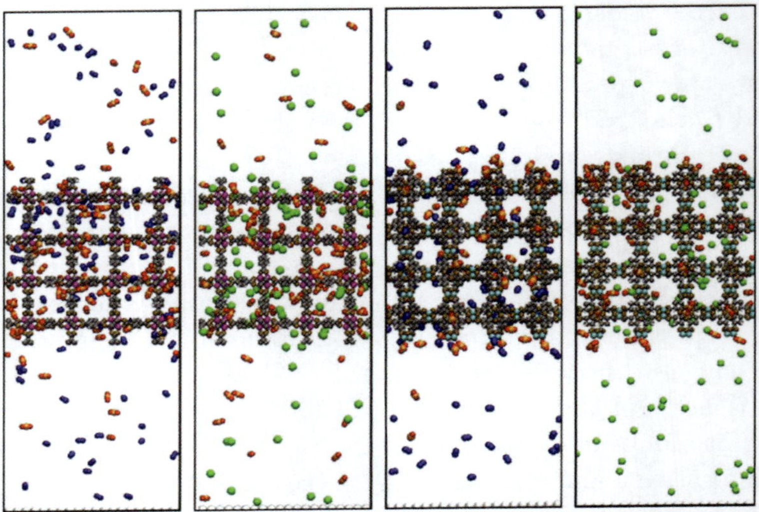

Fig. 8. MD snapshots of final configurations (at 3 ns) (a) IRMOF-1 and CO_2–N_2, (b) IRMOF-1 and CO_2–CH_4, (c) CuBTC and CO_2–N_2, and (d) CuBTC and CO_2–CH_4. The color codes: C (CO_2), orange; O (CO_2), red; N, blue; CH_4, green. For the framework atoms: Zn, magenta; Cu, cyan; C, gray; O, ocher. Reprinted from Ref. 100b with permission from the Elsevier.

results showed that MOF membranes show similar separation performance for CO_2–N_2 mixture as shown in Figs. 8(a) and 8(c). However, CuBTC was reported to perform better to separate CO_2 from CH_4 compared to IRMOF-1 membrane. The analysis of MD simulations also showed that diffusion mechanism of the gases through the MOF membranes is dominated by rotational motion for all molecules.[100b]

3.2. *MOF-based MMMs*

The systematic computational methodology which combines molecular simulation of MOF membranes with the permeation models was implemented to predict gas-separation performances of various MOF-based MMMs. The accuracy of the methodology was validated by comparing permeability and selectivity predictions with the experimental results of MOF-based MMMs. Figure 9 shows comparison of 58 experimental data points with the computational predictions for

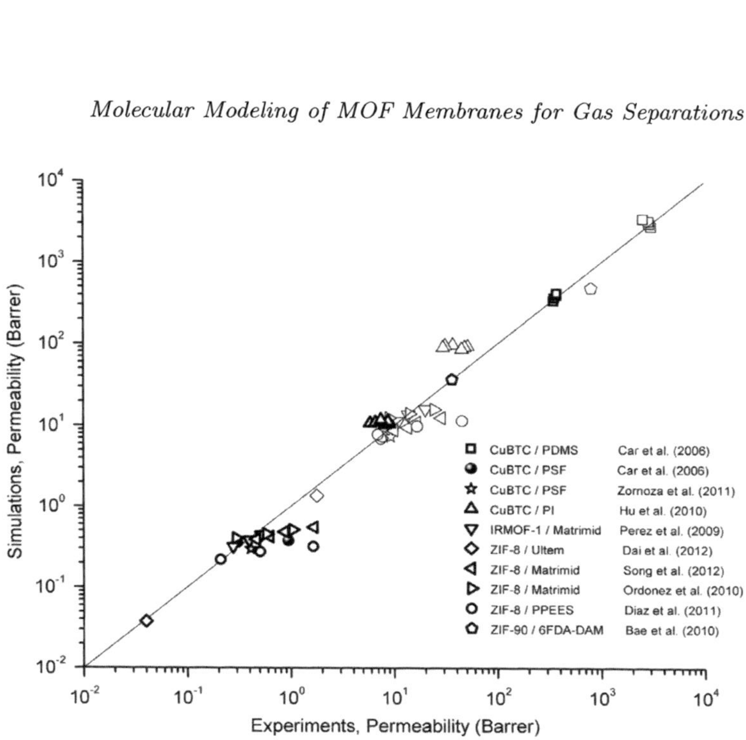

Fig. 9. Comparison of our molecular simulation results with experimentally measured: CO_2 (light symbols) and N_2 (bold symbols) permeability of MOF- and ZIF-based MMMs. Reprinted from Ref. 101 with permission from the Elsevier.

CO_2 and N_2 permeabilities of different types of MOF- and ZIF-based MMMs.[101] In order to estimate the gas permeabilities through MOF membranes, molecular simulations were performed at the same conditions with the experiments, and then the modified Maxwell model was used to calculate the gas permeability through MMMs. Results were in good agreement with the experimentally measured permeability data as shown in Fig. 9.[101] Theoretical predictions slightly underestimated N_2 permeability of ZIF-8-based MMMs because of the flexibility of ZIF-8 which will be discussed later. Following this study, Altintas and Keskin[102] compared more experimental data points with the predictions for CO_2 and N_2 permeabilities of different types of MOF-based MMMs and found a good agreement. They also compared the average absolute relative error (AARE %) values for the predictions of Maxwell, and the modified Maxwell models for MMMs contained different polymers and MOF fillers such as CuBTC, ZIF-8 and CuTPA. They found that the Maxwell model gave lower AARE % values, and using the Maxwell model they predicted gas-separation performances of 200 new PCN-based MMMs which have not been fabricated yet.[102] Results showed that using molecular simulations and the theoretical permeation models, one can predict promising MOF–polymer combinations for gas separations in a time-efficient manner prior to experiments.

ZIFs were also used as fillers in the polymer membranes to investigate gas-separation performance of MMMs. Thornton *et al.*[103] predicted separation performance of ZIF-based MMMs using the Bruggeman model for hydrogen production from coal gasification (H_2–CO_2 and H_2–N_2), postcombustion carbon capture from coal flue streams (CO_2–N_2), natural gas purification (CO_2–CH_4), and air separation (O_2–N_2). Figure 10 shows the predicted selectivity/permeability properties of pure ZIF membranes and ZIF-based MMMs for natural gas purification. Experimental data were also shown for comparison. ZIF-90 particles were added into three polymers, Ultem, Matrimid, and 6FDA-DAM to obtain ZIF-90-based

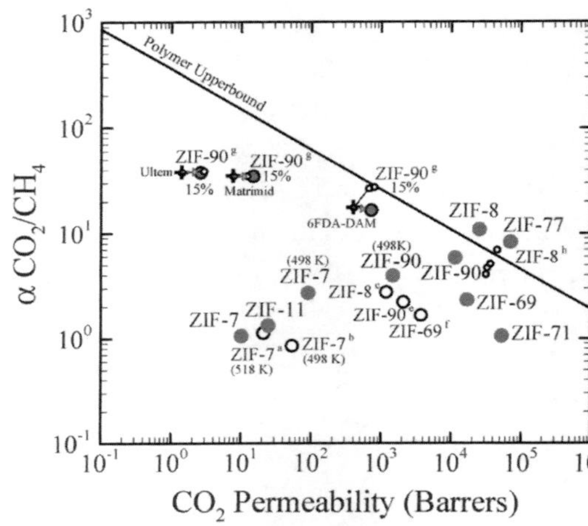

Fig. 10. Comparison of predicted CO_2–CH_4 selectivities and CO_2 permeabilities (filled symbols) with experimental results (empty symbols with superscripts) for ZIF membranes. Polymers (crosses) with connecting lines for MMM predictions (dotted lines) and experimental results (solid line). Reprinted from Ref. 103 with permission form the American Chemical Society.

MMMs. Predictions for ZIF-90–Ultem and ZIF-90–Matrimid membranes were in good agreement with the experimental data. However, deviations were found for ZIF-90–6FDA-DAM membrane which was attributed to the polymer chain rigidification due to the glassy nature of 6FDA-DAM. This result shows the importance of MOF–polymer selection to achieve high gas-separation performance.

Careful selection of MOF–polymer pairs is very important because the performance of MOF-based MMMs depends on the identies of both MOFs and polymers. If the MOF filler has higher permeability and selectivity compared to polymer, then separation performance of MMM can be enhanced. If the MOF filler has a mediocre selectivity or permeability, performance enhancement of MMMs is dependent on the identity of polymers. To illustrate this, Fig. 11 shows selectivity and permeability of MMMs where MAB-JOB with mediocre CO_2–CH_4 selectivity and high CO_2 permeability

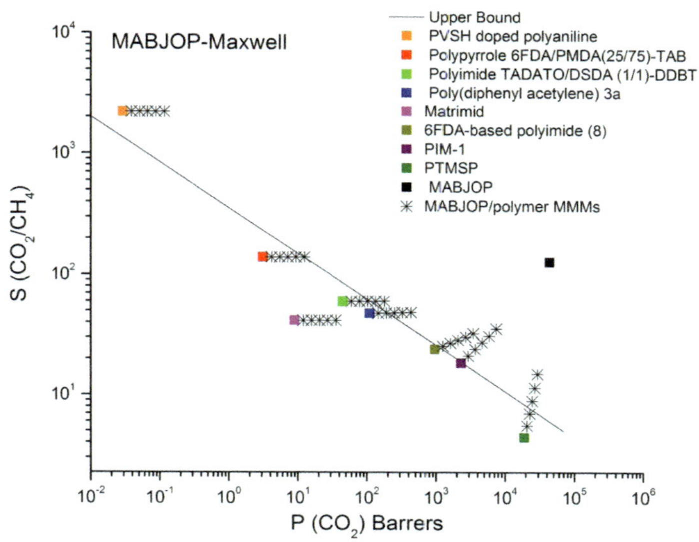

Fig. 11. Maxwell model predictions for CO_2 selectivity and permeability of MOF-based MMMs. Squares represent the performance of pure polymers and pure MOF, MABJOP, and stars represent the performance of MMMs with filler particles having volume fractions of 0.1, 0.2, 0.3, 0.4, and 0.5. Reprinted from the Ref. 83 with permission of the American Chemical Society.

was used as filler particles.[83] The Maxwell model was used to estimate CO_2 permeabilities of MOF-based MMMs. When MABJOP was added into the polymers, the CO_2 permeabilities of all polymers increased since MOF has higher CO_2 permeability compared to the pure polymers. Selectivity of 6FDA-based polyimide, PIM-1, and PTMSP increased, and these MMMs exceeded the upper bound. As shown in Fig. 11, polymers have generally moderate selectivity and permeability, and adding an MOF with medium selectivity/high permeability such as MABJOP can easily enhance separation performances of pure polymers and carry them above the upper bound.[83] More detailed discussions for careful selection of MOF–polymer pairs to achieve high separation efficiency in MMM applications can be found in the literature.[104]

Computational studies that we discussed so far assumed rigid framework in molecular simulations of MOFs. The effect of filler flexibility on the separation performance of MOF-based MMMs was investigated in the literature.[101,102] Yilmaz *et al.*[101] considered different scenarios to examine the effect of MOF's flexibility on CO_2–N_2 separation and concluded that flexibility has a significant effect on the performance of pure membrane but a negligible effect on the performance of MOF-based MMMs. This was attributed to the nature of permeation models which estimate the performance of MMMs based on the polymer phase. The same analysis was also done for PCN-based MMMs, and similar results were reported.[102] As a result, it was concluded that if the MOF's permeability is much higher than the polymer's permeability ($P_{\mathrm{MOF}}/P_{\mathrm{POLYMER}} > 100$) and the volume fraction of the filler is low ($\phi \leq 0.3$), rigid simulations of MOFs can be safely used to calculate gas permeabilities of MOF-based MMMs. Performing flexible simulations is highly costly; therefore, a rigid-framework assumption saves significant time for simulation of MOFs.

Computing electrostatic interactions for quadrupolar gas molecules such as CO_2 and N_2 is also challenging due to the large computation time, especially for gas mixtures. Effects of electrostatic interactions on the performance of MOF-based MMMs were also examined for CO_2–CH_4, CO_2–N_2, and H_2–CO_2 separations.[105] Similar to flexibility discussion, the electrostatic interactions were reported to have negligible effect on the performance of MOF-based MMMs. However, the separation performance of pure MOF membranes changed significantly when the electrostatic interactions between gas molecules and MOF atoms were neglected. It can be concluded that neglecting framework flexibility of MOFs and electrostatic interactions between MOFs and gases is a reasonable assumption to provide an initial estimate about the separation performances of MOF-based MMMs in which the volume fraction of the MOF is low. However, both flexibility and electrostatic interactions must be considered to assess separation performances of pure MOF membranes.

Studies of MMMs that we discussed so far are based on two-phase description: one filler phase and one polymer phase. Ternary systems including two fillers–one polymer and one filler–two polymers were also modeled in the literature. We predicted gas permeability of MMMs having two different types of fillers, a MOF and a zeolite.[101] Permeability of MOF–zeolite–polymer MMMs was calculated by using the permeability of zeolite–polymer membrane and the permeability of MOF–polymer membrane. Results showed that predictions of the Maxwell model are in good agreement with the experimental data for ZIF-8/SC1(silicalite-1)–PSF and CuBTC–SC1–PSF MMMs for CO_2–N_2 and CO_2–CH_4 mixtures. MMMs composed of one type of filler and two different types of polymers were also considered.[101] N_2, CO_2, CH_4, and O_2 permeabilities of ZSM-5–polyimide (PI)–PSF and CuBTC–PI–PSF MMMs were predicted. These predictions agreed well with the experimental data, showing that ternary MMM systems can be also modeled using molecular simulations and theoretical permeation models.

Computational studies of MOF-based MMMs that we discussed so far only simulate MOF phase at the atomic level without simulating the polymer phase. The first fully atomistic approach was reported by Zhang *et al.*[106] who simulated both polymer and MOF phases and calculated solubility and diffusivity of H_2 and CO_2 in ZIF-7–PBI (polybenzimidazole) MMMs using GCMC and EMD simulations, respectively. They reported that both gas permeabilities and mechanical properties of pure PBI membranes are enhanced with the addition of ZIF-7 fillers.[106] In another study,[107] a fully atomistic methodology was developed to study MOF–polymer interface. The surfaces of both MOF and polymer were initially created and then both rigid and flexible MD simulations were performed to characterize the ZIF-8/PIM-1 (polymer of intrinsic microporosity) interface.[107] Results showed that the surface interactions are dependent on the rigidity of the polymer, the energetically favored interaction sites, and the distribution of the MOF pockets. The methodology introduced in that study[107] can be also used to investigate the potential affinity between different polymers and MOF surfaces. This type of modeling studies is highly useful to understand the experimental feasibility of MMMs for further applications.

4. Outlook

We reviewed the current research on computational modeling of thin-film MOF membranes and MOF-based MMMs. We showed that computational predictions agreed well with the experimental measurements for different systems including thin-film MOF membranes, MOF-based MMMs, and ternary MMM systems and proposed that molecular simulations and theoretical permeation models can be used to accurately predict gas-separation performances of MOF membranes. Considering the quick development of the computational techniques, much progress in this exciting field is expected in the future. The major challenges and opportunities in this research area can be discussed as follows:

- Current computational studies on MOF membranes focus on predicting gas separation performances. One important issue is to predict long-term stability of MOF membranes. Long-term industrial use of MOF membranes strongly depends on their stabilities under different chemical environments. Therefore, computational studies that investigate mechanical and thermal stabilities of MOF membranes under realistic operation conditions will be very useful.
- Most computational studies on MOF membranes examine binary gas separations. For example, molecular simulations consider a binary mixture of CH_4 and CO_2 for natural gas purification, N_2

and CO_2 for flue-gas separations. However, other gas species such as water, SO_x, and NO_x also exist in the real mixtures, and these impurities may affect predicted separation performance of MOF membranes. It will be helpful to study multicomponent gas mixtures in molecular simulations to evaluate the real separation performance of MOF membranes in practical applications.

- Considering the large number of available MOF structures, it is challenging to identify the most promising MOF membranes for targeted gas separations. Most of the large-scale computational screening studies examined MOF membranes for CO_2 separation applications. Studies for other industrially important gas separations such as air separation (N_2–O_2), hydrogen recovery (H_2–N_2, H_2–CO, H_2–CH_4), and hydrocarbon separations (C_2H_6–CH_4) will be useful to assess the potential of MOF membranes. The potential of MOF membranes was recently investigated for the separation of olefin–paraffin mixtures which are important for the petrochemical industry. For example, Rungta *et al.*[108] assessed membrane-based gas-separation performance of polymeric membranes, carbon membranes, zeolite membranes, and MOF-based MMMs for C_2H_4–C_2H_6 separations. They showed that ZIF-8-based MMM exhibited much higher C_2H_4 permeability compared to zeolite 4A membrane due to its narrow pore openings. Altintas and Keskin[109] computationally screened a large number of MOFs for adsorption-based separation of C_2H_4–C_2H_6 mixtures and investigated the top performing MOFs for membrane-based separation of the same gas mixture. Their results showed that MOF membranes can outperform zeolites and polymeric membranes in terms of C_2H_4 permeability. More computational and experimental studies are highly required to investigate the separation performance of MOF membranes for olefin–paraffin mixtures.
- Current studies on MOF-based MMMs use the atomistic approach for simulating MOFs and experimentally available data for polymers. There is only one study, to the best of our knowledge, which uses a fully atomistic approach to model MOF–polymer system.[106] Separation performance of MMMs is also dependent on the compatibility between the filler and polymer phases. For

example, improvement of the interface compatibility has been recently observed with the addition of two-dimensional ultra thin MOF nanosheets into the polymer phase which led to a promising MMM for CO_2 separation.[110] Therefore, the fully atomistic approach can be used to better understand the compatibility between the MOF and polymer and to design high quality MOF–polymer membranes.

- Quantitative structure–property relationship (QSPR) analysis correlates structural properties of materials to their functional properties. This type of analyses is very useful to quickly predict the performance of a material using its structural characteristics before performing extensive experiments. Large-scale QSPR analyses have been done for adsorption-based gas separations and gas storage using MOFs.[111] A similar analysis was performed to predict membrane selectivity of 20 PCNs using simple structural parameters such as pore volume, surface area, and pore diameters of materials.[92] Developing QSPR models for large number of MOF membranes will be useful to unlock structure–performance relations and to provide guidelines to experimentalists for the design of high-performance MOF membranes.
- Generic force fields, Dreiding, and UFF are commonly used in molecular simulations of MOFs. However, these force fields are not capable of accurately representing flexible structures because they only consider nonbonded interactions. Bonded force field parameters must be defined for MOF atoms to consider the flexibility of frameworks. In the literature, time-consuming parameterization of the force fields for specific MOFs is performed.[112] However, this type of force fields are material-specific and generally not transferable to different materials. Constructing a flexible force field database which is applicable to all MOFs will be recommended for future studies in this research area.

References

1. Kosinov, N., Gascon, J., Kapteijn, F., Hensen, E. J. Recent developments in zeolite membranes for gas separation. *J. Membr. Sci.* 2016 , 499, 65–79.

2. Baker, R. W. Future directions of membrane gas separation technology. *Ind. Eng. Chem. Res.* 2002, 41, 1393–1411.
3. Li, J.-R., Sculley, J., Zhou, H.-C. Metal-organic frameworks for separations. *Chem. Rev.* 2012, 112, 869–932.
4. Pandey, P., Chauhan, R. S. Membranes for gas separation. *Prog. Polym. Sci.* 2001, 26, 853–893.
5. Qiao, Z.-A., Chai, S.-H., Nelson, K., Bi, Z., Chen, J., Mahurin, S. M., Zhu, X., Dai, S. polymeric molecular sieve membranes via in situ cross-linking of non-porous polymer membrane templates. *Nat. Commun.* 2014, 5, 1–8.
6. Graham, T. On the law of the diffusion of gases. *J. Membr. Sci.* 1995, 100, 17–21.
7. Ulbricht, M. Advanced functional polymer membranes. *Polymer* 2006 47, 2217–2262.
8. (a) Baker, R. W., Low, B. T. Gas separation membrane materials: A perspective. *Macromolecules* 2014, 47, 6999–7013; (b) Sanders, D. F., Smith, Z. P., Guo, R., Robeson, L. M., McGrath, J. E., Paul, D. R., Freeman, B. D. Energy-efficient polymeric gas separation membranes for a sustainable future: A review. *Polymer* 2013, 54, 4729–4761.
9. Robeson, L. M. Correlation of separation factor versus permeability for polymeric membranes. *J. Memb. Sci.* 1991, 62, 165–185.
10. Robeson, L. M. The upper bound revisited. *J. Memb. Sci.* 2008, 320, 390–400.
11. Morooka, S., Kusakabe, K. Microporous inorganic membranes for gas separation. *MRS Bull.* 1999, 24, 25–29.
12. de Vos, R. M., Verweij, H. Improved performance of silica membranes for gas separation. *J. Memb. Sci.* 1998, 143, 37–51.
13. Ismail, A. F., David, L. I. B. A review on the latest development of carbon membranes for gas separation. *J. Memb. Sci.* 2001 193, 1–18.
14. (a) Vu, D. Q., Koros, W. J., Miller, S. J. Mixed matrix membranes using carbon molecular sieves: I. Preparation and experimental results. *J. Memb. Sci.* 2003, 211, 311–334; (b) Das, M., Perry, J. D., Koros, W. J. Gas-transport-property performance of hybrid carbon molecular sieve-polymer materials. *Ind. Eng. Chem. Res.* 2010, 49, 9310–9321.
15. (a) Moore, T. T., Koros, W. J. Sorption in zeolites modified for use in organic-inorganic hybrid membranes. *Ind. Eng. Chem. Res.* 2008, 47, 591–598; (b) Vankelecom, I., Dotremont, C., Morobe, M., Uytterhoeven, J.-B., Vandecasteele, C. Zeolite-filled PDMS membranes. 1. Sorption of halogenated hydrocarbons. *J. Phys. Chem. B* 1997, 101, 2154–2159.
16. Qiu, S., Wu, L., Shi, G., Zhang, L., Chen, H., Gao, C. Preparation and pervaporation property of chitosan membrane with functionalized multiwalled carbon nanotubes. *Ind. Eng. Chem. Res.* 2010, 49, 11667–11675.
17. Car, A., Stropnik, C., Peinemann, K.-V. Hybrid membrane materials with different metal-organic frameworks (MOFs) for gas separation. *Desalination.* 2006, 200, 424–426.

18. Chung, T.-S., Jiang, L. Y., Li, Y., Kulprathipanja, S. mixed matrix membranes (MMMs) comprising organic polymers with dispersed inorganic fillers for gas separation. *Prog. Polym. Sci.* 2007, 32, 483–507.
19. (a) Jeazet, H. B. T., Staudt, C., Janiak, C. Metal-organic frameworks in mixed-matrix membranes for gas separation. *Dalton Trans.* 2012, 41, 14003–14027; (b) Aroon, M. A., Ismail, A. F., Matsuura, T., Montazer-Rahmati, M. M. Performance studies of mixed matrix membranes for gas separation: A Review. *Sep. Purif. Technol.* 2010, 75, 229–242; (c) Bastani, D., Esmaeili, N., Asadollahi, M. Polymeric mixed matrix membranes containing zeolites as a filler for gas separation applications: A review. *J. Ind. Eng. Chem.* 2013, 19, 375–393; (d) Jusoh, N., Yeong, Y. F., Chew, T. L., Lau, K. K., Shariff, A. M. Current development and challenges of mixed matrix membranes for CO_2/CH_4 separation. *Sep. Purif. Rev.* 2016, 45, 321–344.
20. Li, H., Eddaoudi, M., O'Keeffe, M., Yaghi, O. M. Design and synthesis of an exceptionally stable and highly porous metal-organic framework. *Nature.* 1999, 402, 276–279.
21. Frost, H., Düren, T., Snurr, R. Q. Effects of surface area, free volume, and heat of adsorption on hydrogen uptake in metal-organic frameworks. *J. Phys. Chem. B* 2006, 110, 9565–9570.
22. Sethia, G., Sayari, A. Activated carbon with optimum pore size distribution for hydrogen storage. *Carbon.* 2016, 99, 289–294.
23. Eddaoudi, M., Kim, J., Rosi, N., Vodak, D., Wachter, J., O'Keeffe, M., Yaghi, O. M. Systematic design of pore size and functionality in isoreticular MOFs and their application in methane storage. *Science.* 2002, 295, 469–472.
24. Allen, F. H. The Cambridge structural database: A quarter of a million crystal structures and rising. *Acta Crystallogr. Sect. B: Struct. Sci.* 2002 58, 380–388.
25. Wilmer, C. E., Kim, K. C., Snurr, R. Q. An extended charge equilibration method. *J. Phys. Chem. Lett.* 2012, 3, 2506–2511.
26. Eddaoudi, M., Li, H., Yaghi, O. Highly porous and stable metal-organic frameworks: Structure design and sorption properties. *J. Am. Chem. Soc.* 2000, 122, 1391–1397.
27. Ma, S., Zhou, H.-C. Gas storage in porous metal-organic frameworks for clean energy applications. *Chem. Commun.* 2010, 46, 44–53.
28. Keskin, S., van Heest, T. M., Sholl, D. S. Can Metal-organic framework materials play a useful role in large-scale carbon dioxide separations? *Chem Sus Chem.* 2010, 3, 879–891.
29. Erucar, I., Keskin, S. Efficient storage of drug and cosmetic molecules in bio-compatible mOFs: A molecular simulation study. *Ind. Eng. Chem. Res.* 2016, 55, 1929–1939.
30. Della Rocca, J., Liu, D., Lin, W. Nanoscale metal-organic frameworks for biomedical imaging and drug delivery. *Acc. Chem. Res.* 2011, *44*, 957–968.
31. Kuppler, R. J., Timmons, D. J., Fang, Q.-R., Li, J.-R., Makal, T. A., Young, M. D., Yuan, D., Zhao, D., Zhuang, W., Zhou, H.-C. Potential applications of metal-organic frameworks. *Coord. Chem. Rev.* 2009, 253, 3042–3066.

32. Allendorf, M. D., Stavila, V. Crystal engineering, structure-function relationships, and the future of metal-organic frameworks. *Cryst Eng Comm* 2015, 17, 229–246.
33. Liu, Y., Ng, Z., Khan, E. A., Jeong, H.-K., Ching, C.-B., Lai, Z. Synthesis of continuous MOF-5 membranes on porous α-alumina substrates. *Microporous Mesoporous Mater.* 2009, 118, 296–301.
34. (a) Yoo, Y., Lai, Z., Jeong, H.-K. Fabrication of MOF-5 membranes using microwave-induced rapid seeding and solvothermal secondary growth. *Microporous and Mesoporous Mater.* 2009, *123*, 100–106, (b) Zhao, Z., Ma, X., Kasik, A., Li, Z., Lin, Y. S. Gas separation properties of metal organic framework (MOF-5) membranes. *Ind. Eng. Chem. Res.* 2013, 52, 1102–1108, (c) Zhao, Z., Ma, X., Li, Z., Lin, Y. S. Synthesis, Characterization and gas transport properties of MOF-5 membranes. *J. Membr. Sci.* 2011, 382, 82–90.
35. Guo, H., Zhu, G., Hewitt, I. J., Qiu, S. "Twin Copper Source" Growth of metal-organic framework membrane: $Cu_3(BTC)_2$ with high permeability and selectivity for recycling H_2. *J. Am. Chem. Soc.* 2009, 131, 1646–1647.
36. (a) Nan, J., Dong, X., Wang, W., Jin, W., Xu, N. Step-by-step seeding procedure for preparing HKUST-1 membrane on porous α-alumina support. *Langmuir* 2011, 27, 4309–4312, (b) Guerrero, V. V., Yoo, Y., McCarthy, M. C., Jeong, H.-K. HKUST-1 Membranes on porous supports using secondary growth. *J. Mater. Chem.* 2010, 20, 3938–3943, (c) Shah, M. N., Gonzalez, M. A., McCarthy, M. C., Jeong, H.-K. An unconventional rapid synthesis of high performance metal-organic framework membranes. *Langmuir.* 2013, 29, 7896–7902, (d) Mao, Y., Cao, W., Li, J., Liu, Y., Ying, Y., Sun, L., Peng, X. Enhanced gas separation through well-intergrown MOF membranes: Seed Morphology and Crystal Growth Effects. *J. Mater. Chem. A* 2013, 1, 11711–11716, (e) Cao, F., Zhang, C., Xiao, Y., Huang, H., Zhang, W., Liu, D., Zhong, C., Yang, Q., Yang, Z., Lu, X. Helium recovery by a Cu-BTC metal–organic-framework membrane. *Ind. Eng. Chem. Res.* 2012, *51*, 11274–11278, (f) Ben, T., Lu, C., Pei, C., Xu, S., Qiu, S. Polymer-Supported and free-standing metal-organic framework membrane. *Chem. — Eur. J.* 2012, 18, 10250–10253, (g) Nagaraju, D., Bhagat, D. G., Banerjee, R., Kharul, U. K. In situ growth of metal-organic frameworks on a porous ultrafiltration membrane for gas separation. *J. Mater. Chem. A* 2013, 1, 8828–8835, (h) Hara, N., Yoshimune, M., Negishi, H., Haraya, K., Hara, S., Yamaguchi, T. Metal-Organic framework membranes with layered structure prepared within the porous support. *RSC Adv.* 2013, 3, 14233–14236.
37. Adatoz, E., Avci, A. K., Keskin, S. Opportunities and challenges of MOF-based membranes in gas separations. *Sep. Purif. Technol.* 2015, 152, 207–237.
38. Shah, M., McCarthy, M. C., Sachdeva, S., Lee, A. K., Jeong, H.-K. Current status of metal-organic framework membranes for gas separations: promises and challenges. *Ind. Eng. Chem. Res.* 2012, 51, 2179–2199.
39. Venna, S. R., Carreon, M. A. Metal organic framework membranes for carbon dioxide separation. *Chem. Eng. Sci.* 2015, 124, 3–19.

40. (a) Qiu, S., Xue, M., Zhu, G. Metal-organic framework membranes: From synthesis to separation application. *Chem. Soc. Rev.* 2014, *43*, 6116–6140, (b) Yao, J., Wang, H. Zeolitic imidazolate framework composite membranes and thin films: synthesis and applications. *Chem. Soc. Rev.* 2014, 43, 4470–4493, (c) Zhang, C., Koros, W. J. Zeolitic imidazolate framework-enabled membranes: Challenges and opportunities. *J. Phys. Chem. Lett.* 2015, 6, 3841–3849, (d) Caro, J. Are MOF membranes better in gas separation than those made of zeolites? *Curr. Opin. Chem. Eng.* 2011, 1, 77–83.
41. Li, W., Zhang, Y., Li, Q., Zhang, G. Metal-Organic framework composite membranes: Synthesis and separation applications. *Chem. Eng. Sci.* 2015, 135, 232–257.
42. Meek, S. T., Greathouse, J. A., Allendorf, M. D. Metal-organic frameworks: A rapidly growing class of versatile nanoporous materials. *Adv. Mater.* 2011, 23, 249–267.
43. Beckner, M., Dailly, A. Hydrogen and methane storage in adsorbent materials for automotive applications. *Int. J. Energy Res.* 2016, 40, 91–99.
44. Yehia, H., Pisklak, T., Ferraris, J., Balkus, K., Musselman, I. Methane facilitated transport using copper (II) biphenyl dicarboxylate-triethylenediamine/poly (3-acetoxyethylthiophene) mixed matrix membranes. *Polym. Prepr.* 2004, 45, 35–36.
45. Perez, E. V., Balkus, K. J., Ferraris, J. P., Musselman, I. H. Mixed-matrix membranes containing MOF-5 for gas separations. *J. Memb. Sci.* 2009, 328, 165–173.
46. Dong, X., Liu, Q., Huang, A. Highly Permselective MIL-68 (Al)/Matrimid mixed matrix membranes for CO_2/CH_4 separation. *J. Appl. Polym. Sci.* 2016, 133, 43485–43493.
47. (a) Zornoza, B., Tellez, C., Coronas, J., Gascon, J., Kapteijn, F. Metal organic framework based mixed matrix membranes: An increasingly important field of research with a large application potential. *Microporous Mesoporous Mater.* 2013, 166, 67–78, (b) Goh, P. S., Ismail, A. F., Sanip, S. M., Ng, B. C., Aziz, M. Recent Advances of inorganic fillers in mixed matrix membrane for gas separation. *Sep. Purif. Technol.* 2011, 81, 243–264, (c) Rezakazemi, M., Amooghin, A. E., Montazer-Rahmati, M. M., Ismail, A. F., Matsuura, T. State-of-the-art membrane based CO_2 separation using mixed matrix membranes (MMMs): An overview on current status and future directions. *Prog. Polym. Sci.* 2014, 39, 817–861, (d) Seoane, B., Coronas, J., Gascon, I., Benavides, M. E., Karvan, O., Caro, J., Kapteijn, F., Gascon, J. Metal-organic framework based mixed matrix membranes: A solution for highly efficient CO_2 Capture? *Chem. Soc. Rev.* 2015, 44, 2421–2454.
48. Frenkel, D., Smit, B. *Understanding Molecular Simulation: From Algorithms to Applications*, 2nd ed. Academic Press: London, 2002.
49. Sarkisov, L., Martin, R. L., Haranczyk, M., Smit, B. On the flexibility of metal-organic frameworks. *J. Am. Chem. Soc.* 2014, 136, 2228–2231.
50. Smit, B. Phase diagrams of Lennard-Jones fluids. *J. Chem. Phys.* 1992, 96, 8639–8640.

51. Coudert, F.-X., Fuchs, A. H. Computational characterization and prediction of metal-organic framework properties. *Coord. Chem. Rev.* 2016, 307, 211–236.
52. Rappé, A. K., Casewit, C. J., Colwell, K., Goddard Iii, W., Skiff, W. UFF, a full periodic table force field for molecular mechanics and molecular dynamics simulations. *J. Am. Chem. Soc.* 1992, 114, 10024–10035.
53. Mayo, S. L., Olafson, B. D., Goddard, W. A. Dreiding: A generic force field for molecular simulations. *J. Phys. Chem.* 1990, 94, 8897–8909.
54. (a) Keskin, S., Liu, J., Johnson, J. K., Sholl, D. S. Atomically detailed models of gas mixture diffusion through CuBTC membranes. *Microporous and Mesoporous Mater.* 2009, 125, 101–106, (b) Guo, H.-c., Shi, F., Ma, Z.-f., Liu, X.-q. Molecular simulation for adsorption and separation of CH_4/H_2 in zeolitic imidazolate frameworks. *J. Phys. Chem. C* 2010, 114, 12158–12165.
55. (a) Lorentz, H. Ueber die Anwendung Des Satzes vom Virial in der Kinetischen Theorie der Gase. *Ann. Phys. (Berlin, Ger.)* 1881, 248, 127–136, (b) Berthelot, D. Sur le Melange des Gaz. *C. R. Hebd. Seanc. Acad. Sci. (Paris)* 1898, 126, 1703–1855.
56. Xu, Q., Zhong, C. A general approach for estimating framework charges in metal-organic frameworks. *J. Phys. Chem. C* 2010, 114, 5035–5042.
57. Yang, Q., Liu, D., Zhong, C., Li, J.-R. Development of computational methodologies for metal-organic frameworks and their application in gas separations. *Chem. Rev.* 2013, 113, 8261–8323.
58. Buch, V. Path integral simulations of mixed para-D2 and ortho-D2 clusters: The orientational effects. *J. Chem. Phys.* 1994, 100, 7610–7629.
59. Martin, M. G., Siepmann, J. I. Transferable potentials for phase equilibria. 1. United-atom description of n-alkanes. *J. Phys. Chem. B* 1998, 102, 2569–2577.
60. Potoff, J. J., Siepmann, J. I. Vapor-liquid equilibria of mixtures containing alkanes, carbon dioxide, and nitrogen. *AIChE J.* 2001, 47, 1676–1682.
61. Makrodimitris, K., Papadopoulos, G. K., Theodorou, D. N. prediction of permeation properties Of CO_2 and N_2 through silicalite via molecular simulations. *J. Phys. Chem. B* 2001, 105, 777–788.
62. Smit, B., Maesen, T. L. M. Molecular simulations of zeolites: Adsorption, diffusion, and shape selectivity. *Chem. Rev.* 2008, 108, 4125–4184.
63. Einstein, A. On the Theory of the Brownian movement. *Ann. Phys. (Berlin, Ger.)* 1906, 4, 371–381.
64. Darken, L. S. Diffusion, mobility and their interrelation through free energy in binary metallic systems. *Trans. Aime* 1948, 175, 184–201.
65. Keil, F. J., Krishna, R., Coppens, M.-O. Modeling of diffusion in zeolites. *Rev. Chem. Eng.* 2000, 16, 71–197.
66. Skoulidas, A. I., Sholl, D. S. Molecular dynamics simulations of self-diffusivities, corrected diffusivities, and transport diffusivities of light gases in four silica zeolites to assess influences of pore shape and connectivity. *J. Phys. Chem. A* 2003, 107, 10132–10141.

67. Ackerman, D. M., Skoulidas, A. I., Sholl, D. S., Karl Johnson, J. Diffusivities of Ar and Ne in Carbon Nanotubes. *Mol. Simul.* 2003, 29, 677–684.
68. Keskin, S., Sholl, D. S. Efficient methods for screening of metal organic framework membranes for gas separations using atomically detailed models. *Langmuir* 2009, 25, 11786–11795.
69. Basdogan, Y., Sezginel, K. B., Keskin, S. Identifying highly selective metal organic frameworks for CH_4/H_2 separations using computational tools. *Ind. Eng. Chem. Res.* 2015, 54, 8479–8491.
70. Yilmaz, G., Ozcan, A., Keskin, S. Computational screening of ZIFs for CO_2 separations. *Mol. Simul.* 2015, 41, 713–726.
71. Keskin, S., Sholl, D. S. Assessment of a metal-organic framework membrane for gas separations using atomically detailed calculations: CO_2, CH_4, N_2, H_2 mixtures in MOF-5. *Ind. Eng. Chem. Res.* 2009, 48, 914–922.
72. Wesselingh, J., Krishna, R. *Mass Transfer in Multicomponent Mixtures*, 1st ed. Delft University Press: Delft, 2000.
73. Fick, A. V. On liquid diffusion. *The London, Edinburgh, and Dublin Philos. Mag. (1841–1944)* 1855, 10, 30–39.
74. Kesting, R. E., Fritzsche, A. *Polymeric Gas Separation Membranes.* John Wiley & Sons, Inc.: New York, 1993.
75. Krishna, R., van Baten, J. M. In silico screening of zeolite membranes for CO_2 capture. *J. Memb. Sci.* 2010, 360, 323–333.
76. (a) Adatoz, E., Keskin, S. Application of MD simulations to predict membrane properties of MOFs. *J. Nanomater.* 2015, 2015, 1–9, (b) Atci, E., Keskin, S. Understanding the potential of zeolite imidazolate framework membranes in gas separations using atomically detailed calculations. *J. Phys. Chem. C* 2012, 116, 15525–15537.
77. Maxwell, J. C. *A Treatise on Electricity and Magnetism.* Dover Publications: New York, 1954.
78. Bruggeman, V. D. Berechnung Verschiedener Physikalischer Konstanten von Heterogenen Substanzen. I. Dielektrizitätskonstanten und Leitfähigkeiten der Mischkörper aus Isotropen Substanzen. *Ann. Phys.(Leipzig)* 1935, 416, 636–664.
79. Lewis, T., Nielsen, L. Dynamic mechanical properties of particulate-filled composites. *J. Appl. Polym. Sci.* 1970, 14, 1449–1471.
80. Pal, R. New Models for thermal conductivity of particulate composites. *J. Reinf. Plast. Compos.* 2007, 26, 643–651.
81. Moore, T. T., Mahajan, R., Vu, D. Q., Koros, W. J. Hybrid membrane materials comprising organic polymers with rigid dispersed phases. *AIChE J.* 2004, 50, 311–321.
82. Felske, J. Effective thermal conductivity of composite spheres in a continuous medium with contact resistance. *Int. J. Heat Mass Transfer* 2004, 47, 3453–3461.
83. Erucar, I., Keskin, S. Screening metal-organic framework-based mixed-matrix membranes for CO_2/CH_4 separations. *Ind. Eng. Chem. Res.* 2011, 50, 12606–12616.

84. (a) Erucar, I., Keskin, S. Molecular modeling of MOF-based mixed matrix membranes. *Curr. Org. Chem.* 2014, 18, 2364–2380, (b) Erucar, I., Yilmaz, G., Keskin, S. Recent advances in metal-organic framework-based mixed matrix membranes. *Chem. Asian J.* 2013, 8, 1692–1704, (c) Vinh-Thang, H., Kaliaguine, S. Predictive models for mixed-matrix membrane performance: A Review. *Chem. Rev.* 2013, 113, 4980–5028.
85. Bouma, R., Checchetti, A., Chidichimo, G., Drioli, E. Permeation through a heterogeneous membrane: The effect of the dispersed phase. *J. Memb. Sci.* 1997, 128, 141–149.
86. Keskin, S., Sholl, D. S. Selecting metal organic frameworks as enabling materials in mixed matrix membranes for high efficiency natural gas purification. *Energy Environ. Sci.* 2010, 3, 343–351.
87. Atci, E., Keskin, S. Atomically detailed models for transport of gas mixtures in ZIF membranes and ZIF/Polymer composite membranes. *Ind. Eng. Chem. Res.* 2012, 51, 3091–3100.
88. Huang, A., Dou, W., Caro, J. Steam-stable zeolitic imidazolate framework ZIF-90 membrane with hydrogen selectivity through covalent functionalization. *J. Am. Chem. Soc.* 2010, 132, 15562–15564.
89. Krishna, R., van Baten, J. M. In silico screening of metal-organic frameworks in separation applications. *Phys. Chem. Chem. Phys.* 2011, 13, 10593–10616.
90. Krishna, R., van Baten, J. M. Investigating the potential of MgMOF-74 membranes for CO_2 capture. *J. Membr. Sci.* 2011, 377, 249–260.
91. Park, K. S., Ni, Z., Côté, A. P., Choi, J. Y., Huang, R., Uribe-Romo, F. J., Chae, H. K., O'Keeffe, M., Yaghi, O. M. Exceptional chemical and thermal stability of zeolitic imidazolate frameworks. *Proc. Natl. Acad. Sci.* 2006, 103, 10186–10191.
92. Ozturk, T. N., Keskin, S. Computational screening of porous coordination networks for adsorption and membrane-based gas separations. *J. Phys. Chem. C* 2014, 118, 13988–13997.
93. Qiao, Z., Wang, N., Jiang, J., Zhou, J. Design of amine-functionalized metal-organic frameworks for CO2 separation: the more amine, the better? *Chem. Commun.* 2016, 52, 974–977.
94. Zhang, L., Wu, G., Jiang, J. Adsorption and Diffusion of CO_2 and CH_4 in zeolitic imidazolate framework-8: effect of structural flexibility. *J. Phys. Chem. C* 2014, 118, 8788–8794.
95. (a) Bohrman, J. A., Carreon, M. A. Synthesis and CO_2/CH_4 separation performance of Bio-MOF-1 membranes. *Chem. Commun.* 2012, 48, 5130–5132, (b) Xie, Z., Li, T., Rosi, N. L., Carreon, M. A. Alumina-supported cobalt–adeninate MOF membranes for CO_2/CH_4 Separation. *J. Mater. Chem. A* 2014, 2, 1239–1241, (c) Erucar, I., Keskin, S. Computational modeling of bio-MOFs for CO_2/CH_4 separations. *Chem. Eng. Sci.* 2015, 130, 120–128.
96. (a) Ozcan, A., Keskin, S. Effects of molecular simulation parameters on predicting gas separation performance of ZIFs. *J. Chem. Technol. Biotechnol.* 2015, 90, 1707–1718, (b) Liu, B., Smit, B. molecular simulation studies of

separation of CO_2/N_2, CO_2/CH_4, and CH_4/N_2 by ZIFs. *J. Phys. Chem. C* 2010, 114, 8515–8522.

97. Haldoupis, E., Nair, S., Sholl, D. S. Efficient calculation of diffusion limitations in metal organic framework materials: A tool for identifying materials for kinetic separations. *J. Am. Chem. Soc.* 2010, 132, 7528–7539.
98. Watanabe, T., Sholl, D. S. Accelerating applications of metal-organic frameworks for gas adsorption and separation by computational screening of materials. *Langmuir* 2012, 28, 14114–14128.
99. Sitprasert, C., Wang, F., Rudolph, V., Zhu, Z. Ideal and mixture permeation selectivity of flexible prototypical zeolitic imidazolate framework-8 membranes. *Chem. Eng. Sci.* 2014, 108, 23–32.
100. (a) Tong, M., Yang, Q., Ma, Q., Liu, D., Zhong, C. Few-layered ultrathin covalent organic framework membranes for gas separation: A computational study. *J. Mater. Chem. A* 2016, 4, 124–131, (b) Cabrales-Navarro, F. A., Gómez-Ballesteros, J. L., Balbuena, P. B. Molecular Dynamics simulations of metal-organic frameworks as membranes for gas mixtures separation. *J. Memb. Sci.* 2013, 428, 241–250.
101. Yilmaz, G., Keskin, S. Molecular modeling of MOF and ZIF-Filled MMMs for CO_2/N_2 Separations. *J. Memb. Sci.* 2014, 454, 407–417.
102. Altintas, C., Keskin, S. Molecular simulations of porous coordination network-based mixed matrix membranes for CO_2/N_2 separations. *Mol. Simul.* 2015, 41, 1396–1408.
103. Thornton, A. W., Dubbeldam, D., Liu, M. S., Ladewig, B. P., Hill, A. J., Hill, M. R. feasibility of zeolitic imidazolate framework membranes for clean energy applications. *Energy Environ. Sci.* 2012, 5, 7637–7646.
104. Erucar, I., Keskin, S. Computational screening of metal organic frameworks for mixed matrix membrane applications. *J. Memb. Sci.* 2012, 407-408, 221–230.
105. Erucar, I., Manz, T. A., Keskin, S. Effects of electrostatic interactions on gas adsorption and permeability of MOF membranes. *Mol. Sim.* 2014, 40, 557–570.
106. Zhang, L., Hu, Z., Jiang, J. Metal-organic framework/polymer mixed-matrix membranes for H_2/CO_2 separation: A fully atomistic simulation study. *J. Phys. Chem. C* 2012, 116, 19268–19277.
107. Semino, R., Ramsahye, N. A., Ghoufi, A., Maurin, G. Microscopic model of the MOF/Polymer Interface: A First step towards understanding the compatibility in mixed matrix membranes. *ACS Appl. Mater. Interfaces* 2016, 8, 809–819.
108. Rungta, M., Zhang, C., Koros, W. J., Xu, L. Membrane-based ethylene/ethane separation: The upper bound and beyond. *AIChE J.* 2013, 59, 3475–3489.
109. Altintas, C., Keskin, S. Computational screening of MOFs for C_2H_6/C_2H_4 and C_2H_6/CH_4 separations. *Chem. Eng. Sci.* 2016, 139, 49–60.
110. Rodenas, T., Luz, I., Prieto, G., Seoane, B., Miro, H., Corma, A., Kapteijn, F., Llabrés i Xamena, F. X., Gascon, J. Metal-Organic framework

nanosheets in polymer composite mfor gas separation. *Nat. Mater.* 2015, 14, 48–55.
111. Fernandez, M., Woo, T. K., Wilmer, C. E., Snurr, R. Q. Large-scale quantitative structure-property relationship (QSPR) analysis of methane storage in metal-organic frameworks. *J. Phys. Chem. C* 2013, 117, 7681–7689.
112. Tafipolsky, M., Schmid, R. Systematic first principles parameterization of force fields for metal-organic frameworks using a genetic algorithm approach. *J. Phys. Chem. B* 2009, 113, 1341–1352.

Chapter 4

Membrane Processes for N_2–CH_4 Separation

Shiguang Li[*,§], Zhaowang Zong[†], Miao Yu[‡,¶] and Moises A. Carreon[†,||]

[*] *Gas Technology Institute, 1700 S. Mount Prospect Road, Des Plaines, IL 60018*

[†] *Chemical and Biological Engineering Department, Colorado School of Mines, Golden, CO 80401*

[‡] *Department of Chemical Engineering, University of South Carolina, Columbia, SC 29208*

[§] *shiguang.li@gastechnology.org*

[¶] *yumiao@cec.sc.edu*

[||] *mcarreon@mines.edu*

An increasing demand for clean energy has led to a search for alternative energies with less environmental impact, of which natural gas is an important option. The large consumption of natural gas requires an effective technology to improve the purity of natural gas, in which inert gases such as nitrogen play a large part. However, current technologies cannot provide economic solutions to remove nitrogen from natural gas, which leads to a decrease in the heat value of the natural gas. Membrane separation by zeolites represents a viable energy-saving method that can potentially offer an effective way to remove nitrogen from natural gas. In particular, SAPO-34 zeolite membranes have been extensively studied for carbon dioxide separation because of their exceptional molecular-sieving effects, higher thermal and chemical resistances. These membranes can also be effective for N_2–CH_4 separation due to their molecular-sieving effects and adsorption properties. In this mini review, we discuss the different technologies employed for nitrogen rejection from natural gas, emphasizing zeolite membranes as an effective and promising technology to economically separate N_2 from CH_4.

1. Natural Gas Generalities

An increasing demand for clean energy has led to a search for alternative energies with less environmental impact, of which natural gas is an important option. Natural gas represents a good source of supply for a number of reasons, including economy, operation, and environment.[1] According to US Energy Information Administration (EIA),[2] the natural gas consumption has grown from 5.0 trillion cubic feet (TCF) in 1949 to 27.0 TCF in 2014 and is expected to rise to 31.6 TCF in 2040,[3] as shown in Fig. 1. This large demand requires an effective technology to process and purify natural gas.

Natural gas is a fossil fuel formed when buried plants, gases, and animals are exposed to high temperature and pressure over thousands of years. However, a large part of the world's natural gas to date is currently unavailable for production because of the limitations in separation technology.[4] The natural gas consists primarily of methane (CH_4) and also includes varying amount of other higher alkanes, carbon dioxide (CO_2), nitrogen (N_2), and hydrogen sulfide, as shown in Table 1. These inert gases with large mole fraction, such as CO_2 and N_2, could decrease the heat value of the natural

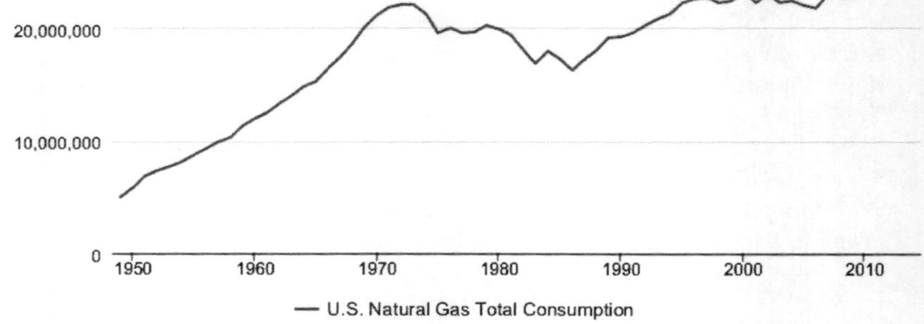

Fig. 1. U.S. natural gas total consumption.[2]

Table 1. Typical composition of natural gas.[5]

Component	Formula	Mole fraction
Methane	CH_4	70–90%
Ethane	C_2H_6	
Propane	C_3H_8	0–20%
Butane	C_4H_{10}	
Carbon dioxide	CO_2	0–8%
Oxygen	O_2	0–0.2%
Nitrogen	N_2	0–5%
Hydrogen sulfide	H_2S	0–5%
Rare gases	Ar, He, Ne, Xe	Trace

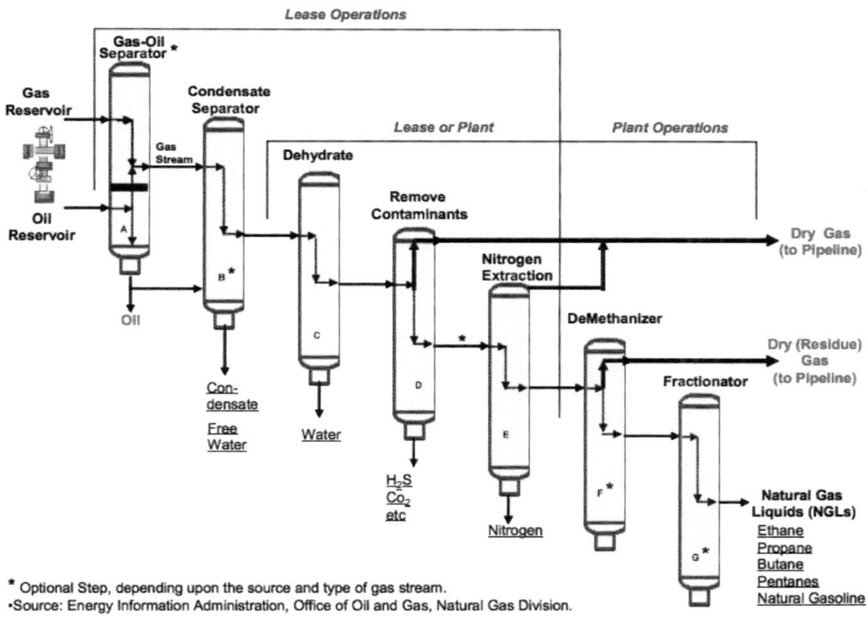

Fig. 2. Generalized natural gas processing schematic.[6]

gas. Therefore, it is necessary to purify the natural gas in order to improve its heat content and to avoid the erosion of pipelines from CO_2 in the presence of moisture.

Figure 2 shows a generalized schematic to process natural gas. The number of steps and the types of techniques used depend upon the source and composition of the wellhead production stream. In the first step, gas–oil separators are employed to separate gas hydrocarbons from the heavier oil by gravity. The resulting gas flows into a condensate separator to remove free water. However, this step cannot remove all the free water. The residue free water may form hydrates under specific temperature/pressure conditions. So a dehydration unit is included to further remove the free water. A contaminant-removal unit follows to eliminate hydrogen sulfide, CO_2, and water vapor to maintain acceptable levels. Once processed, the stream is routed to a N_2 extraction unit to remove the inert N_2 to maintain allowed concentration in pipeline specifications. A CH_4 separation unit is incorporated to demethanize the gas stream. This is used to separate CH_4 from natural gas liquids. These liquids undergo a fractionation process by varying boiling points of the individual hydrocarbons in the stream. Not all these units are required for a specific gas reservoir.

2. Overview of Natural Gas N_2 Rejection Technologies

The typical U.S. natural gas pipeline specification for inert gases is less than 4%. Approximately 14% of U.S. natural gas contains >4% nitrogen, and thus nitrogen-rejection technologies are required.[4] Most of the nitrogen-rejection plants in the U.S. use a cryogenic distillation process. The basic principle is to liquefy the natural gas to the point where most of the gas is in the liquid phase and then distill the resulting liquid..

Cryogenic plants are most suited to large natural gas wells that can deliver 50–500 million standard cubic feet per day (MMSCFD) for 10–20 years.[4] These large wells allow the high capital cost of the cryogenic plant to be defrayed over several years. Principal variations in the cryogenic process involve whether to use one or two columns (at different pressures).

The cost of cryogenic process is high and sensitive to the plant size. For a plant capacity of 20 MMSCFD assuming $4/MSCF (thousand standard cubic feet) gas price for the plant and 15 vol.%

N_2 in feed, the cost is approximately \$0.80/MSCF, whereas for a plant capacity of 1 MMSCF/day, the cost is as high as about \$2.40/MSCF.[7]

In addition to the cryogenic processing plants, other technologies are highlighted below:

2.1. *Nitrotec Energy's PSA Process*

Nitrotec Energy has installed three plants of essentially the same size and design in West Texas. Nitrotec Energy previously installed helium recovery and purification plants in Colorado using this technology. Their process utilizes a specially treated carbon molecular sieve (CMS) to remove hydrocarbons from the stream containing N_2. The process operates at an optimum pressure between 30 and 50 psig, making it more suitable for the low pressure-associated gas typical of the Permian Basin. The stream flows through the CMS bed with the hydrocarbons being trapped, and the N_2 is vented. The CMS bed is depressurized, with about 95% of the hydrocarbons being recovered. Some gas-containing hydrocarbons must be compressed for recycling, and the recovered hydrocarbon stream must be pumped from a vacuum pressure and then further compressed to pipeline delivery pressure.

UOP has also provided a PSA process for a high N_2 gas stream in North Central Texas. Their process appears to be similar to Nitrotec's process although they indicate a pressure range of 50–300 psig, which is somewhat greater than Nitrotec's.

2.2. *Engelhard Molecular Gate*

The Engelhard Molecular Gate absorbent is a new type of molecular sieve that has the unique ability to adjust pore size openings within an accuracy of 0.1 Å. The pore size is precisely adjusted in the manufacturing process to allow the production of a molecular sieve with a pore size tailored to size-selection separations.

Nitrogen and methane molecular diameters are approximately 3.6 and 3.8 Å, respectively. In a Molecular Gate adsorbent-based system for upgrading nitrogen-contaminated natural gas, a pore size of 3.7 Å is used. This adsorbent permits the nitrogen (as well as carbon dioxide, hydrogen sulfide, water, and oxygen) to enter the pore and be adsorbed while excluding the methane, which passes through the fixed bed of adsorbent at essentially the same pressure as the feed.

The adsorbent is applied in a pressure swing adsorption system where the system operates by swinging the pressure from a high pressure for the absorption step to a low pressure for the regeneration step to remove the previously adsorbed nitrogen. The swings between the high and low pressure levels are completed in the order of minutes rather than hours to minimize adsorbent inventory.

In general, the system consists of four to seven adsorber vessels, with four being the most typical number. Buffer tanks will smoothen flow fluctuations. These are designed to operate unattended, with daily monitoring by a well pumper or equivalent.

2.3. *Mehra Process^{SM}'s Solvent Process*

The Mehra Process^{SM} technology uses a refrigerated solvent to absorb methane and heavy hydrocarbons from a stream with high N_2. The process has been tested on a natural gas stream containing 16% N_2 producing a sales gas stream containing 2% N_2. However, the N_2 vent stream contained 14.1% hydrocarbons. Fuel is required for recompression of the hydrocarbon streams produced at various let-down pressures during the separation of the hydrocarbons from the solvent. Fuel is also required for a refrigeration system used to cool the inlet gas and recycle solvent prior to entering the absorber. Fuel is also required for regeneration of the glycol used for injection to prevent hydrates. Electrical energy is consumed primarily for the large solvent pumps.

This process may be economical in situations where part or the entire produced hydrocarbon can be used at low pressure such as for fuel. Otherwise, a significant part of the hydrocarbon stream would require compression from near-atmospheric pressure. It is indicated solvent circulation would have to be significant to assure minimum hydrocarbon content in the N_2 vent. The Mehra Process^{SM} should also be more efficient where the N_2 is to be recovered and reinjected for field pressure maintenance since the N_2 stream can be available at a higher pressure.

2.4. *Bend Research Process*

Bend Research was developing a metal–liquid transition absorbent process in the 1990s but has abandoned the development effort. The use of a solvent in this process to separate N_2 from natural gas is similar to the Mehra Process^{SM}. However, the Mehra Process^{SM} absorbs the hydrocarbon selectively away from the N_2, and the Bend process absorbs the N_2 away from the hydrocarbon.

In the Bend process, therefore, the N_2 which is normally vented would be separated from the solvent at very low pressures. The Bend process should be much more energy efficient than both the cryogenic processes, which required significant refrigerant and recompression horsepower, and the Mehra Process^{SM}, which also requires some refrigerant horsepower and a significant amount of recompression horsepower for the low-pressure hydrocarbon stream. Bend expected this process to result in processing economics close to the amine solvent type CO_2 removal process. In the Bend process, the N_2 is "complexed" by the transition metal compound. The advantage of this process would have been that the hydrocarbon gas does not have to be reduced in pressure and then recompressed like the other NRU processes discussed earlier. A small amount of methane is physically dissolved by the solution but most can be recovered in an intermediate pressure flash tank. The process is anticipated to operate very similar to the amine-treatment process for CO_2 removal, which has a contactor and a regeneration process for the amine. Bend anticipated this process to result in the processing economics for N_2 removal to be similar to the economics for amine solvent CO_2 removal.

2.5. *Membrane Process*

Membrane separation offers the advantage of low energy cost relative to the more established gas-separation processes such as adsorption

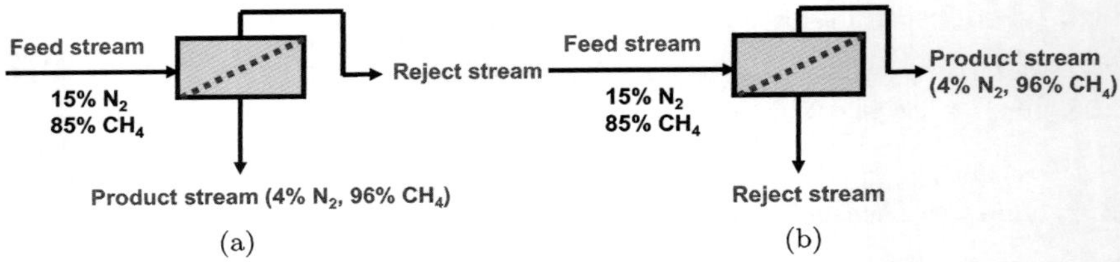

Fig. 3. Membrane processes for N_2 rejection using (a) CH_4-selective membranes and (b) N_2-selective membranes. Reprinted from Ref. 8 with permission of the American Chemical Society.

and cryogenic distillation. For membrane separation, the challenge is to develop membranes with the necessary nitrogen–methane separation characteristics. Two types of membranes can be used in N_2 rejection: CH_4-selective or N_2-selective membranes (Fig. 3).

CH_4-selective membranes: For CH_4-selective membranes, CH_4 is more permeable than N_2, so it is concentrated in the low-pressure permeate side and needs to be recompressed to pipeline pressure before entering the pipeline (Fig. 3(a)). Meyer and Henson[9] evaluated the economics for CH_4-selective membranes. The effect of selectivity on membrane area, percentage of CH_4 recovery, and processing cost for a single-stage system for a highly permeable CH_4 selective membrane is shown in Fig. 4. Increasing selectivity increases the methane recovery but has a very small effect on membrane area as expected. The processing cost drops over the range but is not in the range of economic interest well beyond selectivities of 7.

Various two- and three-stage systems were analyzed. In multiple stage configurations, recycling and compression are used to minimize hydrocarbon loses. Figure 5 shows that the two-stage system with residual recycle had the lowest area and compression costs of the two-stage systems, while the three-stage system had the overall lowest operating cost. The estimated cost is, however, too high to be of commercial interest.

N_2-selective membranes. For a process using N_2-selective membranes, CH_4 remains in the retentate stream (Fig. 3b) at high pressures, and thus there is a significant saving in recompression cost as compared to CH_4-selective membranes. However, the

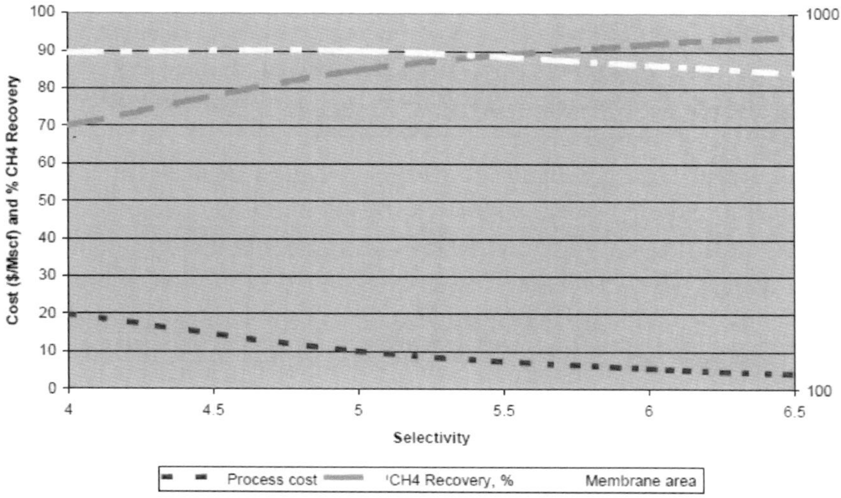

Fig. 4. Effect of selectivity for single-stage system using CH_4 selective membrane.

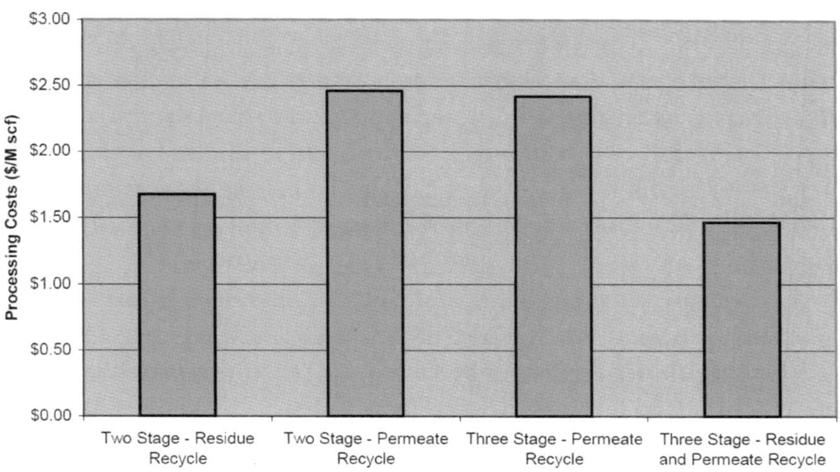

Fig. 5. Processing costs for multiple stage systems.

state-of-the-art N_2-selective polymeric membranes display low N_2 permeabilities, typically less than 10 barrers (1 barrer = 10^{-10} cm^3 (STP) $\cdot$ cm/(cm$^2 \cdot$ s$\cdot$ cmHg)) at a selectivity of approximately 3, as shown in the "Robeson" plot.[10]

3. Nitrogen-Selective Membranes

Recently, some inorganic materials have been made into N_2-selective membranes and shown N_2–CH_4 separation performances above the so-called upper bound observed for dense polymers.[10]

Ning and Koros,[11] using a carbon molecular sieve (CMS) membrane, obtained a N_2/CH_4 selectivity of 7.7 and a N_2 permeability of 6.8 barrers at 24°C, which surpasses the polymer membrane upper bound line. CMS membranes are pyrolyzed from polymer membrane cursors under controlled atmosphere and temperature protocol. In this process, scission and reorganization of polymer main chains occur, with small gas molecules released and free radicals formed. Those free radicals can form interchain or intrachain cross-linking to obtain conjugated aromatic 3D network with graphene-like structure.[12] With increasing temperature, those can form ordered structures and trubostratic structures, which are amorphous and isotropic. The packing imperfections between ordered regions form "slit-like" pore structure, combined with a bimodal pore distribution, which are described CMS structures-"micropores" with size of 0.7–2 nm connected by molecular-sieving "ultramicropores" with size <0.7 nm. This morphology enables high permeance and selectivity. This membrane was pyrolyzed from a commercial polyimide, Matrimid® 5218 as a precursor, but can show significantly better separation performance than the polymer precursor dense film. However, this CMS membrane only possesses a moderate N_2 permeance.

Wu *et al.*[13] prepared SSZ-13 zeolite membranes for N_2–CH_4 separation and obtained selectivity of 13 and N_2 permeance of 2.2×10^{-8} mol/($m^2 \cdot s \cdot Pa$) (=66 GPU) at 20°C for a feed pressure of 270 kPa. This SSZ-13 zeolite displays chabazite topology. Its framework contains a tridimensional pore system made up of 0.38 nm, eight-membered oxygen rings. Unlike the SAPO-34 zeolite, the SSZ-13 does not contain phosphorous, but it has a high Si/Al ratio, which possesses the increased ability to water. The SSZ-13 zeolite membrane was prepared by secondary seeded growth on the outer surface of the tubes, which is different from other zeolite membranes. However, the SSZ-13 still cannot provide sufficient N_2 permeance to economically reject N_2 during separation.

Of all reported membranes for N_2–CH_4 separation, SAPO-34 membranes have displayed the best separation performance for this binary gas mixture. Therefore, the rest of this mini-review focuses on SAPO-34 membranes for N_2–CH_4 separation.

4. SAPO-34 Zeolites

Zeolites are hydrated aluminosilicate minerals made from interlinked tetrahedral of alumina (AlO_4) and silica (SiO_4). These tetrahedrally coordinated atoms (T-atoms) are coordinated to four oxygen atoms and linked to other T-atoms by sharing oxygen with a neighboring T-atom tetrahedron.[14] Zeolites are also defined as microporous solids with a relatively open, 3D crystal structure built from the elements aluminum, oxygen, and, silicon, with alkali or alkaline-earth metals plus water molecules trapped in the gaps between them. Figure 6 is a typical schematic of zeolites. In this figure, the "–" sign on the framework means a negative charge and the "x" sign means exchangeable cations which can balance the charges on the framework to form neutral charge framework.[15]

Zeolites are one of the most important porous materials for a series of applications, including adsorption, catalysis, and gas separations.[17–27] Zeolites exhibit very uniform pore size distribution, high

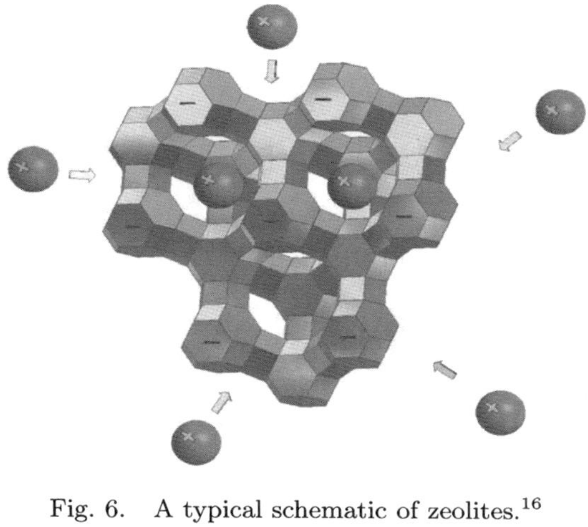

Fig. 6. A typical schematic of zeolites.[16]

specific surface area, high porosity, variable chemical composition, and controllable acid–base properties.[14] These characteristics allow zeolites to be promising functional materials.

Silicoaluminophosphates (SAPOs) are crystalline nanoporous zeolites formed by silicon, aluminum, phosphorus, and oxygen atoms in tetrahedral coordination and arranged in an orderly structure. The first reported SAPO was disclosed in 1984.[28] This family of SAPO-*n* exhibits structure diversity with different frameworks, having properties both of aluminophosphates and low silica to alumina ratio zeolites.[15]

SAPO-34 belongs to SAPO-n family having the composition $Si_xAl_yP_zO_2$ where x, y, and z represent the mole fractions of silicon, aluminum, and phosphorus and range from 0.01 to 0.98, 0.01 to 0.60, and 0.01 to 0.52, respectively, with $x + y + z = 1$.[15] It has framework characteristics of natural zeolite chabazite as shown in Fig. 7. This molecular sieve includes eight-ring apertures with a pore diameter of ~0.38 nm that permit access to 3D channel and cage systems. This geometry allows small molecules to diffuse easily

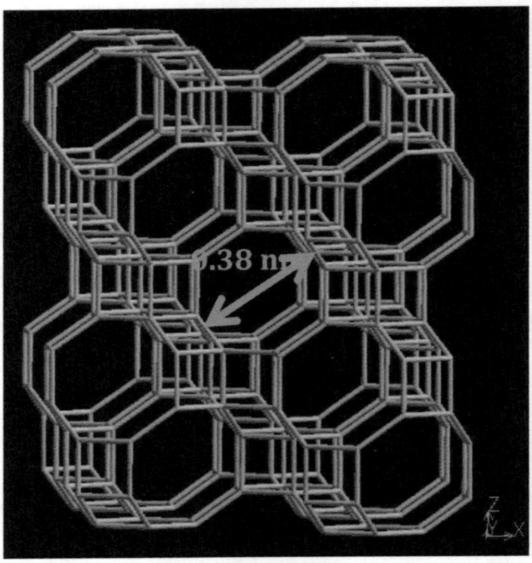

Fig. 7. SAPO-34 crystal structure and pore opening.[29]

through the crystal structure. These have anionic frameworks with a net negative charge depending upon the silicon substitution into the frameworks.[15]

SAPO-34

was first discovered in Union Carbide Corporation.[28] SAPO-34 has been prepared by different templates, such as tetraethylammonium hydroxide (TEAOH),[15] morpholine,[30] piperidine,[31] diethylamine (DEA),[32,33] triethylamine (TEA),[34,35] isopropylamine,[36] TEAOH-dipropylamine (DPA),[37] and TEAOH-cyclohexylamine (CHA).[37] The synthetic approaches include different heat treatment methods, such as hydrothermal treatment, two-stage temperature varying hydrothermal treatment synthesis,[30] microwave treatment,[38] vapor phase transport method,[39–41] ultrasound irradiation synthesis.[42] However, the control of crystal size of SAPO-34 with high surface area, high crystallinity, and preferential adsorption capacity is still challenging.

Microwave treatment is a useful approach, which has been successfully used in zeolite and zeolite membrane's synthesis. The first published article related to microwave synthesis of microporous zeolite appeared in 1993,[43] in which microwave-assisted crystallization of Y-type and ZSM-5 zeolites could be achieved in much shorter time and without undesired phases compared with conventional heating.

Conventional heating relies on a heat source on the outside and conducts the heat to the center from the outside. Instead, the microwave treatment has the following advantages compared to conventional heating[44]: (1) the microwave heating has much higher heating rates than that of conventional heating; (2) the microwaves treat chemical reactors remotely without direct contact between energy source and the reacting chemicals; (3) there is no wall or heat-diffusion effects; (4) it can selectively heat because the chemicals and the contaminants do not interact equally with microwaves; (5) there are no "hot spots." Microwave heating leads to more uniform crystals, which can potentially form thinner continuous membranes. However, this method can lead to a compositional difference of the synthesized zeolites.

5. SAPO-34 Membranes for N_2–CH_4 Separation

In the following paragraphs, we describe and discuss the state-of-the-art work related to SAPO-34 membranes for N_2–CH_4 separation.

Wu *et al.*[13] prepared SAPO-34 membranes towards N_2–CH_4 separation for the first time with the same recipe as the membranes for CO_2–CH_4 separation. The authors also studied the effects of adding propane in the gas mixture, which decreased the N_2 permeance.

Li *et al.*[8] explored synthesis parameters for the development of SAPO-34 membranes. These parameters including water content in the gel, crystallization time, support pore size, and aluminum source were investigated and optimized. N_2 and CH_4 adsorption isotherms were measured on SAPO-34 crystals collected from the best SAPO-34 membrane. The heats of adsorption for N_2 and CH_4 on SAPO-34 were also measured to explain the effect of competitive adsorption in mixture separation. The dependences of the N_2–CH_4 separation performance on temperature and feed pressure were also studied.

In Zong *et al.*,[45] the synthesis conditions for SAPO-34 membranes were further optimized towards N_2–CH_4 separation. Pervious SAPO-34 membranes have been prepared with water mole contents of 55–150. Even with a water content of 150, phase separation of gel occurred during the synthesis of the tubular membranes, and settling of crystals to the bottom of the autoclave was observed. These could decrease the quality of the membrane. To avoid gel-phase separation and settling of crystals to the bottom of the autoclave, water contents of 150, 200, 250, 300, and 400 were investigated, and two membranes were prepared at each water content.

Huang *et al.*[46] prepared slice-shaped SAPO-34 zeolite seeds and used these seeds with high aspect ratio (AR) for preparing high quality thin SAPO-34 membranes towards high N_2 permeance and improving N_2–CH_4 selectivity. SAPO-34 seeds with larger ARs facilitated the formation of thinner continuous membranes upon secondary crystallization. The exceptional separation performance for equimolar N_2–CH_4 mixture (50–50) was achieved with the thinnest membranes (~2 μm) prepared using seeds with an average AR of 10.

5.1. *SAPO-34 Membranes Synthesis*

SAPO-34 membranes have been synthesized via secondary seeded growth onto the inside surfaces of porous supports.[8,13,45,46] These porous tubes were typically 6-cm-long with effective areas between 6.5 and 7.3 cm^2. The success of synthesizing continuous and defect-free SAPO-34 membranes depends to a great extent on the quality of the crystals used as "seeds" to promote membrane growth. In the following paragraphs, we describe the reported research efforts on the synthesis of SAPO-34 seeds.

5.1.1. *Synthesis of SAPO-34 Seeds*

Wu *et al.*[13] prepared SAPO-34 seeds by microwave heating at 180°C for 7 h with the gel composition of 1.0 Al_2O_3: 1.0 P_2O_5: 0.6 SiO_2: 2.0 TEAOH: 75 H_2O. No results were shown in the study.

Li *et al.*[8] and Zong *et al.*[45] prepared the SAPO-34 seeds with a molar composition of 1.0 Al_2O_3: 2.0 P_2O_5: 0.6 SiO_2: 4.0 TEAOH: 150 H_2O. In a typical synthesis, aluminum isopropoxide ($Al(i\text{-}C_3H_7O)_3$, Aldrich 99.99%) and deionized water were stirred for 1 h to form a homogeneous solution followed by addition of phosphoric acid (H_3PO_4, Sigma-Aldrich 85 wt.% aqueous solution) for another 2 h of stirring. This mixing order was different from previous studies reported by Wu *et al.*[13] and Funke *et al.*[47] where these three chemicals were added at the same time. Next, Ludox AS-40 colloidal silica (40 wt.% suspension in water) was added, and the resulting solution was stirred for another 3 h. Tetraethylammonium hydroxide (TEAOH, Aldrich 35 wt.% aqueous solution) was then added, and the solution was stirred overnight at room temperature (~22°C). The gel was placed in an autoclave and heated in a microwave oven (CEM Mars 5 Microwave Reaction System with XP-1500 plus control vessel) at 180°C for 7 h. After the solution was cooled to room temperature, it was centrifuged at 3300 rpm for 10 minutes to collect the seeds, which were followed by deionized water washing. The process was repeated three times, and the resulting precipitate was dried overnight at 100°C.

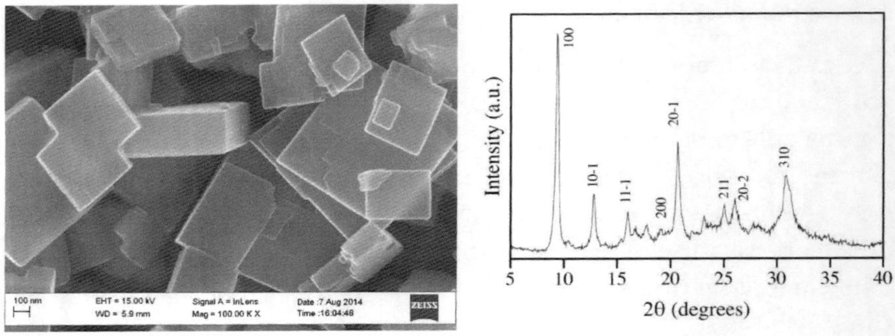

(a) (b)

Fig. 8. (a) SEM image of the seeding crystals synthesized by microwave treatment; (b) XRD pattern of the SAPO-34 seeds. Reprinted from Ref. 8 with permission from Elsevier.

Figure 8(a) shows an SEM image of the seeding crystals prepared by microwave heating. These crystals are composed of nanosheets with lengths and widths between 200 and 800 nm and thicknesses between 50 and 220 nm.[8] XRD confirms the CHA topology typical of SAPO-34 (Fig. 8(b)).

Layered zeolitic materials in nanosheet form have been used for preparing ultrathin zeolite membranes by combining filtration deposition of a thin zeolite coating with subsequent short-period hydrothermal growth.[46] For instance, MFI zeolite nanosheets (~3 nm in thickness) have been applied to the growth of thin MFI zeolite films which showed superior performance in p-xylene/o-xylene separation.[48] Although it is highly desirable to use thin zeolite nanosheets for growing zeolite membranes, only MWW and MFI zeolites were currently available in crystalline nanosheet form. Attempts to exfoliate other layered zeolites, including certain layered silicates and aluminophosphates with microporous layers, did not preserve the crystallographic order of the layers.[49] Thus, preparation of zeolite nanosheets or high-aspect-ratio slice-shaped zeolites is still highly challenging, yet desirable, for preparing thin and high quality zeolite membranes.[46]

Huang *et al.*[46] prepared highly crystalline SAPO-34 seeds with average AR of ~1, 3, 10, and 20, and submicron lateral sizes were

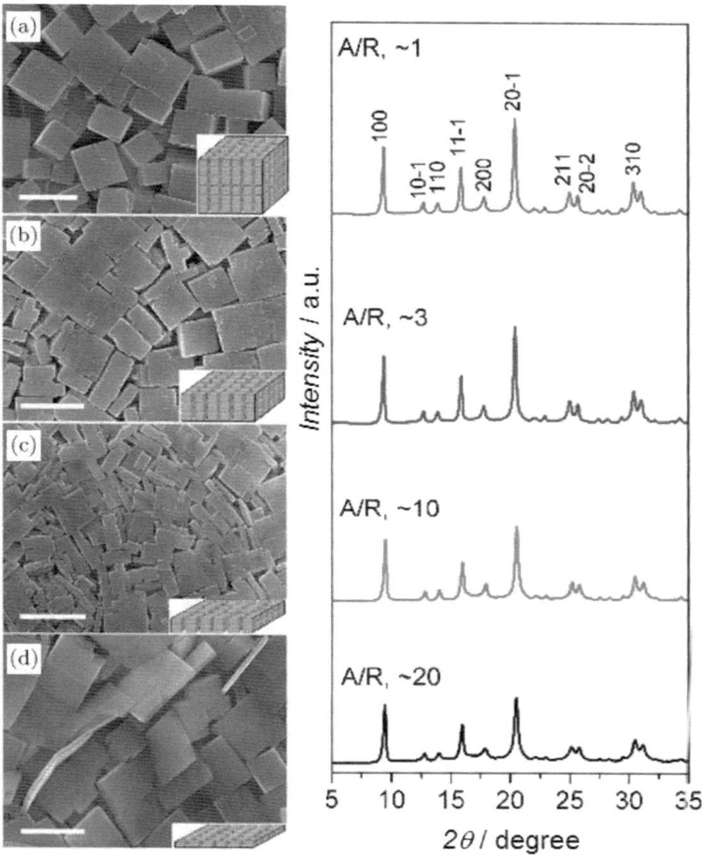

Fig. 9. SEM images (left column) and XRD patterns (right) of SAPO-34 seeds with average AR of (a) 1, (b) 3, (c) 10, and (d) 20. The inset of each SEM image provides a schematic drawing of the typical morphology of a seed. Scale bar in (a), (b), and (c) is 1 μm; in (d), 0.5 μm. Reprinted from Ref. 46 with permission of Wiley-VCH.

synthesized hydrothermally (Fig. 9) through careful and extensive recipe constructions and fine tuning of synthesis conditions. XRD patterns of all as-synthesized seeds can be easily indexed to the characteristic CHA structure, indicating that they were phase pure and highly crystalline (Fig. 9, XRD).

SAPO-34 cubes with AR of ~1 (denoted as S1, Fig. 9(a)) and particle size distribution in the range of 0.5~1.2 μm (average size = 0.8 μm) were prepared at 200°C for 24 h using a gel composition of 1.0 Al_2O_3: 0.6 SiO_2: 1.0 P_2O_5: 60.0 H_2O: 2.0 TEAOH.[46] Thinner seeds with AR of ~3 (denoted as S2, Fig. 9(b)) and ~10 (denoted as S3, Fig. 9(c)) were prepared by using a modified gel composition (1.0 Al_2O_3: 0.6 SiO_2: 2.0 P_2O_5: 75.0 H_2O: 4.0 TEAOH) and crystallization at 200°C and 210°C, respectively, for 24 h. S2 seeds with a thickness of 0.3 ± 0.075 μm had a fairly uniform lateral size (0.85 ± 0.2 μm), whereas much thinner S3 seeds had a broader lateral size distribution (0.4–1.2 μm). This may be because of the shorter gel mixture aging time (72 h) for S3 than that for S2 (144 h). In zeolite synthesis, extended aging of gel mixture favors the formation of more nuclei and thus might affect the size, morphology, and uniformity of the final product. However, the majority of S3 seeds were over 0.6 μm in lateral size and less than 0.1 μm in thickness, which yielded an average AR of ~10 (based on size examinations of more than 5000 seeds). It is also worthwhile mentioning that in this sample the AR could reach as high as 20 for a small fraction, 6–9%) of crystals with a thickness of ~0.05 μm and lateral sizes over 1 μm. Moreover, the as-synthesized S3 seeds were highly dispersible in water and easily deposited as oriented layers on flat surfaces, indicative of their good structural stability in thin slice form.

SAPO-34 seeds with further reduced thickness (denoted as S4) were also synthesized at 170°C for 24 h with a gel composition of 1.0 Al_2O_3: 0.6 SiO_2: 1.0 P_2O_5: 55.0 H_2O: 1.5 TEAOH.[46] Their lateral sizes were in the range of 0.1–0.7 μm (0.5 μm in average) and smaller than those of S1–S3. However, most of the seeds were thinner than 0.05 μm with an average AR of ~20. Some SAPO-34 crystals (1–3% in quantity) observed under microscope were as thin as 0.015 μm (15 nm) which corresponds to approximately 10 CHA unit cells (unit cell dimensions: $a = 13.68$ Å; $b = 13.68$ Å; and $c = 14.77$ Å). Moreover, those thin seeds looked semi transparent under the scanning electron microscope (SEM) observation with an acceleration voltage of 15 kV, again indicating very thin thickness of the seeds. More importantly, these very low SAPO-34 seeds still preserved high crystallinity according to XRD study (Fig. 9). Authors claimed that S4 contains the thinnest SAPO-34 crystals reported thus far which corresponds to ~10 unit cells in thickness.[46]

5.1.2. *Preparation of SAPO-34 Membranes*

SAPO-34 zeolite membranes have been extensively studied because of their exceptional molecular-sieving effects, higher thermal and chemical resistance. SAPO-34 membranes have been successfully used mainly for CO_2–CH_4[23,47,50–53] and CO_2–N_2 separations.[27,54,55] Typically, a continuous polycrystalline layer of SAPO-34 membrane is prepared on a porous support to ensure mechanical strength.[56] The membranes are mainly prepared by *in situ* synthesis and secondary seeded growth. For *in situ* crystallization, the porous support is immersed into the synthesis solution, and then the membrane is formed by direct crystallization. "Seeds" or crystals can be used to promote the formation of a well-intergrown membrane.[57] This method gives a better control over the membrane formation by separating the crystal nucleation and growth with a shortened crystallization time.[58] In addition, it ensures the formation of the pure-phase zeolite crystals on the support.[59] Secondary seeded growth is preferred during the SAPO-34 membrane synthesis.

In recent reports,[8,13,45,46] porous alumina tubes were used as porous supports to grow SAPO-34 membranes. The inside of these supports was seeded by rubbing with SAPO-34 seeds. The outside of the supports was wrapped with Teflon tape to avoid growth of membranes on the outside surface. After seeding, the supports were placed in an autoclave filled with the synthesis gel. Usually two seeded supports were placed in one autoclave, which was then filled with the synthesis gel to about 0.5 cm above the top of the supports. Hydrothermal treatment was carried out in a conventional oven at 220°C for 6–24 h. After this step, the membranes were washed for 15 min with flowing tap water and dried at 100°C for ~2 h. The membranes were calcined at 400°C for 4 h with heating and cooling rates of 0.8°C/min under weak vacuum (pressure of 300 Pa).

Wu *et al.*[13] prepared SAPO-34 membranes by secondary seeded growth on the inside of seeded, porous Al_2O_3 tubes (1 cm O.D., 100 nm average pore size, 40–55% porosity). The membrane synthesis was carried out at 180°C for 6 h using a gel composition of 0.85 Al_2O_3: 1.0 P_2O_5: 0.3 SiO_2: 2.0 TEAOH: 155 H_2O. The template was removed by vacuum calcination at 450°C for 6 h with heating and cooling rates of 1°C/min. The author reported a N_2–CH_4 selectivity 5–7 for a feed pressure of 350 kPa, and the N2 permeance was ~300 GPU.

Li *et al.*[8] prepared SAPO-34 membranes with several gel compositions containing Al_2O_3, P_2O_5, and SiO_2 sources, templates, and H_2O. Al(i-C_3H_7O)$_3$ and $Al(OH)_3$ were used as Al_2O_3 sources, whereas H_3PO_4 and Ludox AS-40 were used as P_2O_5 and SiO_2 sources, respectively. TEAOH was used as pivotal template for the formation of SAPO-34 structure, and DPA was employed as secondary structure directing agent. For example, a typical synthesis using Al(*i*-C_3H_7O)$_3$ as Al_2O_3 source and dual templates had a gel composition of 1.0 Al_2O_3: 1.0 P_2O_5: 0.3 SiO_2: 1.0 TEAOH: 1.6 DPA: 150 H_2O. To prepare the gel, Al(*i*-C_3H_7O)$_3$ and deionized water were stirred for 1 h to form a homogeneous solution, followed by addition of H_3PO_4 into the gel for another 2 h of stirring. Then Ludox AS-40 colloidal silica was added, and the resulting solution was stirred for another 3 h. The TEAOH (Aldrich 35 wt.% aqueous solution) was then added, followed by addition of DPA (Acros 99%) after 1 h. The gel was stirred at 45–50°C for 4 days before it was used in membrane growth.

Zong *et al.*[45] optimized synthesis conditions for SAPO-34 membranes toward N_2–CH_4 separation by diluting the membrane gels. The gel composition was 1.0 Al_2O_3: 1.0 P_2O_5: 0.3 SiO_2: 1.0 TEAOH: 1.6 DPA: $x H_2O$. Figure 10 shows XRD patterns of the crystals collected from membrane synthesis with water contents of 150, 200, 250, 300, and 400. The patterns were essentially identical as water mole content increased from 150 to 400, and the peaks match those for simulated SAPO-34. Moreover, the crystallinity did not show any decrease due to a 2.7-fold increase in water content.

Some membranes were broken and analyzed by EDX and SEM. The Si/Al and Si/P ratios of the surfaces of the membranes prepared by different water contents were essentially identical (Fig. 11). These ratios (Si/Al = 0.18-0.19; Si/P = 0.23-0.25) were higher in the membranes than in the gel (Si/Al = 0.15; Si/P = 0.15), indicating

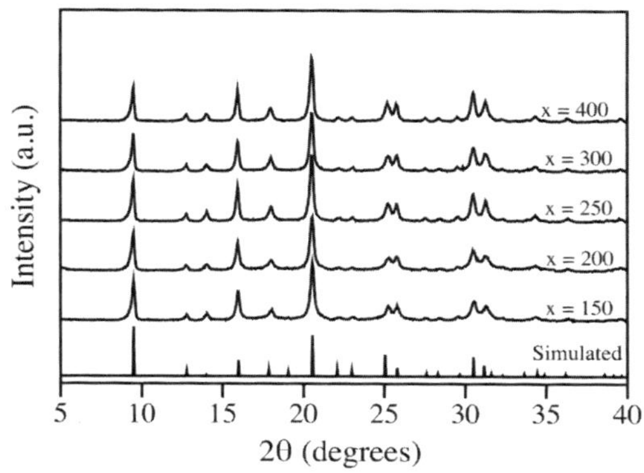

Fig. 10. XRD patterns of powders collected from the bottom of the membranes with gel composition of 1.0 Al_2O_3: 1.0 P_2O_5: 0.3 SiO_2: 1.0 TEAOH: 1.6 DPA: x H_2O are shown at the top, and simulated XRD pattern for pure SAPO-34 crystals is shown at the bottom for comparison. Reprinted from Ref. 45 with permission from Elsevier.

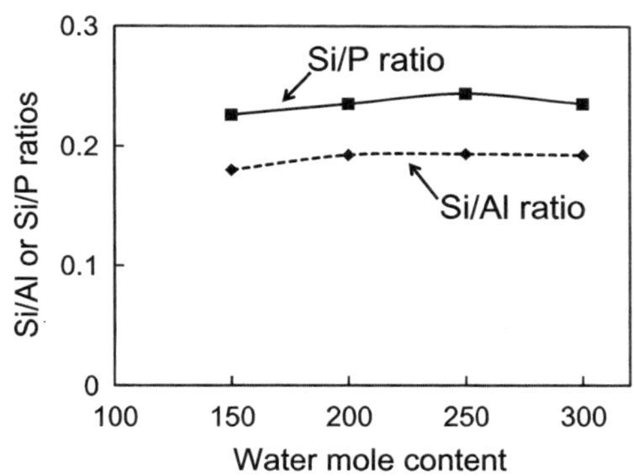

Fig. 11. Si/Al and Si/P ratios for the membranes prepared with different water mole contents. Reprinted from Ref. 45 with permission from Elsevier.

that under the current synthesis conditions, Si in the gels was preferentially incorporated into the SAPO framework. In contrast, when using single template of TEAOH and a crystallization temperature of 200°C for SAPO-34 membrane synthesis, the Si/Al ratio was lower in the crystal (Si/Al = 0.13) than in the gel (Si/Al = 0.15). The differences in Si/Al and S/P ratios would affect specific retention volumes and adsorption enthalpies of N_2 and CH_4 in SAPO-34 crystals as reported by Maple and Williams,[60] which would further affect the N_2/CH_4 separation performance of the SAPO-34 membranes.

Huang *et al.*[46] also synthesized SAPO-34 membranes by secondary growth on porous tubular α-Al_2O_3 supports with 100-nm pores (Inopor® Micro, 100 nm). Before synthesis, the supports were boiled in deionized water for 4 h and then dried at 100°C overnight. SAPO-34 seeds in Fig. 9 were used to provide nucleation for membrane growth. The inside of the supports was seeded by rubbing with SAPO-34 seeds (S1–S4). The uncalcined SAPO-34 seeds were first dried at 100°C for 12 h and then stored in a series of 5–10 mL glass vials for future use. The inside support of treated tubular α-Al_2O_3 supports was then repeatedly rubbed using a Q-tips cotton swab with uncalcined SAPO-34 powder on the cotton-tipped head. The rubbing process was performed for at least 3 min, ensuring that the inside surface was completely rubbed with seeds. Sparsely distributed seeds on the support surface can be seen, and it seems most of the seeds laid flatly on the support surface when high aspect ratio seeds (A–R = 10) were used. The outside of the supports was then wrapped with Teflon tape to avoid growth of membranes on the outside surface. After seeding, the supports were placed in an autoclave filled with the synthesis gel. Two gel compositions (1.0 Al_2O_3: 1.0 P_2O_5: x SiO_2: 1.0 TEAOH: 1.6 DPA: 150 H_2O, $x = 0.3$ and 0.6) for the secondary growth of SAPO-34 seeds were used in this study. Usually, 2–3 seeded supports were placed in an autoclave. The synthesis gel solution was about 0.5 cm above the top of the supports. Secondary growth of the seeds was carried out in a mechanical convection oven at 210°C for 6–6.5 h. After that, the membranes were cooled and washed for 15 minutes with flowing tap water. SAPO-34 membranes were then soaked in DI water for 18 h and dried at 100°C for ~2 h. The membranes were calcined at 450°C for 6 h with heating and cooling rates of 0.5 and 1.0°C/min, respectively, under weak vacuum (pressure of 6.8 kPa) to remove templates. All membranes were stored at 220°C before gas permeation measurement.

To understand the effect of seed AR on membrane growth, for each seed 6–9 membranes were prepared at 210°C for 6.5 h using the same gel, 1.0 Al_2O_3: 1.0 P_2O_5: 0.3 SiO_2: 1.0 TEAOH: 1.6 DPA: 75 H_2O.[46] M*n* (n = 1–4) corresponded to batch number of membranes prepared using seeds S1–S4, respectively, while other synthetic parameters were kept the same. Average membrane thicknesses were 6.6 ± 0.3, 4.5 ± 0.4, 2.4 ± 0.25, and 1.8 ± 0.35 μm when using S1–S4 seeds, respectively. As shown in Fig. 12(a), membrane thicknesses decreased monotonically (almost linearly) with the increase of the logarithm of the average seed ARs, indicating that the higher-aspect-ratio seeds allowed the formation of thinner membranes. Propane with a kinetic diameter of 0.43 nm has big difficulty in penetrating into SAPO-34 pores. Therefore, its permeation test can be used to identify the flow through defects larger than 0.50 nm.[13]

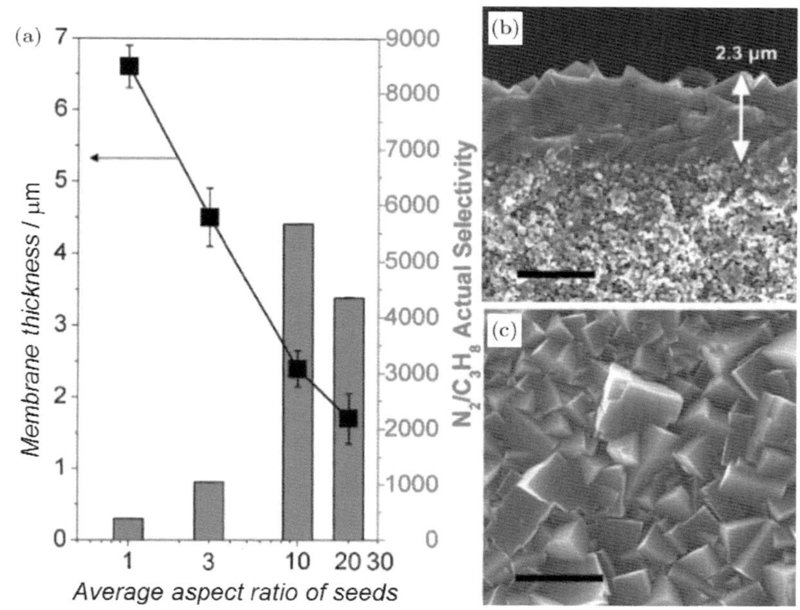

Fig. 12. (a) Effect of the seed AR on membrane thickness (solid squares) and the N_2–C_3H_8 ideal selectivity (red bars) for corresponding membranes; (b) cross-sectional and (c) top-view SEM images of a representative SAPO-34 membrane prepared using S3 seeds. Scale bars in (b) and (c): 2 μm. Reprinted from Ref. 46 with permission from Wiley-VCH.

Single-gas permeation was measured for N_2 and propane on SAPO-34 membranes (Fig. 23(a)). Interestingly, the thickest M1 (~6.5 μm) displayed the lowest N_2–C_3H_8 ideal selectivity of 379 ± 53. The number increased to 1046 ± 120 for M2 (~4.6 μm) followed by a dramatic rise to 5665 ± 523 for thinner M3 (~2.3 μm). Apparently, the seed AR not only affected the thickness of membranes but also played a crucial role on the control of defect formation. High-aspect-ratio crystals may pack much better and form a much thinner seeding layer than cubic crystals. These may help minimize the amount of non-zeolite pores/defects while maintaining a thin membrane thickness. Figures 12(b) and 12(c) presented cross-sectional and top-view SEM images of a membrane prepared using S3 seeds. SAPO-34 crystals with grain size of approximately 2 μm were well intergrown to form an ~2.3 μm thick continuous membrane free of visible defects (e.g., cracks and pinholes).

The same authors reported[46] SAPO-34 membranes prepared with S4 seeds (AR: ~20) displayed slightly lower N_2–C_3H_8 selectivity of 4355 ± 205. The effective membrane thickness was actually much thicker than the top layer thickness (1.8 ± 0.35 μm) observed under SEM. Partial seed inclusion in the support pores (~0.1 μm) may happen during rubbing process. Note, according to SEM observation, S4 contained ~5% seeds smaller than 0.1 μm in lateral size. Seed inclusion could lead to the formation of thicker membranes with degraded separation performance. We expect thinner membranes with improved selectivity would be obtained if supports with smaller pores were used to avoid seed inclusion in support pores.

5.2. *N_2–CH_4 Separation Performance of SAPO-34 Membranes*

In a previous study,[8] the compositions of the feed, retentate, and permeate streams were measured, after attaining the steady state, using a gas chromatograph (SRI instruments, 8610C) equipped with a thermal-conductivity detector and HAYESEP-D packed column. The oven, injector, and detector temperatures in the GC were kept at 40°C, 50°C, and 150°C, respectively.

The permeance of the component i, P_i, is defined as

$$P_i = \frac{J_i}{\Delta p_{\ln,i}}, \tag{1}$$

where J_i is the flux through the membrane for component i. For the cross-flow configuration, because one component preferentially permeates through the membrane, the partial pressures in the feed and retentate are quite different. Therefore, a log-mean pressure drop, $\Delta p_{\ln,i}$, was calculated by:

$$\Delta p_{\ln,i} = \frac{(p_{f,i} - p_{r,i})}{\ln[(p_{f,i} - p_{p,i})/(p_{r,i} - p_{p,i})]}, \tag{2}$$

where $p_{f,i}$, $p_{r,i}$, and $p_{p,i}$ are partial pressures for component i in feed, retentate, and permeate sides, respectively. The permeability is the permeance multiplied by membrane thickness. The ideal selectivity is the ratio of the single-gas permeances, and the separation selectivity, $\alpha_{i/j}^{\text{sep}}$, is the ratio of the permeances for mixtures.

Li *et al.*[8] employed six approaches with synthesis conditions including pore size of the support, gel composition, alumina source, and crystallization time as shown in Table 2. The number of membranes prepared for each approach is also provided.

Water content in the gel: In this research, Li *et al.*[8] also studied the effects of the water content in the membrane gel. Approaches 1 and 2 used the same synthesis conditions except that the water mole ratio was 77 in approach 1, whereas in the approach 2 nearly twice the amount of water (mole ratio of 150) was used. The SEM images of the membrane surface and cross-section for approaches 1 and 2 are shown in Figs. 13 and 14.

The surface SEM images for membranes prepared by both approaches (Figs. 13(a) and 14(a)) show well-intergrown zeolite crystals. The crystals obtained via the approach 2 (Fig. 14(a)) look more uniform than those prepared by approach 1 (Fig. 13(a)). Moreover, as expected, the dilute gel led to a thinner dense zeolite layer (7.3 μm) as shown in the cross-sectional SEM micrographs (Figs. 14(b) versus 13(b)).

Table 2. Membrane preparation conditions.

Approach	Support pore size (nm)	Membrane synthesis conditions: Membrane gel composition	Membrane synthesis conditions: Al source	Crystallization time (h)	Number of membranes prepared
1	100	1.0 Al_2O_3: 1.0 P_2O_5: 0.3 SiO_2: 1.0 TEAOH: 1.6 DPA: 77 H_2O	$Al(i\text{-}C_3H_7O)_3$	24	2
2	100	1.0 Al_2O_3: 1.0 P_2O_5: 0.3 SiO_2: 1.0 TEAOH: 1.6 DPA: 150 H_2O	$Al(i\text{-}C_3H_7O)_3$	24	2
3	100	1.0 Al_2O_3: 1.0 P_2O_5: 0.3 SiO_2: 1.0 TEAOH: 1.6 DPA: 150 H_2O	$Al(i\text{-}C_3H_7O)_3$	6	3
4	50	1.0 Al_2O_3: 1.0 P_2O_5: 0.3 SiO_2: 1.0 TEAOH: 1.6 DPA: 150 H_2O	$Al(i\text{-}C_3H_7O)_3$	6	2
5	200	1.0 Al_2O_3: 1.0 P_2O_5: 0.3 SiO_2: 1.0 TEAOH: 1.6 DPA: 150 H_2O	$Al(i\text{-}C_3H_7O)_3$	6	2
6	100	1.0 Al_2O_3: 1.0 P_2O_5: 0.3 SiO_2: 1.0 TEAOH: 1.6 DPA: 150 H_2O	$Al(OH)_3$	6	1

Reprinted from Ref. 8 with permission from the American Chemical Society.

Mixture (50–50 N_2–CH_4) permeations were measured for the membranes prepared by approaches 1 and 2. Note that the N_2 (kinetic diameter: 0.364 nm) and CH_4 (kinetic diameter: 0.38 nm) molecules can permeate through both SAPO-34 pores (0.38 nm) and defects larger than 0.38 nm. The SAPO-34 pores are selective, whereas the defects are generally not selective for light gas mixtures.[26,53,61] Table 3 shows that as the water mole ratio increased from 77 to 150, the N_2–CH_4 selectivity increased from 2.8 ± 1.1 to 6.0 ± 0.49, indicating that the fraction of the defects of the membranes decreased. A diluted gel formed a thinner membrane layer, and thus the average N_2 permeance obtained for the approach 2 was higher than that for the approach 1.

(a) (b)

Fig. 13. (a) Surface and (b) cross-sectional SEM micrographs of a membrane prepared by approach 1. Membranes were synthesized at 220°C for 24 h onto 100-nm-pore Al_2O_3 supports with a gel composition of 1.0 Al_2O_3: 1.0 P_2O_5: 0.3 SiO_2: 1.0 TEAOH: 1.6 DPA: 77 H_2O. $Al(i\text{-}C_3H_7O)_3$ was used as the Al_2O_3 source. Reprinted from Ref. 8 with permission from Elsevier.

(a) (b)

Fig. 14. (a) Surface and (b) cross-sectional SEM micrographs of a membrane prepared by approach 2. Membranes were synthesized at 220°C for 24 h onto 100-nm-pore Al_2O_3 supports with a gel composition of 1.0 Al_2O_3: 1.0 P_2O_5: 0.3 SiO_2: 1.0 TEAOH: 1.6 DPA: 150 H_2O. $Al(i\text{-}C_3H_7O)_3$ was used as the Al_2O_3 source. Reprinted from Ref. 8 with permission from Elsevier.

Crystallization time: To obtain the effects of the crystallization time, the authors decreased the time to 6 h in approach, from 24 h in approaches 1 and 2. The shortened crystallization time led to a thinner dense zeolite layer (6.2 μm) as indicated by the cross-sectional

Table 3. Comparison of N_2–CH_4 separation performance for membranes prepared by approaches 1 and 2.

				N_2–CH_4 separation*	
Approach	Gel composition	Membrane thickness (μm)	Number of membranes prepared	N_2 permeance (GPU)	Selectivity
1	1.0 Al_2O_3: 1.0 P_2O_5: 0.3 SiO_2: 1.0 TEAOH: 1.6 DPA: 77 H_2O	8.2	2	320 ± 120	2.8 ± 1.1
2	1.0 Al_2O_3: 1.0 P_2O_5: 0.3 SiO_2: 1.0 TEAOH: 1.6 DPA: 150 H_2O	7.3	2	370 ± 52	6.0 ± 0.49

*: All the +/− values are standard deviations.
Reprinted from Ref. 8 with permission from the American Chemical Society.

(a) (b)

Fig. 15. (a) Surface and (b) cross-sectional SEM micrographs of a membrane prepared by approach 3. Membranes were synthesized at 220°C for <u>6 h</u> onto 100-nm-pore Al_2O_3 supports with a gel composition of 1.0 Al_2O_3: 1.0 P_2O_5: 0.3 SiO_2: 1.0 TEAOH: 1.6 DPA: 150 H_2O. $Al(i\text{-}C_3H_7O)_3$ was used as the Al_2O_3 source. Reprinted from Ref. 8 with permission from Elsevier.

SEM image (Fig. 15(b)). The SEM image of the membrane surface (Fig. 15(a)) still shows well-intergrown zeolite crystals.

When the crystallization time was reduced from 24 h to 6 h, the average N_2 permeance increased by 19% (Table 3). This was mainly

Table 4. Comparison of N_2–CH_4 separation performance for membranes prepared by approaches 2 and 3.

Approach	Crystallization time (h)	Membrane thickness (μm)	Number of membranes prepared	N_2–CH_4 separation*	
				N_2 permeance (GPU)	Selectivity
2	24	7.3	2	370 ± 52	6.0 ± 0.49
3	6	6.2	3	440 ± 74	7.3 ± 0.21

*: All the +/− values are standard deviations.
Reprinted from Ref. 8 with permission from the American Chemical Society.

because the membrane thickness decreased by 15%. The average N_2–CH_4 selectivity was also improved from 6 to 7.3 (Table 4).

Pore size of the support: In the same study, the authors[8] also explored the effects of the pore size of the support. Rubbing was used in the current study to seed the SAPO crystals onto the support. Achieving a high-quality seeding layer is important because it affects the property of the resulting membrane. The quality of the seeding layer is usually influenced by the size match between the seeding crystals and the support pores as well as the roughness of the inside surface of the support. Approaches 3, 4, and 5 used the same preparation conditions except that membranes were synthesized on alumina supports with pore size of 100 nm, 50 nm, and 200 nm, respectively. Intergrown zeolite crystals were formed for all these supports (Figs. 15(a), 16(a), and 17(a)). The membrane thicknesses estimated from SEM analysis of the membrane cross-sections decreased with increasing pore size of the support. As shown in Table 5, among the three types of supports, the seeding crystals have been implanted the best onto the 100-nm-pore support for secondary growth of a membrane as the highest N_2–CH_4 selectivity was observed for the 100-nm-pore supported membranes (Table 5).

Alumina source: To study the effects of the Al_2O_3, $Al(OH)_3$ was used instead of $Al(i\text{-}C_3H_7O)_3$ as the Al_2O_3 source in approach 6. Other synthesis conditions were the same as those in approach 3. The

(a) (b)

Fig. 16. (a) Surface and (b) cross-sectional SEM micrographs of a membrane prepared by approach 4. Membranes were synthesized at 220°C for 6 h onto 50-nm-pore Al_2O_3 supports with a gel composition of 1.0 Al_2O_3: 1.0 P_2O_5: 0.3 SiO_2: 1.0 TEAOH: 1.6 DPA: 150 H_2O. Al(i-$C_3H_7O)_3$ was used as the Al_2O_3 source. Reprinted from Ref. 8 with permission from Elsevier.

(a) (b)

Fig. 17. (a) Surface and (b) cross-sectional SEM micrographs of a membrane prepared by approach 5. Membranes were synthesized at 220°C for 6 h onto 200-nm-pore Al_2O_3 supports with a gel composition of 1.0 Al_2O_3: 1.0 P_2O_5: 0.3 SiO_2: 1.0 TEAOH: 1.6 DPA: 150 H_2O. Al(i-$C_3H_7O)_3$ was used as the Al_2O_3 source. Reprinted from Ref. 8 with permission from Elsevier.

SEM images of the membrane surface show poorly intergrown zeolite crystals (Fig. 18(a)). The membrane thicknesses estimated from SEM analysis of the membrane cross-sections (5.6 μm, Fig. 18(b)) was close to that of the membrane prepared by approach 3. N_2 permeance as

Table 5. Comparison of N_2–CH_4 separation performance for membranes prepared by approaches 3, 4, and 5.

Approach	Pore size of the support (nm)	Membrane thickness (μm)	Number of membranes prepared	N_2–CH_4 separation* N_2 permeance (GPU)	Selectivity
3	100	6.2	3	440 ± 74	7.3 ± 0.21
4	50	7.7	2	580 ± 42	2.4 ± 0.92
5	200	3.6	2	610 ± 25	3.2 ± 1.2

*: All the +/− values are standard deviations.
Reprinted from Ref. 8 with permission from the American Chemical Society.

(a) (b)

Fig. 18. (a) Surface and (b) cross-sectional SEM micrographs of a membrane prepared by approach 6. Membranes were synthesized at 220°C for 6 h onto 100-nm-pore Al_2O_3 supports with a gel composition of 1.0 Al_2O_3: 1.0 P_2O_5: 0.3 SiO_2: 1.0 TEAOH: 1.6 DPA: 150 H_2O. $Al(OH)_3$ was used as the Al_2O_3 source. Reprinted from Ref. 8 with permission from Elsevier.

high as 2600 GPU was obtained for this membrane, but the selectivity was only 2.3 (Table 6), indicating a poor quality of the membrane when using $Al(OH)_3$ as the alumina source. During the synthesis, we found that $Al(i\text{-}C_3H_7O)_3$ was more soluble in aqueous solution than $Al(OH)_3$, which led to a more homogenous gel solution for membrane synthesis and thus better membrane quality.

Because the membranes prepared by approach 3 had the highest N_2–CH_4 separation selectivity and moderate N_2 permeance, they were used for further study to investigate the separation mechanisms.

Table 6. Comparison of N_2–CH_4 separation performance for membranes prepared by approaches 3 and 6.

				N_2–CH_4 separation*	
Approach	Al source	Membrane thickness (μm)	Number of membranes prepared	N_2 permeance (GPU)	Selectivity
3	$Al(i\text{-}C_3H_7O)_3$	6.2	3	440 ± 74	7.3 ± 0.21
6	$Al(OH)_3$	5.6	1	2,600	2.3

*: All the +/− values are standard deviations.
Reprinted from Ref. 8 with permission from the American Chemical Society.

SEM images of the membrane surfaces (Fig. 19 left column) show well-intergrown micron range zeolite crystals for SAPO-34 membranes prepared with different water contents. Cross-sectional SEM images of the membranes (Fig. 19 right column) show zeolite layers. The thickness of the layer decreases from $\sim$6.8 to $\sim$2.7 μm as water content increases from 150 to 400; a diluted gel forms a thinner membrane layer.

Membranes prepared with different water contents were used to separate a premixed 50–50 N_2–CH_4 mixture. The feed pressure was 223 kPa, and the pressure in the permeate side was 85 kPa. The separation results at 23°C for these membranes are shown in Table 7.

N_2–CH_4 separation selectivities were greater than 5.4 for membranes with water molar contents of 150, 200, 250, and 300. The maximum average N_2–CH_4 selectivity was 7.2 obtained for the water content of 300. The maximum average N_2 permeance was 1200 GPU obtained for the water content of 250. When the water content was increased to 400, the resulting membranes could not hold pressure for a trans membrane drop permeation measurement, indicating the membranes had high concentration of defects. Although the SEM surface image of the membrane prepared with water content of 400 (Fig. 19(e) left column) shows intergrown zeolite crystals, some crystals were as large as 2.7 μm, which was about the SEM thickness of the membrane (Fig. 19(e) right column). Intercrystalline boundaries

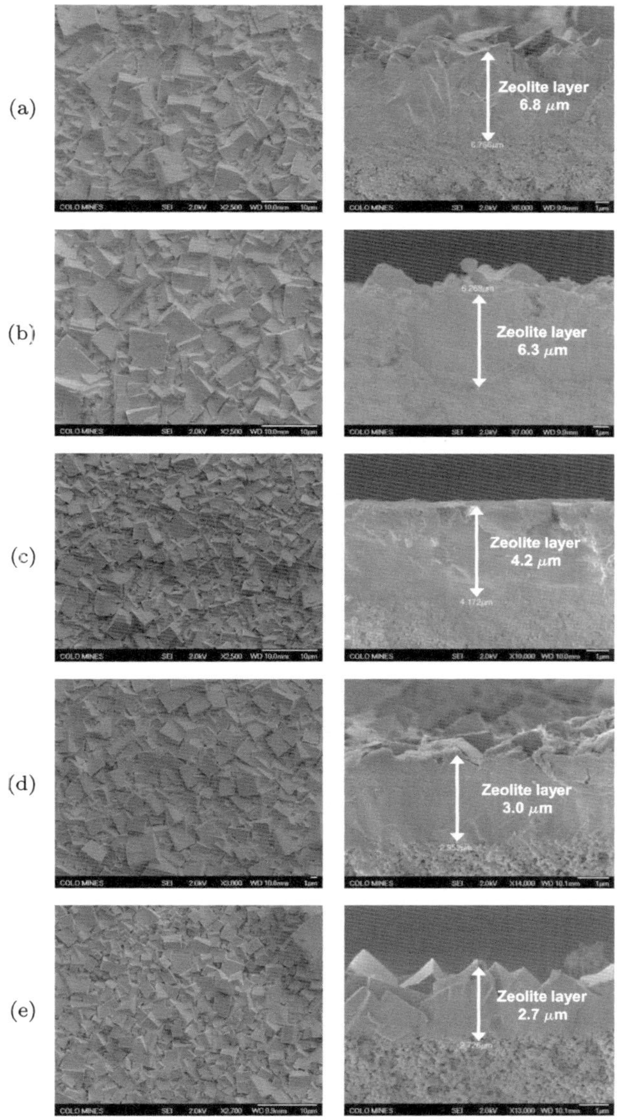

Fig. 19. Surface (left column) and cross-sectional SEM (right column) micrographs of SAPO-34 membranes prepared with gel compositions of 1.0 Al_2O_3: 1.0 P_2O_5: 0.3 SiO_2: 1.0 TEAOH: 1.6 DPA: x H_2O: (a) $x = 150$, (b) $x = 200$, (c) $x = 250$, (d) $x = 300$, and (e) $x = 400$. Reprinted from Ref. 45 with permission from Elsevier.

Table 7. Membrane thickness and N_2–CH_4 separation performance.

Water content	Membrane SEM thickness (μm)	N_2/CH_4 separation	
		N_2 permeance (GPU)	Selectivity
$x = 150$	6.8	670 ± 31	5.4 ± 0.92
$x = 200$	6.3	670 ± 40	6.2 ± 0.71
$x = 250$	4.2	1200 ± 240	6.5 ± 0.28
$x = 300$	3.0	880 ± 47	7.2 ± 0.070
$x = 400$	2.7	NA	NA

*: Two membranes were prepared at each water content. Average permeances and selectivities are presented. All the +/− values are standard deviations. Reprinted from Ref. 45 with permission from Elsevier.

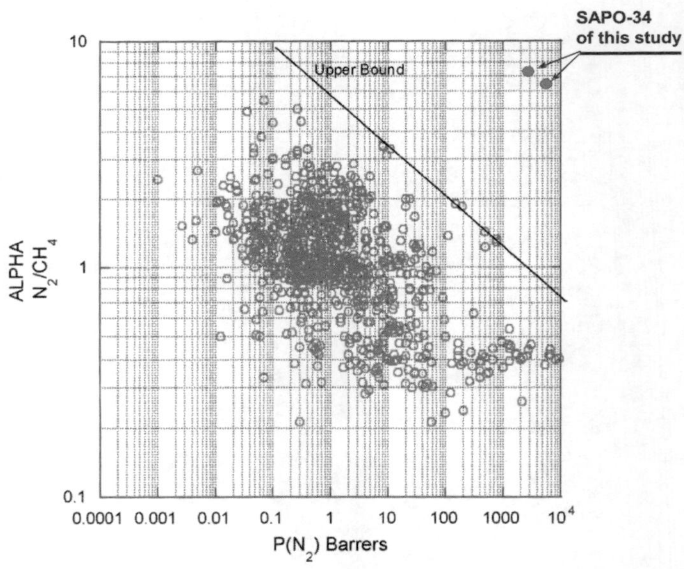

Fig. 20. Comparison of membranes presented in Robeson plot for N_2–CH_4.[10] Reprinted from Ref. 45 with permission from Elsevier.

around these large crystals might have resulted in the high concentration of defects.

Figure 20 compares our SAPO-34 membranes to polymer membranes presented in "Robeson" plot[10] for the N_2–CH_4 separation.

The SEM thicknesses were used in the calculation of N_2 permeabilities (permeance × membrane thickness) for our SAPO-34 membranes. The data points for the SAPO-34 membranes are significantly above the upper bound.

Huang *et al.*[46] tested SAPO-34 membranes for separation of equimolar N_2–CH_4 mixture at 22°C, and a fixed feed, and permeate pressures of 275 kPa and 101 kPa, respectively. Separation performance of representative membranes presented in Table 8. Usually, thicker zeolite membranes were prepared to minimize flux contribution of nonselective defects (non zeolite pores). This typically leads to increased selectivity and decreased permeance. However, we found an inversed relationship between N_2 permeance and N_2–CH_4 selectivity for the membranes M1–M3 and achieved a simultaneous increase in permeance and selectivity by using seeds with increased ARs (Table 8). Specifically, N_2 flux increased linearly as a function of the logarithm of seed ARs from $(0.941 \pm 0.5) \times 10^{-7}$ to $(1.93 \pm 0.1) \times 10^{-7}$ and then to $(2.87 \pm 0.15) \times 10^{-7}$ mol/($m^2 \cdot s \cdot Pa$) for membranes M1, M2, and M3, respectively. N_2–CH_4 selectivity slightly increased from 5.7 ± 0.42 to 6.0 ± 0.36 when increasing the AR of seeds from 1 to 3. Maximum N_2–CH_4 selectivity, 11.1 ± 0.46, was obtained for membrane M3 by further increasing AR to 10. Separation performance of M4 was similar to that for M2. As discussed earlier, the degraded separation performance can be ascribed to a thicker effective membrane thickness. Therefore, the N_2 permeance

Table 8. N_2–CH_4 Separation Properties of SAPO-34 Membranes at 22°C and a Permeate Pressure of 101 kPa and Pressure Drop of ~175 kPa.

Membrane ID	Average AR of seeds	Permeance, $\times 10^7$ mol/($m^2 \cdot s \cdot Pa$) N_2	CH_4	N_2/CH_4 mixture selectivity
M1	1	0.94 ± 0.50	0.164 ± 0.07	5.7 ± 0.42
M2	3	1.93 ± 0.10	0.324 ± 0.02	6.0 ± 0.36
M3	10	2.87 ± 0.15	0.258 ± 0.02	11.1 ± 0.46
M4	20	2.13 ± 0.21	0.246 ± 0.01	8.7 ± 0.38
M5*	10	4.02 ± 0.22	0.356 ± 0.02	11.3 ± 0.31

*M5 was prepared with double amount of silicon in the gel.

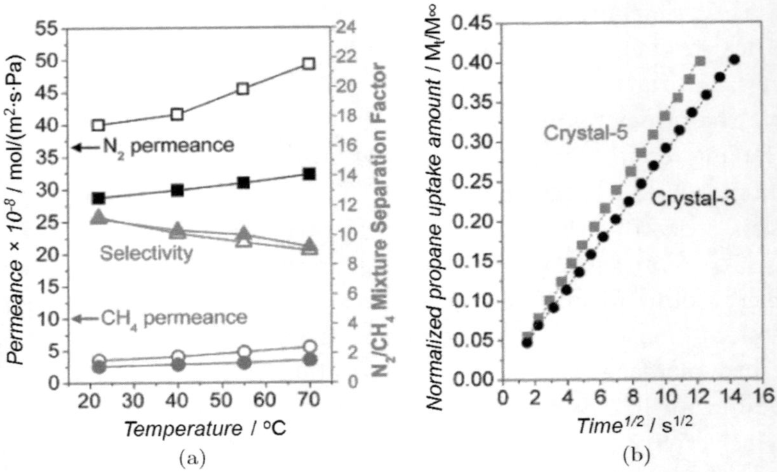

Fig. 21. Equimolar N_2–CH_4 mixture separation at different temperatures by M3 (solid symbols) and M5 (open symbols) (a). The pressure drop was ~175 kPa with a permeate side pressure at 101 kPa. (b) Propane adsorption kinetics at 22°C on Crystal-3 and Crystal-5 collected in M3 and M5 preparation, respectively. M_t is the adsorbed amount of propane at time t, and M_∞ is adsorbed amount at equilibrium. Reprinted from Ref. 46 with permission from Wiley-VCH.

and selectivity over CH_4 are expected to increase when an appropriate support is used.

N_2–CH_4 mixture separations at elevated temperatures (40, 55, and 70°C) were also performed for SAPO-34 membranes (Fig. 21(a)). For M3, both N_2 and CH_4 permeances increased with the increase of temperature, but N_2 permeance increased more slowly than that of CH_4, resulting in slightly decreased N_2–CH_4 selectivity. The highest N_2 permeance for this membrane was $(3.23 \pm 0.32) \times 10^{-7}$ mol/(m$^2 \cdot$ s $\cdot$ Pa) at 70°C, and the highest N_2–CH_4 separation selectivity was 11.1 ± 0.46 at 22°C. Importantly, we also observed a substantial increase in N_2 permeance without sacrificing N_2–CH_4 selectivity for membranes prepared with doubled silicon content in the gel. For example, M5 with higher Si/(Si+Al+P) ratio showed the highest N_2 permeance of $(4.93 \pm 0.25) \times 10^{-7}$ mol/(m$^2 \cdot$ s $\cdot$ Pa) at 70°C and N_2–CH_4 selectivity of 11.3±0.31 at 22°C (Fig. 21(a), open symbols). SEM results confirmed that M5 was well intergrown and had a similar low thickness, ~2 μm as M3. Increased N_2 permeance may result from the different framework compositions that may slightly change crystal structure and thus affect diffusion in SAPO-34 crystals.[62,63] To prove this assumption, we selected a tight-fit molecule to SAPO-34 pores, propane and used its adsorption uptake kinetics to explore the small potential change of the SAPO-34 structure upon composition change. Propane adsorbs slowly onto SAPO-34 crystals, allowing us to precisely monitor its diffusivity change that may reveal to slight structure variation upon framework composition change. SAPO-34 crystals with different compositions were collected from the gels used for M3 and M5 syntheses and denoted as Crystal-3 and Crystal-5, respectively. Adsorption-kinetics profiles (M_t/M_∞ versus time) a shown in Fig. 21(b). SAPO-34 with higher Si content (Crystal-5) showed a larger slope which corresponds to a faster diffusion of propane in SAPO-34 pores. This indicates that incorporating more Si into the structure slightly increases SAPO pore size. Therefore, fine tuning of framework composition may influence diffusion of molecules in SAPO-34 crystals and thus may be a viable way of further improving membrane-separation performance if membrane quality is high and thus flux contribution from defects are negligible.

Figure 22 compared our SAPO-34 membranes to other reported inorganic membranes and polymer membranes presented in "Robeson" plot[10] for the N_2–CH_4 separation. The SEM thicknesses were used in the calculation of N_2 permeabilities (permeance × membrane thickness) for our SAPO-34 membranes. SAPO-34 membranes from this work were significantly above the upper bound and also showed superior performance to other inorganic membranes. Moreover, our membranes were highly reproducible using the reported synthetic and secondary growth technologies. We also found that the SAPO-34 membranes demanded a sufficient drying period to achieve and reproduce its high permeance and selectivity due to the moisture sensitivity of SAPO-34 structure. To apply our approach for fabrication of commercial-sized membrane tubes, uniformity of the seeding layer prepared by rubbing might be a key challenge. The mitigation strategy is to use a more scalable dip-coating technique instead of rubbing; a previous study has shown that both rubbing and

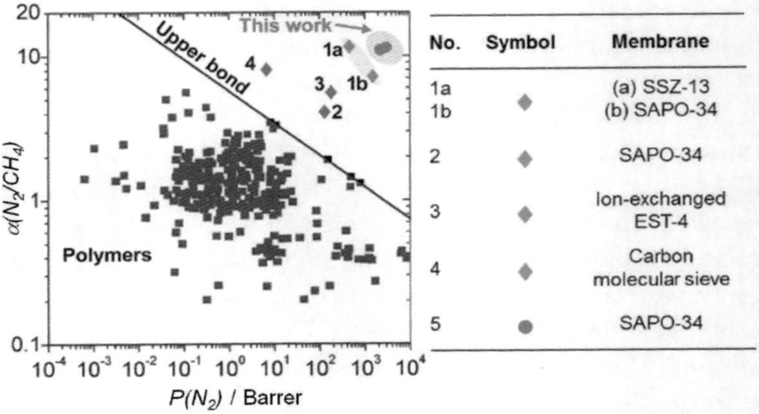

Fig. 22. Comparison of polymer membranes presented in Robeson plot and other inorganic membranes for N_2–CH_4 separation. Reprinted from Ref. 46 with permission from Wiley-VCH.

Table 9. Comparison of inorganic membranes employed for N_2–CH_4 separation.

Material	Separation selectivity	N_2 permeance (GPUs)	Thickness (μm)	Refs.
Carbon molecular sieve	7.7	~0.1	70 ± 15	65
SSZ-13	13	66	7.8	13
SAPO-34	5–7	300	2.0–3.0	13
SAPO-34	8	500	6.2	8
SAPO-34	5.7–11.3	281–1,200	2.3	46
SAPO-34	6.5–7.4	880–1,300	3.0–4.2	45

dip-coating seeding methods had yielded SAPO-34 membranes with similar CO_2–CH_4 separation properties and membrane performances were slightly more reproducible when dip-coating was used.[64]

Table 9 compares N_2–CH_4 selectivities and N_2 permeances for the SAPO-34 membranes prepared in this study vs. other N_2-selective inorganic membranes reported in the literature.

5.3. *Adsorption Isotherms and Heats of Adsorption*

Li *et al.*[8] collected SAPO-34 powder from the bottom of the membranes prepared by approach 3. The powder was calcined and used

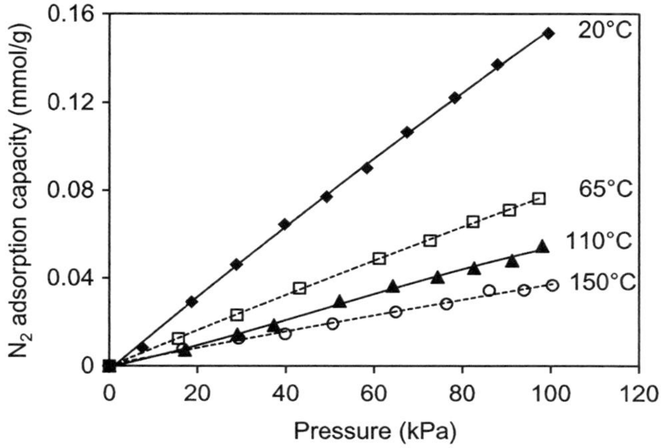

Fig. 23. Adsorption isotherms for N_2 on SAPO-34 powder at various temperatures. Reprinted from Ref. 8 with permission from Elsevier.

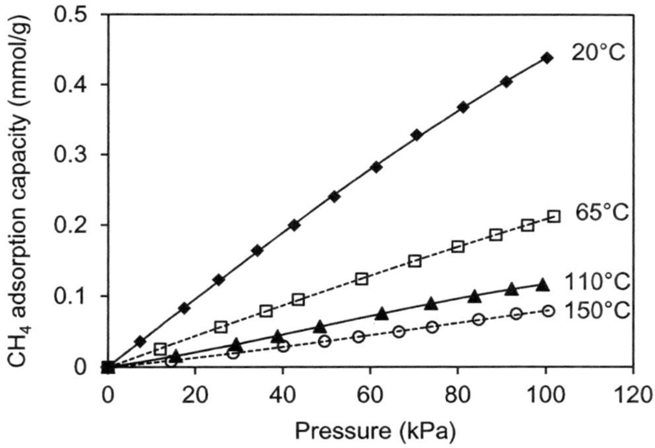

Fig. 24. Adsorption isotherms for CH_4 on SAPO-34 powder at various temperatures. Reprinted from Ref. 8 with permission from Elsevier.

for adsorption study. Adsorption isotherms for N_2 and CH_4 at various temperatures are shown in Figs. 23 and 24. The amounts of CH_4 adsorbed at high pressures were 2.1–2.9 times the amounts of N_2 adsorbed.

For light gases on zeolites, Langmuir single-site adsorption can be assumed:

$$\theta = \frac{q}{q_{\text{sat}}} = \frac{bp}{1+bp}, \tag{3}$$

where θ is the fractional coverage, q is the coverage, q_{sat} is the saturation coverage, p is the pressure. Equation (3) can be converted to

$$\frac{1}{q} = \frac{1}{bq_{\text{sat}}} \cdot \frac{1}{p} + \frac{1}{q_{\text{sat}}}. \tag{4}$$

Through which bq_{sat} can be obtained by plotting $\frac{1}{q}$ vs. $\frac{1}{p}$ using the adsorption isotherm data. Note that the saturation coverage is the maximum number of molecules that can fit in the zeolite pores, and thus it should be a constant for a certain molecule absorbed onto a certain zeolite. The adsorption equilibrium constant, b, can be written in terms of the heat and entropy of adsorption:

$$b = \exp\left(\frac{\Delta S}{R} - \frac{\Delta H_{\text{ads}}}{RT}\right) \tag{5}$$

and thus,

$$bq_{\text{sat}} = \exp\left(\frac{\Delta S}{R} - \frac{\Delta H_{\text{ads}}}{RT}\right) \cdot q_{\text{sat}}. \tag{6}$$

Equation (6) can be further converted to

$$\ln bq_{\text{sat}} = -\frac{\Delta H_{\text{ads}}}{RT} + \frac{\Delta S}{R} + \ln q_{\text{sat}}. \tag{7}$$

The heats of adsorption ($-\Delta H_{\text{ads}}$) determined from Arrhenius plots (Fig. 25) of the adsorption equilibrium constants are 11 kJ/mol for N_2 and 15 kJ/mol for CH_4, indicating CH_4 adsorbs more strongly than N_2 onto SAPO-34 crystals.

Zong *et al.*[45] studied the transport mechanism through zeolite membranes based on adsorption–diffusion. Depending upon the type of zeolite, the mixture system, and the operating conditions, mixtures are separated by at least one of the following three mechanisms:

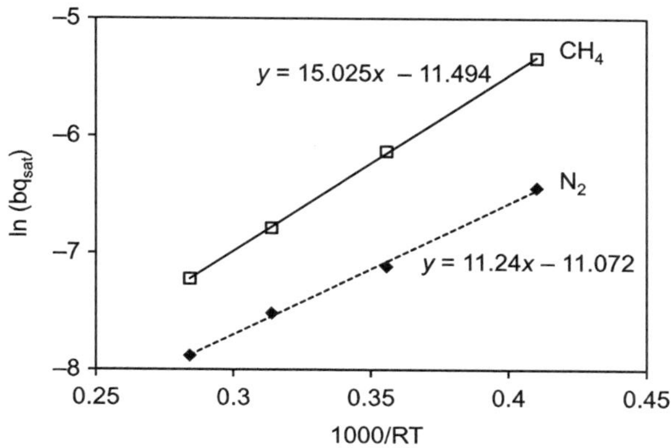

Fig. 25. Adsorption equilibrium constants as a function of inverse temperature for N_2 and CH_4 adsorption onto SAPO-34. Reprinted from Ref. 8 with permission from Elsevier.

(1) molecular sieving — larger molecules cannot fit into the pores, and thus the smaller molecules preferentially permeate; (2) differences in diffusivity — the smaller, less hindered type of molecule in a mixture diffuses faster than the larger ones; and (3) competitive adsorption — one type of molecule is more strongly adsorbed on the zeolite and thus can dramatically inhibit permeation of the second molecule.

Both N_2 (kinetic diameter: 0.364 nm) and CH_4 (kinetic diameter: 0.38 nm) can fit into the SAPO-34 pores (0.38 nm), which rules out the separation by molecular sieving. The CH_4 adsorbs more strongly than N_2 on SAPO-34 crystals based on the heats of adsorption obtained from Fig. 25. Thus, the preferential adsorption of CH_4 would favor separating CH_4 over N_2 in the mixture. On the other hand, the smaller molecule N_2 diffuses faster than the larger molecule CH_4. For example, Li *et al.*[55] reported that the diffusivity of the N_2 was 24 times that of the CH_4 through a SAPO-34 membrane prepared by the single-template method. The differences in diffusivity would favor separating N_2 over CH_4 in the mixture. Single-gas and mixture permeations were thus measured to determine the roles of differences in diffusivity and competitive adsorption.

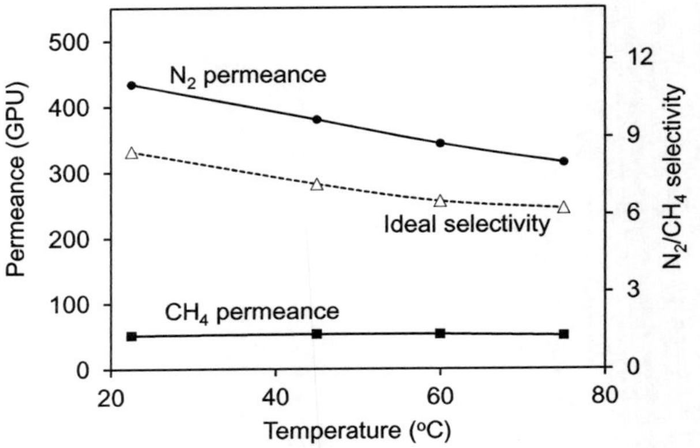

Fig. 26. Single-gas permeances for N_2 and CH_4 and N_2–CH_4 ideal selectivity through SAPO-34 membrane M-100nm-6h-1 as a function of temperature. The feed pressure was 223 kPa, and the pressure drop was 138 kPa. Reprinted from Ref. 8 with permission from Elsevier.

Figure 26 shows single-gas permeances of N_2 and CH_4 through membrane M-100nm-6h-1 as a function of temperature. M-100nm-6h-1 was one of the membranes prepared by the approach 3. The N_2 single-gas permeance decreased with increasing temperature, whereas CH_4 permeance was almost constant. The highest N_2–CH_4 ideal selectivity was 8.4 (observed at 23°C), and it decreased to 6.2 as temperature increased to 75°C.

Similar to the single-gas behavior, the N_2 permeance for a N_2–CH_4 mixture (50–50) decreased with temperature, whereas the CH_4 permeance was essentially independent of temperature for the membrane (Fig. 27). Thus, the N_2–CH_4 separation selectivity decreased with increasing temperature but was lower than the ideal selectivity. The highest N_2 permeance for this membrane was 480 GPU, and its highest N_2–CH_4 separation selectivity was 7.5.

The N_2 molecule permeates faster than CH_4 molecule in both single-gas and mixture permeation because N_2 is smaller (diffuses faster). The higher heat of adsorption for CH_4 than N_2 causes the preferential adsorption of CH_4 in the N_2–CH_4 mixture. Thus, CH_4 inhibits the permeation of N_2 in the mixture. That is why the N_2–CH_4 separation selectivity was lower than the ideal selectivity

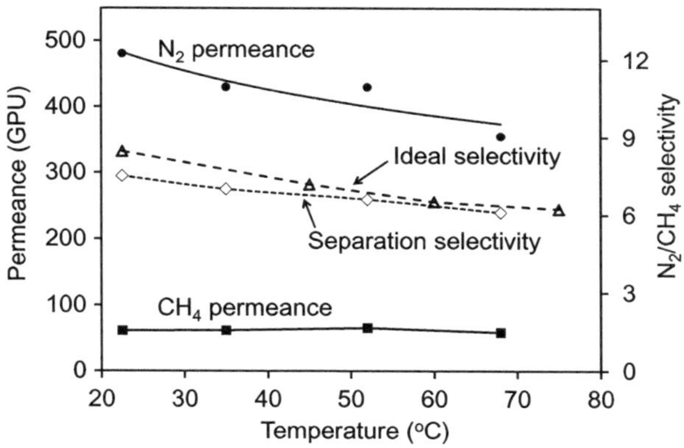

Fig. 27. Permeances and selectivity of a CO_2–CH_4 mixture (50–50) as a function of temperature for SAPO-34 membrane M-100nm-6h-1. The feed pressure was 223 kPa, and the trans membrane pressure drop was 138 kPa. The N_2–CH_4 ideal selectivity is shown for comparison. Reprinted from Ref. 8 with permission from Elsevier.

(Fig. 27). However, the gaps between these two selectivities were within 11%, indicating the preferential adsorption of CH_4 in the N_2–CH_4 mixture does not dominate the separation. Differences in diffusivity played a more critical role than the competitive adsorption in the mixture.

Figure 28 shows permeances and selectivity of a N_2–CH_4 mixture (50–50) at 23°C as a function of trans membrane pressure drop for the SAPO-34 membrane M-100nm-6h-1. The permeate side pressure was kept at 85 kPa. The N_2 permeance decreased with pressure drop, whereas the CH_4 permeance was essentially independent of pressure drop. As a result, the N_2–CH_4 separation selectivity decreased with increasing trans membrane pressure drop.

Note that these permeation tests mentioned above were conducted at Colorado School of Mines (Golden, Colorado, USA). The atmospheric pressure at that elevation is approximately 85 kPa. Mixture separation was also carried out at a permeate side pressure of the standard atmosphere (101 kPa). The N_2 permeance was 500 GPU, and the selectivity was 8 at 23°C for a trans membrane pressure drop of 138 kPa.

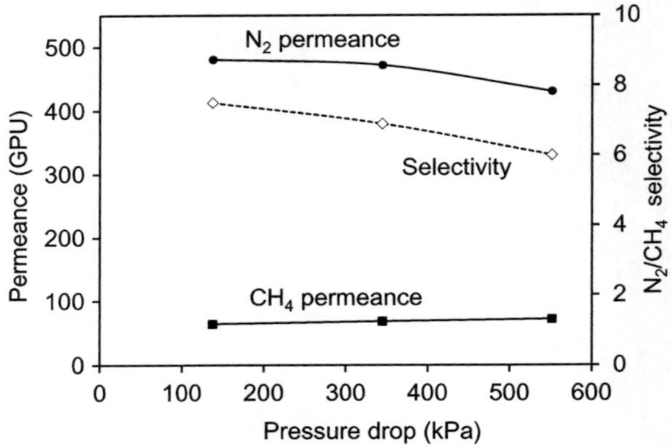

Fig. 28. Permeances and selectivity of a N_2–CH_4 mixture (50–50) at 23°C as a function of trans membrane pressure drop for SAPO-34 membrane M-100nm-6h-1. The permeate pressure was kept at 85 kPa. Reprinted from Ref. 8 with permission from Elsevier.

6. Conclusions

- Membrane separation offers the advantage of low energy cost relative to the more established gas-separation processes such as adsorption and cryogenic distillation. For membrane separation, the challenge is to develop membranes with the necessary nitrogen–methane separation characteristics.
- For a process using N_2-selective membranes, CH_4 remains in the retentate stream at high pressures, and thus there is a significant saving in recompression cost as compared to CH_4-selective membranes.
- SAPO-34 membranes have potential for N_2 rejection in natural gas processing. The heat of adsorption was higher for CH_4 (15 kJ/mol) than N_2 (11 kJ/mol), leading to a preferential adsorption of CH_4 over N_2 in the N_2–CH_4 mixture. On the other hand, the N_2 molecule diffused much faster than the CH_4 molecule, and differences in diffusivity played a more critical role than the competitive adsorption. As such, the SAPO-34 membranes were selective for N_2 over CH_4 in the mixture.
- Diluted gels and increased crystallization temperature grew thinner SAPO-34 membranes and thus higher permeances for N_2–CH_4

separation. The membranes separated N_2 from CH_4 with N_2 permeance as high as 1200 GPU and selectivity of 7.2 at 23°C for a 50–50 N_2–CH_4 mixture. This separation performance is superior to those of the state-of-the-art membranes.

- High-aspect-ratio SAPO-34 seeds have been successfully used to grow thinner and higher quality SAPO-34 membranes to achieve simultaneous increase of N_2 permeance and N_2–CH_4 mixture selectivity. By incorporating more Si into crystal structure, a significant increase in N_2 permeance was obtained, while maintaining N_2–CH_4 selectivity. N_2 permeance as high as ~1500 GPU at 70°C and N_2–CH_4 selectivity of 11.3 ± 0.31 at 22°C for a 50–50 N_2–CH_4 mixture have been obtained. Their separation performance is superior to those of the state-of-the-art membranes.

The current studies[8,13,45,46] have demonstrated the feasibility of preparing continuous and reproducible SAPO-34 membranes for N_2–CH_4 separation. Although SAPO-34 membranes can be synthesized with good separation performance, an improved performance is needed for their potential use at industrial scale. The selectivity and N_2 permeance of current best membranes[66] have not met the targeted value of separation selectivity.[8] Preliminary economic evaluation indicated that the selectivity should increase to 15 to maintain a CH_4 loss below 10% in a one-stage membrane process to compete with the benchmark technology, cryogenic distillation.[8] To improve the separation performance for N_2–CH_4 mixtures, separation selectivity needs to reach the targeted value of at least 15. Seeds' synthesis improvement, membrane-gel improvement, and ion exchange (to decrease the preferential CH4 adsorption over N2) may be useful to increase the N_2–CH_4 selectivity needed for economical used of these membranes that potentially could be used at industrial scale.

References

1. *Natural Gas* (2015). Available at: http://www.iea.org/topics/naturalgas/ (Accessed: May 1, 2015).
2. *U.S. Natural Gas Total Consumption* (2015). Available at: http://www.eia.gov/dnav/ng/hist/n9140us2a.htm (Accessed: May 1, 2015).

3. Sieminski, A (2014). Annual Energy Outlook: Energy Information Administration.
4. Baker, R. W., Lokhandwala, K. Natural gas processing with membranes: An overview. *Ind. Eng. Chem. Res.* 2008, 47(7), 2109–2121.
5. *Overview of Natural Gas* (2015). Available at: http://naturalgas.org/overview/background/ (Accessed: May 1, 2015).
6. Office of Oil and Gas. (2006). *Natural Gas Processing: The crucial link between natural gas production and its transportation to market.* Energy Information Administration. pp. 1–11.
7. Echterhoff, L. W., Pathak, V. K. *Evaluation of Process Costs for Small Scale Nitrogen Removal from Natural Gas.* Gas Research Institute: Arizona, 1991.
8. Li, S., Zong, Z., Zhou, S. J., Huang, Y., Song, Z., Feng, X., Zhou, R., Meyer, H. S., Yu, M., Carreon, M. A. SAPO-34 Membranes for N_2/CH_4 separation: Preparation, characterization, separation performance and economic evaluation. *J. Memb. Sci.* 2015, 487, 141–151.
9. Meyer, H., Henson, M. Methane selective membranes for nitrogen removal from low quality natural gas — high permeation is not enough. *Proceedings of the Natural Gas Technologies II: Ingenuity and Innovation*, Phoenix, AZ, 2004.
10. Robeson, L. M. The upper bound revisited. *J. Memb. Sci.* 2008, 320(1–2), 390–400.
11. Ning, X., Koros, W. J. Carbon molecular sieve membranes derived from Matrimid (R) polyimide for nitrogen/methane separation. *Carbon.* 2014, 66, 511–522.
12. Pierson, H. O. *Handbook of Carbon, Graphite, Diamonds and Fullerenes: Processing, Properties and Applications*: William Andrew Publishing: New York, 2012.
13. Wu, T., Diaz, M. C., Zheng, Y., Zhou, R., Funke, H. H., Falconer, J. L., Noble, R. D. Influence of propane on CO_2/CH_4 and N_2/CH_4 separations in CHA zeolite membranes. *J. Memb. Sci.* 2015, 473, 201–209.
14. Xu, R., Pang, W., Yu, J., Huo, Q., Chen, J. *Chemistry of Zeolites and Related Porous Materials: Synthesis and Structure*: John Wiley & Sons, New York, 2009.
15. Lok, B. M., Messina, C. A., Patton, R. L., Gajek, R. T., Cannan, T. R., Flanigen, E. M. Silicoaluminophosphate molecular sieves — another new class of microporous crystalline inorganic solids. *J. Am. Chem. Soc.* 1984, 106(20), 6092–6093.
16. *What is Zeolite*? Available at: http://www.cmzeolites.com.au/what-is-zeolite (Accessed: May 25, 2015).
17. Lutz, W., Bulow, M., Grossmann, A., Schirmer, W. On pure gas and mixture adsorption of hydrogen-sulfide and normal-butane on NACAA zeolite. 3. separation behavior of hydrogen sulfide-normal-butane mixtures on NACAA zeolite. *Chemische Technik.* 1981, 33(3), 136–138.

18. Frankiewicz, T. C., Donnelly, R. G. Methane/nitrogen gas separation over the zeolite clinoptilolite by the selective adsorption of nitrogen. *ACS Sym. Ser.* 1983, 223, 213–233.
19. Buss, E., Schollner, R. Binary adsorption equilibria of nitrogen, methane and carbon-monoxide on zeolite adsorbents. *Chemische Technik.* 1991, 43(1), 10–14.
20. Jasra, R. V., Choudary, N. V., Bhat, S. G. T. Separation of gases by pressure swing adsorption. *Separ. Sci. Technol.* 1991, 26(7), 885–930.
21. Niwa, M., Yamazaki, K., Murakami, Y. Separation of oxygen and nitrogen due to the controlled pore-opening size of chemical vapor-deposited zeolite A. *Ind. Eng. Chem. Res.* 1991, 30(1), 38–42.
22. Geus, E. R., Denexter, M. J., Vanbekkum, H. Synthesis and characterization of zeolite (MFI) membranes on porous ceramic supports. *J. Chem. Soc. Faraday T.* 1992, 88(20), 3101–3109.
23. Zhang, L. X., Jia, M. D., Min, E. Z. (1997). Synthesis of SAPO-34/ceramic composite membranes. In *Progress in Zeolite and Microporous Materials*, Pts a-C, Vol. 105. Amsterdam: Elsevier, pp. 2211–2216.
24. Poshusta, J. C., Tuan, V. A., Falconer, J. L., Noble, R. D. Synthesis and permeation properties of SAPO-34 tubular membranes. *Ind. Eng. Chem. Res.* 1998, 37(10), 3924–3929.
25. Poshusta, J. C., Noble, R. D., Falconer, J. L. Characterization of SAPO-34 membranes by water adsorption. *J. Memb. Sci.* 2001, 186(1): 25–40.
26. Li, S. G., Martinek, J. G., Falconer, J. L., Noble, R. D., Gardner, T. Q. High-pressure CO_2/CH_4 separation using SAPO-34 membranes. *Ind. Eng. Chem. Res.* 2005, 44(9), 3220–3228.
27. Li, S. G., Fan, C. Q. High-flux SAPO-34 membrane for CO_2/N_2 separation. *Ind. Eng. Chem. Res.* 2010, 49(9), 4399–4404.
28. Lok, B. M., Messina, C. A., Patton, R. L., Gajek, R. T., Cannan, T. R., Flanigen, E. M. *Crystalline Silicoaluminophosphates.* USA. Patent 4440871. Isssued April 3, 1984.
29. Baerlocher, C., McCusker, L. B. Database of Zeolite Structures. Available at: http://www.iza-structure.org/databases.
30. Vistad, O. B., Akporiaye, D. E., Taulelle, F., Lillerud, K. P. Morpholine, an *in situ* C-13 NMR pH meter for hydrothermal crystallogenesis of SAPO-34. *Chem. Mater.* 2003, 15(8), 1650–1654.
31. Dumitriu, E., Azzouz, A., Hulea, V., Lutic, D., Kessler, H. Synthesis, characterization and catalytic activity of SAPO-34 obtained with piperidine as templating agent. *Microporous Mater.* 1997, 10(1–3), 1–12.
32. Liu, G. Y., Tian, P., Li, J. Z., Zhang, D. Z., Zhou, F., Liu, Z. M. Synthesis, characterization and catalytic properties of SAPO-34 synthesized using diethylamine as a template. *Micropor. Mesopor. Mat.* 2008, 111(1–3), 143–149.
33. Liu, G. Y., Tian, P., Zhang, Y., Li, J. Z., Xu, L., Meng, S. H., Liu, Z. M. Synthesis of SAPO-34 templated by diethylamine: Crystallization process

and Si distribution in the crystals. *Micropor. Mesopor. Mater.* 2008, 114(1–3), 416–423.

34. Wei, Y. X., He, Y. L., Zhang, D. H., Xu, L., Meng, S. H., Liu, Z. M., Su, B. L. Study of Mn incorporation into SAPO framework: Synthesis, characterization and catalysis in chloromethane conversion to light olefins. *Micropor. Mesopor. Mater.* 2006, 90(1–3), 188–197.
35. Zhou, H. Q., Wang, Y., Wei, F., Wang, D. Z., Wang, Z. W. *In situ* synthesis of SAPO-34 crystals grown onto α-Al_2O_3 sphere supports as the catalyst for the fluidized bed conversion of dimethyl ether to olefins. *Appl. Catal. A-Gen.* 2008, 341(1–2), 112–118.
36. Rajic, N., Stojakovic, D., Hocevar, S., Kaucic, V. On the possibility of incorporating Mn(II) and Cr(III) in SAPO-34 in the presence of isopropylamine as a template. *Zeolites.* 1993, 13(5), 384–387.
37. Carreon, M. A., Li, S. G., Falconer, J. L., Noble, R. D. SAPO-34 seeds and membranes prepared using multiple structure directing agents. *Adv. Mater.* 2008, 20(4), 729–732.
38. van Heyden, H., Mintova, S., Bein, T. Nanosized SAPO-34 synthesized from colloidal solutions. *Chem. Mater.* 2008, 20(9), 2956–2963.
39. Zhang, L. X., Yao, J. F., Zeng, C. F., Xu, N. P. Combinatorial synthesis of SAPO-34 via vapor-phase transport. *Chem. Comm.* 2003, (17), 2232–2233.
40. Yao, J. F., Wang, H. T., Ringer, S. P., Chan, K. Y., Zhang, L. X., Xu, N. P. Growth of SAPO-34 in polymer hydrogels through vapor-phase transport. *Micropor. Mesopor. Mater.* 2005, 85(3), 267–272.
41. Yao, J. F., Zeng, C. F., Zhang, L. X., Xu, N. P. Vapor phase transport synthesis of SAPO-34 films on cordierite honeycombs. *Mater. Chem. Phys.* 2008, 112(2), 637–640.
42. Askari, S., Halladj, R. Effects of ultrasound-related variables on sonochemically synthesized SAPO-34 nanoparticles. *J. Solid State Chem.* 2013, 201, 85–92.
43. Arafat, A., Jansen, J. C., Ebaid, A. R., Vanbekkum, H. Microwave preparation of zeolite Y and ZSM-5. *Zeolites.* 1993, 13(3), 162–165.
44. Li, Y. S., Yang, W. S. Microwave synthesis of zeolite membranes: A review. *J. Memb. Sci.* 2008, 316(1–2), 3–17.
45. Zong, Z., Feng, X., Huang, Y., Song, Z., Zhou, R., Zhou, S. J., Carreon, M. A., Yu, M., Li, S. Highly permeable N_2/CH_4 separation SAPO-34 membranes synthesized by diluted gels and increased crystallization temperature. *Micropor. Mesopor. Mater.* 2016, 224, 36–42.
46. Huang, Y., Wang, L., Song, Z., Li, S., Yu, M. Growth of high-quality, thickness-reduced zeolite membranes towards N_2/CH_4 separation using high-aspect-ratio seeds. *Angew. Chem. Int. Ed.* 2015, 54(37), 10843–10847.
47. Funke, H. H., Chen, M. Z., Prakash, A. N., Falconer, J. L., Noble, R. D. Separating molecules by size in SAPO-34 membranes. *J. Memb. Sci.* 2014, 456, 185–191.

48. Varoon, K., Zhang, X. Y., Elyassi, B., Brewer, D. D., Gettel, M., Kumar, S., Lee, J. A., Maheshwari, S., Mittal, A., Sung, C. Y., *et al.* Dispersible exfoliated zeolite nanosheets and their application as a selective membrane. *Science*. 2011, 334(6052), 72–75.
49. Tsapatsis, M. 2-dimensional zeolites. *Aiche J.* 2014, 60(7), 2374–2381.
50. Zhou, R. F., Ping, E. W., Funke, H. H., Falconer, J. L., Noble, R. D. Improving SAPO-34 membrane synthesis. *J. Memb. Sci.* 2013, 444, 384–393.
51. Zhang, Y. F., Tokay, B., Funke, H. H., Falconer, J. L., Noble, R. D. Template removal from SAPO-34 crystals and membranes. *J. Memb. Sci.* 2010, 363(1–2), 29–35.
52. Li, S. G., Falconer, J. L., Noble, R. D. Improved SAPO-34 membranes for CO_2/CH_4 separations. *Adv. Mater.* 2006, 18(19), 2601–2603.
53. Li, S. G., Falconer, J. L., Noble, R. D. SAPO-34 membranes for CO_2/CH_4 separation. *Journal of Membrane Science*, 2004, 241(1), 121–135.
54. Poshusta, J. C., Tuan, V. A., Pape, E. A., Noble, R. D., Falconer, J. L. Separation of light gas mixtures using SAPO-34 membranes. *AIChE J.* 2000, 46(4), 779–789.
55. Li, S. G., Falconer, J. L., Noble, R. D., Krishna, R. Modeling permeation of CO_2/CH_4, CO_2/N_2, and N_2/CH_4 mixtures across SAPO-34 membrane with the Maxwell-Stefan equations. *Ind. Eng. Chem. Res.* 2007, 46(12), 3904–3911.
56. Choi, J., Jeong, H. K., Snyder, M. A., Stoeger, J. A., Masel, R. I., Tsapatsis, M. Grain boundary defect elimination in a zeolite membrane by rapid thermal processing. *Science*. 2009, 325(5940), 590–593.
57. Lovallo, M. C., Tsapatsis, M. Preferentially oriented submicron silicalite membranes. *AIChE J.* 1996, 42(11), 3020–3029.
58. Tsapatsis, M., Lovallo, M., Okubo, T., Davis, M. E., Sadakata, M. Characterization of zeolite L nanoclusters. *Chem. Mater.* 1995, 7(9), 1734–1741.
59. Das, J. K., Das, N., Roy, S. N., Bandyopadhyay, S. The growth of SAPO-34 membrane layer on support surface for gas permeation application. *Ceram. Int.* 2012, 38(1), 333–340.
60. Maple, M. J., Williams, C. D. Separating nitrogen/methane on zeolite-like molecular sieves. *Micropor. Mesopor. Mater.* 2008, 111(1–3), 627–631.
61. Li, S. G., Alvarado, G., Noble, R. D., Falconer, J. L. Effects of impurities on CO_2/CH_4 separations through SAPO-34 membranes. *J. Memb. Sci.* 2005, 251(1–2), 59–66.
62. Li, S., Falconer, J. L., Noble, R. D. Improved SAPO-34 membranes for CO2/CH4 separations. *Adv. Mater.* 2006, 18(19), 2601-+.
63. Li, S., Falconer, J. L., Noble, R. D. SAPO-34 membranes for CO2/CH4 separations: Effect of Si/Al ratio. *Micropor. Mesopor. Mater.* 2008, 110(2–3), 310–317.
64. Ping, E. W., Zhou, R. F., Funke, H. H., Falconer, J. L., Noble, R. D. Seeded-gel synthesis of SAPO-34 single channel and monolith membranes for CO_2/CH_4 separations. *J. Memb. Sci.* 2012, 415, 770–775.

65. Ning, X., Koyos, W. J. Carbon molecular sieve membranes derived from Matrimid (R) polyimide for nitrogen/methane separation. *Carbon.* 2014, 66, 511–522.
66. Zong, Z., Feng, X., Li, S., Zhou, S. J., Huang, Y., Song, Z., Zhou, R., Meyer, H. S., Yu, M., Carreon, M. A. Highly permeable N_2/CH_4 separation SAPO-34 membranes synthesized by diluted gels and an increased crystallization temperature. *Micropor. Mesopor. Mater.* 2015, 224, 36–42.

Chapter 5

Polymer Blend Membranes for Gas Separations

Charles J. Holt, Juan P. Vizuet, Inga H. Musselman, Kenneth J. Balkus Jr.* and John P. Ferraris

The University of Texas at Dallas, Department of Chemistry and Biochemistry, 800 West Campbell Road, Richardson, Texas 75080, United States

Although polymer blends have been utilized in several industrial processes, they have not yet found wide acceptance as materials for gas separations due to the demanding conditions of industrial separations. Blending different polymers into a single membrane is attractive because of the synergistic combination of properties. Miscible polymer blends display properties that are intermediate of their comprising individual polymers, while immiscible blends can lead to unique properties depending on what morphology is presented. The gas-separation performance of immiscible blends is largely dependent on the internal morphology resulting from poor mixing. In this chapter, morphology and control of morphology is discussed, along with recent gas-separation examples that illustrate each concept. Reviewing the state of polymer blend membranes for gas separation reveals one major factor: the control of morphology is imperative to developing systems with optimized performance.

1. Introduction

Commercially, polymer blends have been employed in numerous fields including automotive, aerospace, and health care[1] but have yet to find success as commercial gas-separation membranes. The benefit of using polymer blends is the ability to combine a mechanically robust polymer with a highly permeable rubbery polymer to create a gas-separation membrane with improved performance. In recent years, membrane research in blends has largely focused on miscible polymer blends and on uncompatibilized immiscible blends, which reflects the need for higher gas throughput. While their preparation is easier, these membrane systems typically do not exhibit remarkable gas-separation performance. Additionally, the internal morphology of the blend is often not correlated with transport properties. For truly miscible systems, a homogenous morphology is expected. Therefore, it is understandable that there is little emphasis placed on the morphology of miscible blends. However, in the case of immiscible, phase-separated blends, it is also uncommon for researchers to utilize microscopy to understand the morphology. As discussed in later sections, enhancement of gas-separation properties is possible for certain immiscible morphologies. For the advancement of polymeric membranes in gas-separation technology, a better understanding and utilization of the internal microstructure of polymer blend membranes is needed. This chapter reviews some of the current efforts to control microstructure in polymer blend membranes for gas separation.

Commercial gas-separation membranes face many challenges such as low selectivity and flux, scalability, as well as instability towards rigorous processing temperatures and pressures.[2] Despite these obstacles, polymeric membranes have been applied in several industrial processes that include enrichment, recovery, and refinement.[3–6] The purification of nitrogen and the recovery of carbon dioxide and hydrogen from gas streams are the most common commercial applications for polymer membranes.[5] As an illustration of some of the challenges polymeric membranes face, in 2007, the U.S. Department of Energy (DOE) disclosed its Multi-Year Research, Development, and Demonstration Plan, which listed several targets for hydrogen separations.[7] These ambitious targets included operating temperatures and pressure differentials between 250 and 500°C and 800 and 1000 psi, respectively, a H_2 flux of $300\,ft^3h^{-1}ft^{-2}$(SCFH), and a final H_2 purity above 99.99%. Other targets that must be considered are stability (5 years) and tolerance to gas impurities, such as carbon monoxide and sulfur. The DOE cost-effectiveness target for membranes was less than \$100/ft^2. Without considering the mechanical strength necessary for industrial applications, overcoming the trade-off between permeability and selectivity is the biggest challenge to utilizing polymeric membranes for gas separations.[8] The relationship between permeability and selectivity for several gas pairs was correlated by Robeson for several homogenous membranes in 1991[9] and again in 2008.[8] As an example, the "upper-bound" relationship for the H_2–CO_2 gas pair is illustrated in Fig. 1, which showcases the trade-off between permeability and selectivity that is inherent to polymeric membranes. Membranes that surpass the "upper bound", especially those with high permeability and high selectivity, are desirable for commercial applications.

Gas permeabilities are most commonly reported in units of Barrer,[10] where 1 Barrer = 10^{-10} cm^3(STP) cm/cm^2s cmHg. The upper bound for various gas pairs was established by plotting the log of the separation factor (α) versus the log of the permeability of the more permeable gas, resulting in the expression in Eq. (1),

$$P_i = k\alpha_{ij}^n, \tag{1}$$

where P_i is the permeability of the more permeable gas, k is a solubility parameter known as front factor, α is the ideal selectivity,

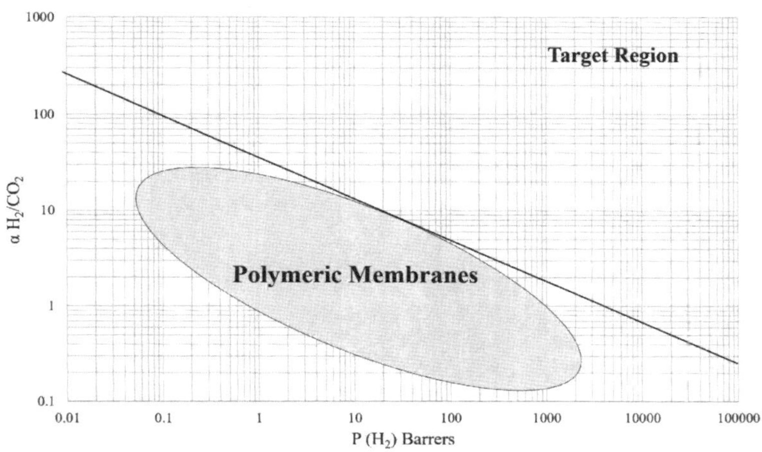

Fig. 1. Robeson plot comparing H_2–CO_2 selectivity (α) to H_2 permeability (P) for polymeric membranes.

and n is the slope of the log–log limit.[8] The trade-off between these properties helps explain the difficulty of developing a membrane with both high selectivity and high permeability.

Polymeric membranes can be conveniently grouped into two types, porous and nonporous. Porous membranes behave similar to filters, where separation is dependent on the size and distribution of any pores or voids in the membrane.[11] Gas transport in nonporous membranes follows the solution-diffusion model, since molecules must diffuse through the dense membrane. Due to their nature, porous membranes typically have high permeabilities and low selectivities, while nonporous membranes are highly selective but have low permeability. This trade-off between permeability and selectivity in dense polymer membranes can be further understood by considering polymer structure. Rigid glassy polymers restrict the movement of gas through the membrane, resulting in low permeability and high selectivity.[12] Rubbery amorphous polymers tend to have low selectivity and high permeability because the chains can rotate and flex more easily, which allows gas to flow at a faster rate.[2]

The solution-diffusion model is defined as the product of gas solubility (S_A) and diffusivity (D_A) (Eq. (2)).

$$P_A = D_A S_A. \tag{2}$$

The diffusion coefficient (D_A) is dependent on the size of the penetrant gas, where smaller gases typically diffuse faster through a membrane than larger ones. Polymer chain flexibility, packing ability, or free-volume defects can be large enough to facilitate diffusion, which, in turn, can affect D_A. The solubility coefficient (S_A) depends on gas condensability and polymer–gas interactions, and is defined as the ratio of the concentration of gas in a polymer (C) to the pressure (P) imparted on a membrane (Eq. (3)).

$$S_A = C/P. \tag{3}$$

Diffusion coefficients are typically expressed in units of cm^2/sec, while solubility is expressed as $cm^3(STP)/(cm^3$ membrane pressure or cmHg). The product of these coefficients results in Barrer units.

For a pure gas, the permeability of a membrane is defined by Eq. (4),

$$P_A = N_A t/(p_2 - p_1), \tag{4}$$

where permeability (P_A) is equal to the product of membrane thickness (t) and normalized steady-state flux (N_A) divided by the pressure difference across the membrane ($p_2 - p_1$). Permeabilities for specific gases are considered a material property, whereas flux is more dependent on the membrane thickness and the difference in pressures.[13] This material property is the basis for gas separations in polymers, since each polymer has a unique permeability coefficient for each gas. Another parameter used to characterize gas-separation ability is ideal selectivity, which is also considered a material property inherent to a polymer, and is defined by Eq. (5) as

$$\alpha = P_A/P_B \quad \text{or} \quad \alpha = D_A S_A/D_A S_B. \tag{5}$$

Common polymers used for gas separations and their permselective properties for various gas pairs are provided in Tables 1 and 2. Table 1 lists some common polymers used in recent years to

Table 1. Common polymers for gas-separation membranes.

Polymer	Repeat Unit
Cellulose acetate (CA)	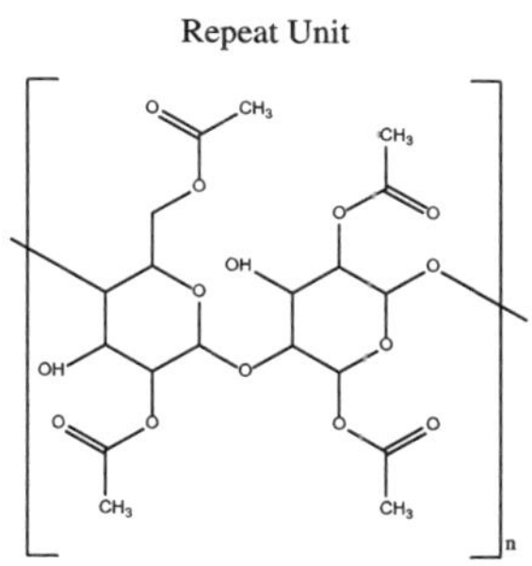
Polyethersulfone (PES)	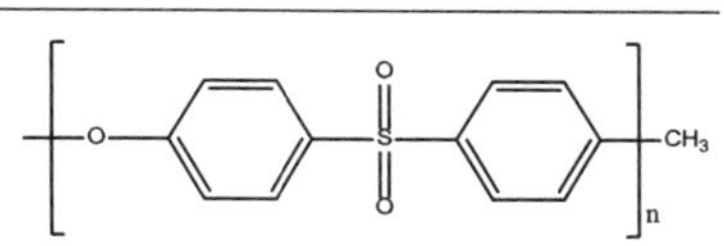

(*Continued*)

Table 1. (*Continued*)

Polymer	Repeat Unit
Polysulfone (PSF)	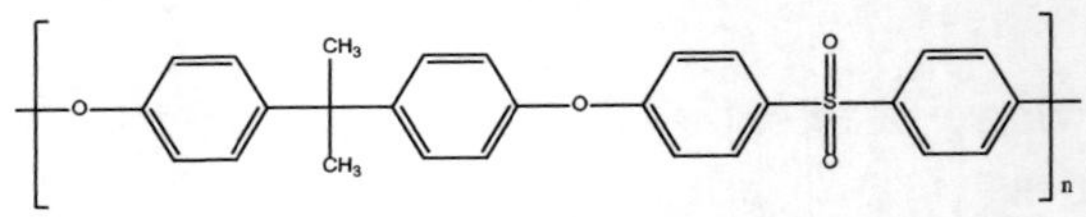
Polydimethylsiloxane (PDMS)	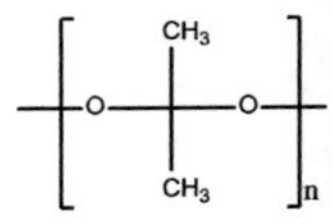
Poly(methyl methacrylate) (PMMA)	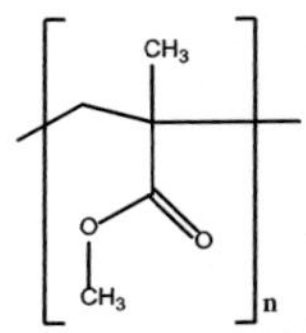
Polyethylene oxide (PEO)	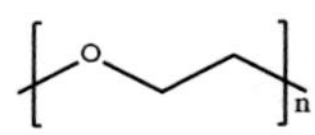
Polymer of Intrinsic Microporosity (PIM-1)	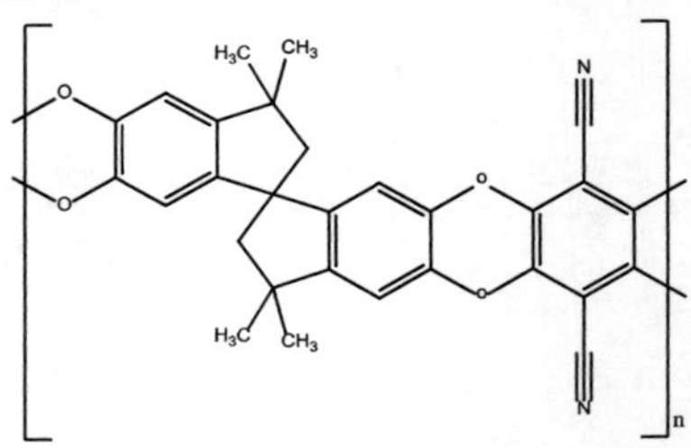
Polyurethane (PU)	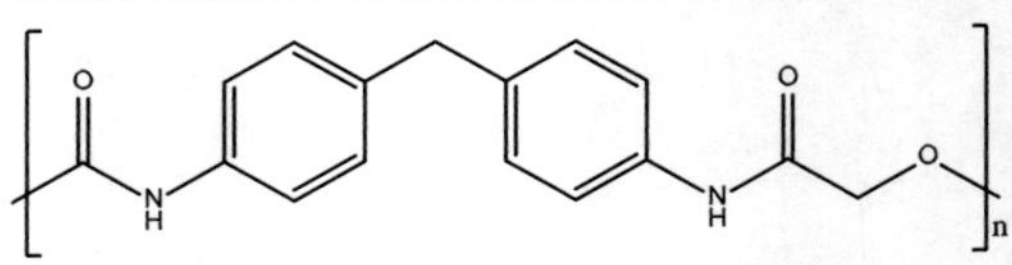
Polybenzimidazole (PBI)	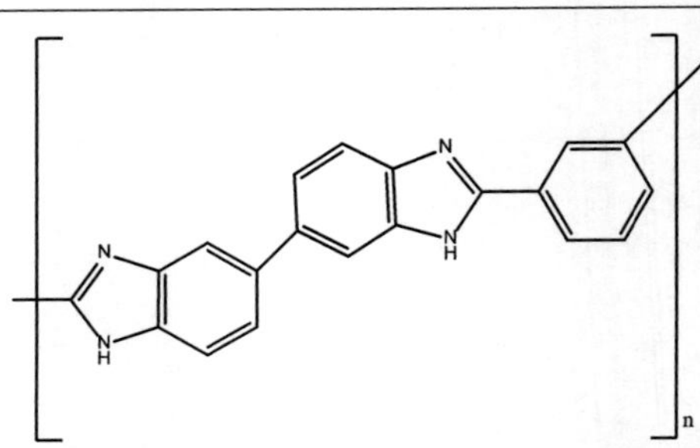

(*Continued*)

Table 1. (*Continued*)

Polymer	Repeat Unit
Matrimid® (MI)	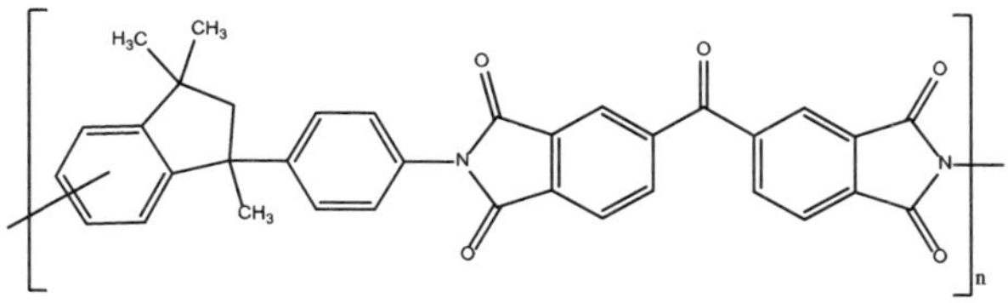
6FDA-DAM:DABA (3:2) (6FDD)	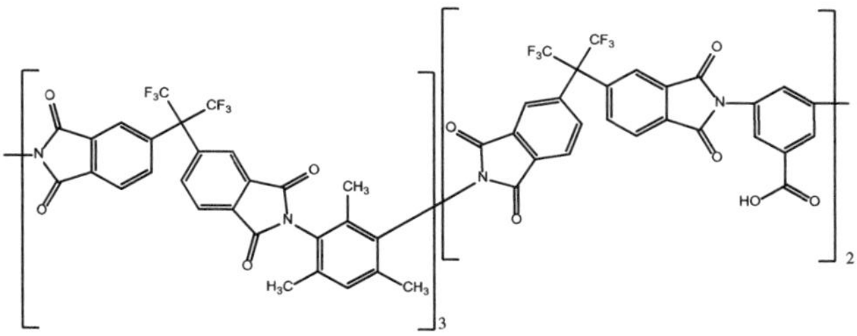
6FDA-Durene (6FDU)	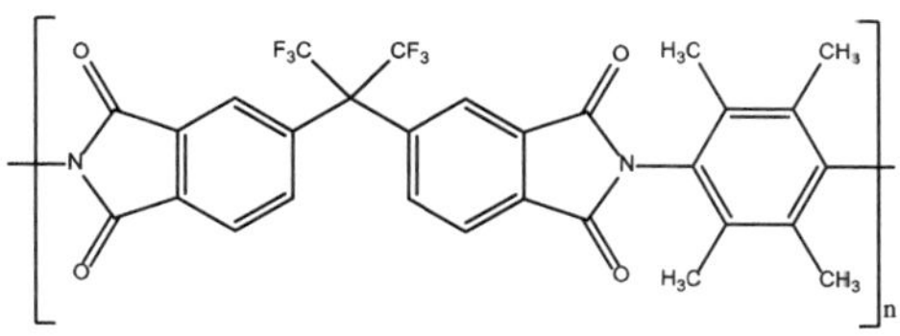
Polyvinyl alcohol (PVA)	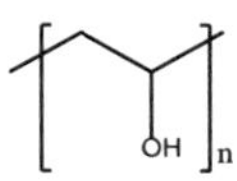
Polyvinyl acetate (PVAc)	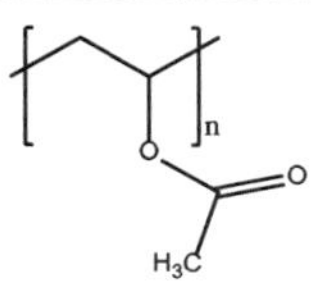
Polyether block amide (Pebax®)	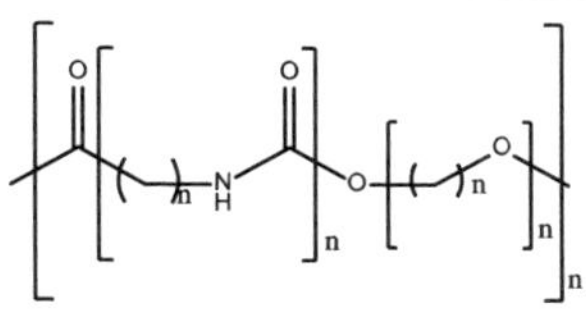

(*Continued*)

Table 1. (*Continued*)

Polymer	Repeat Unit
Poly (p-phenylene oxide) (PPO)	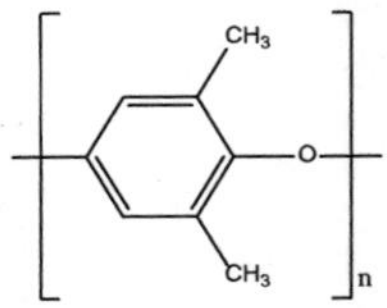
Polypropylene (PP)	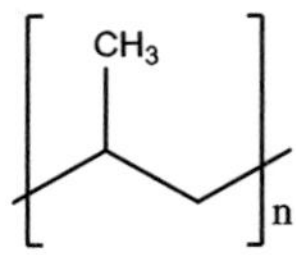
Ethylene vinyl acetate (EVA)	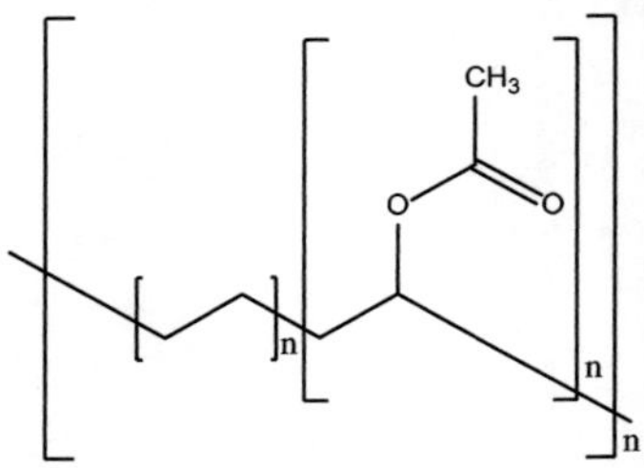
Poly (vinyl methyl ether) (PVME)	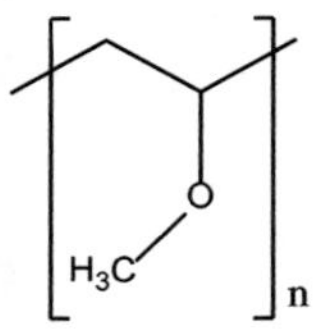
Poly (bisphenol A-co-1,3-phenylene-diamine) (PBNPI)	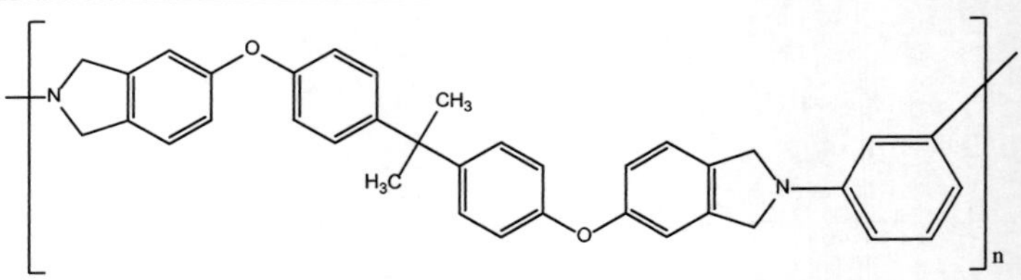
Polyphenyl-sulfone (PPSU)	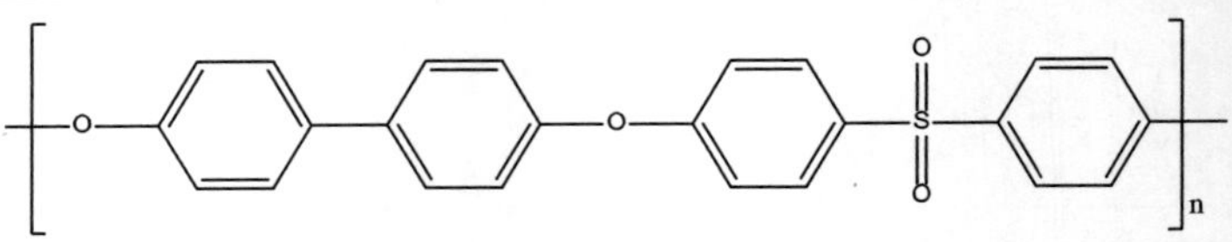

(*Continued*)

Table 1. (*Continued*)

Polymer	Repeat Unit
Acrylonitrile butadiene styrene (ABS)	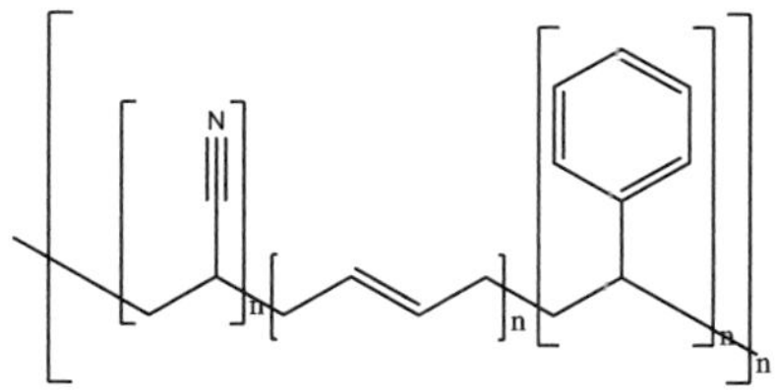
Styrene–butadienerubber (SBR)	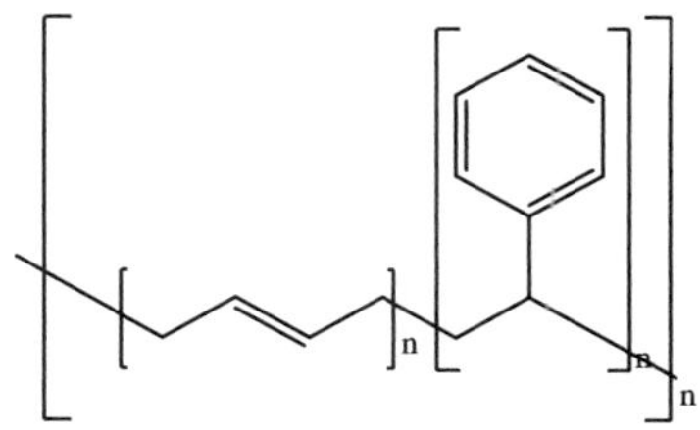
Polyvinylamine (PVAm)	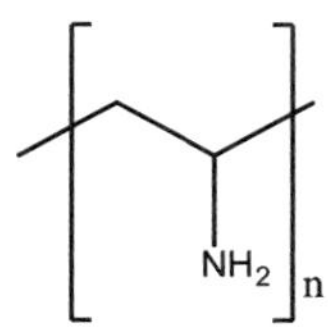
Polyethylenimine (PEI)	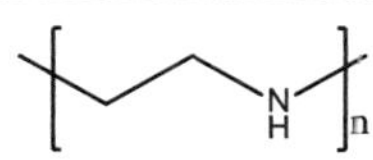
Polyvinylpyrrolidone (PVP)	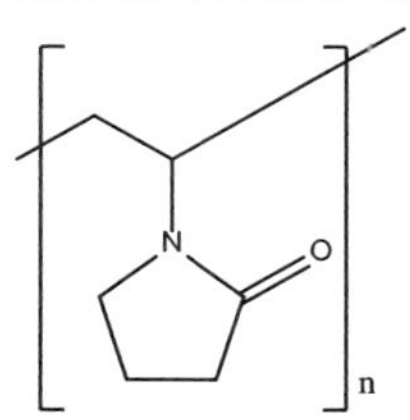

prepare gas-separation membranes, along with the structures of their respective repeating units. This list includes both commercially available polymers, such as cellulose acetate, and newly developed polymers, like 6FDA-durene. Table 2 summarizes the gas permeabilities and ideal selectivities for some of the polymers in Table 1 for

Table 2. Permeability and selectivity data for some of the polymers listed in Table 1, tested at various temperatures and pressures.

Polymer	Permeability (Barrer)					Selectivity				Refs.
	H_2	O_2	N_2	CO_2	CH_4	O_2/N_2	CO_2/N_2	CO_2/CH_4	H_2/CO_2	
CA	14.9	1.05	0.231	5.96	0.205	4.5	25.8	29.1	2.5	14
PES	—	0.29	0.06	1.40	0.05	5.1	24.8	25.6	—	15
PSF	—	1.50	0.257	7.53	0.389	5.84	29.3	19.4	—	16
PDMS	70.57	32.24	11.99	166.02	28.19	2.69	13.85	5.89	0.43	17
PPO	60	7.1	1.3	35	—	5.3	26.3	—	1.7	18
PMMA	3.70	0.0863	0.0111	0.338	0.0058	7.77	30.45	58.28	10.95	19
PIM-1	2918	575	144	2855	173	3.8	19.9	14.3	0.8	20
PU	—	3.71	0.88	25.15	0.91	4.22	28.55	27.63	—	21
PBI	0.6	0.015	0.005	0.16	0.002	3.13	32	88.88	3.75	22
MI	25.7	1.9	0.25	8.4	0.22	7.6	33.6	39.4	3.07	23
6FDD	7.54	—	—	0.73	—	—	—	—	1.9	24
6FDU	518.5	107.7	34.9	468.5	30.0	3.1	13.4	15.6	1.1	25

the most common gases tested (O_2, N_2, H_2, CO_2, and CH_4). The permeability–selectivity trade-off for the polymers is noticeable; the most permeable ones are the less selective and vice versa. There is no universally accepted set of testing conditions for gas-separation membranes, since each membrane can have different requirements for each separation. The values listed in Table 2 were chosen from the results that were reported in units that are easily compared.

While selectivity and permeability are important considerations when developing a membrane for gas separations, a high flux is arguably the most important barrier to bringing membrane technology to the industrial scale, since large amounts of gases are produced and consumed in industrial processes.[2,26] Flux is simply defined as the amount of permeate produced per unit of membrane surface area per unit of time.[2] A membrane's flux is governed by the permeability of the polymer(s) used and the thickness of membrane, with thinner membranes tending to possess higher flux. Logic dictates that decreasing the thickness of the membrane will increase flux, but doing so also increases the chances of structural defects and pinholes, which reduces the structural integrity of the membrane.[2,11] Because of this, developing a mechanically robust polymeric membrane with both high flux and high selectivity is extremely challenging and has kept most polymer systems from being industrially relevant. As an example, in the case of the highly selective polymer polybenzimidazole, the DOE H_2 flux target of 300 SCFH can only be achieved by decreasing the thickness of the membrane to about 30 nm.[27] A membrane this thin is fragile enough to fail through normal handling and, most likely, would not survive the elevated temperatures and pressures of commercial processes. The decreased thickness strategy has been applied to asymmetric and hollow fiber membranes, where the latter relies on a thin selective layer supported by a more permeable polymer that can provide the mechanical strength necessary to withstand elevated pressures. It has been demonstrated that hollow fibers can withstand very high pressure (1000 psi) because of its geometry.[2,28] The hollow fiber geometry possesses an extremely large surface area to volume ratio,

Fig. 2. SEM image of PBI hollow fiber membrane with 150 nm thin, selective skin shown in inset. (Adapted with permission from Ref. 30)

which is a highly desired trait for commercial applications, since a larger ratio increases the efficiency of separations.[29] An example of a PBI hollow fiber membrane with a 150 nm thin, selective layer is shown in Fig. 2.[30]

Developing membranes with high flux and good separation performance is not trivial, and it has been shown that modifying polymers is necessary to overcome the limitations of single polymer membranes.[31] Commonly applied modifications include surface modification techniques,[13,32] cross-linking,[25,33,34] utilizing additives,[3,35,36] and blending polymers.[12,13] These commonly used techniques are employed in order to enhance polymeric membrane performance above the upper bound.[12,31,37] Surface modifications, such as UV initiated grafting,[38] surface carbonization,[39] and surface coating,[26,29,37] are just a few of the methods researchers have used to improve membrane performance. Chemical cross-linking is another modification commonly utilized to enhance the selective properties of the membrane.[25] Cross-linking refers to either covalent or ionic interchain interactions of polymers that result in new bond formation, enhancing mechanical and thermal stability.[40] One of cross-linking's largest benefits is reducing the effect of plasticization.[33,41] Plasticization can be thought of simply as an increase in nonselective interstitial voids in a membrane, due to the movement of polymer chains caused by penetrating species.[41,42] Plasticized polymeric membranes cannot effectively distinguish gas molecules based on their size, which greatly reduces the overall selectivity of the membrane. Cross-linked membranes can have longer lifetimes than unmodified membranes for this reason, since chain mobility is less affected by absorbed species.[41] Unfortunately, since the enhanced selectivity is the result of decreased permeability and flux, cross-linking is not an effective modification on its own. Utilizing cross-linking in polymer blends results in controlled microstructures, as cross-linking limits the available room for phase separation. It has been shown that microphase size is directly related to the amount of cross-linking.[43,44]

One of the most common modifications of polymers employed to improve permselective performance is the incorporation of inorganic materials, resulting in what is known as a mixed-matrix membrane (MMM).[45] Currently, a wide variety of additives are studied as approaches to enhance the separation performance and mechanical stability of the resulting membrane.[3,23,35,36,46] The commercialization of MMMs still faces many challenges, most notably poor selectivity, due to the increase in free void space in the membrane and the aggregation of additives.[47] These approaches to surpass the upper bound come with their own advantages and disadvantages. One of the simpler modifications used to enhance gas-separation performance, and the focus of this chapter, is blending polymers into a single membrane. Modifications have also been applied to polymer blend systems in order to further improve their performance. Examples of surface modification,[48–51] cross-linking,[52–55] and their use as MMMs.[56–59] can be found elsewhere. As mentioned previously, polymer blend membranes display enhanced performances due, in part, to their internal microstructure. Before discussing polymer blends in detail, the possibilities of utilizing different morphologies will be illustrated in the next section.

2. Morphology

Polymer blends can have remarkable gas-separation properties because they can present different morphologies. Some possible morphologies of polymer blend membranes are illustrated in Fig. 3, which showcase both miscible and immiscible compositions.[13]

Polymer blend membranes with uniform microstructures may be the key to maximizing transmembrane gas flux. An increase in flux, without a reduction in selectivity, can be achieved by increasing the surface area accessible to the feed gas, which is maximized with the matrix droplet morphology. As an illustration, the ratio of interfacial surface areas as a function of decreased droplet size for layered and matrix droplet morphologies can be seen in Fig. 4.[60] In the layered morphology, the interfacial surface area is limited to the area between the polymer layers, which can only be increased by increasing the dimensions of the membrane. Reducing the droplet

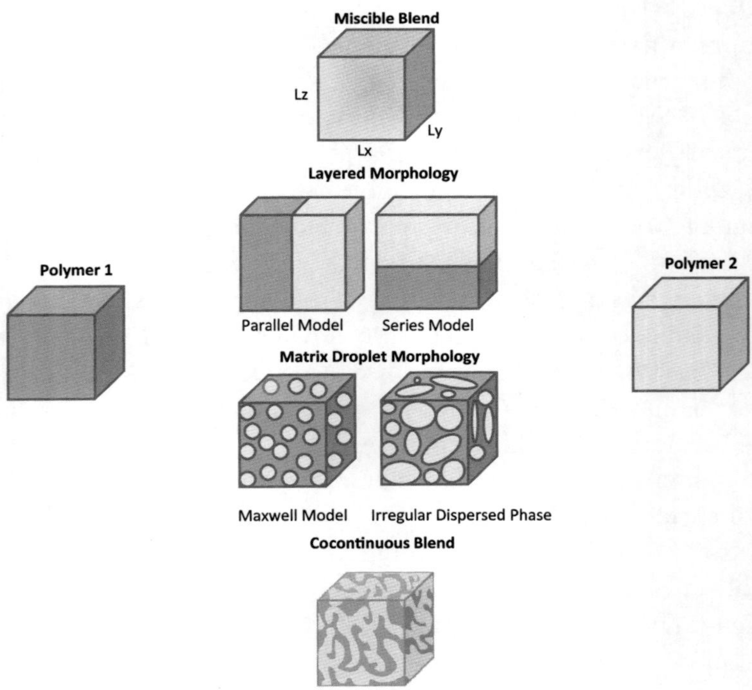

Fig. 3. Possible polymer blend morphologies.

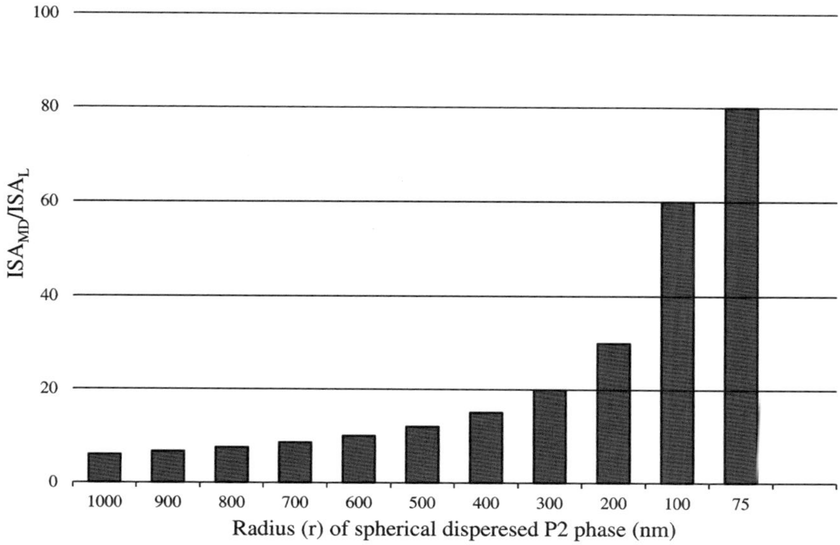

Fig. 4. Ratio of matrix droplet and layered morphology interfacial surface areas in a 1:1 polymer blend as a function of the radius of the dispersed spherical domains. (Adopted with permission from Ref. 60.)

size of the dispersed domain dramatically increases the surface area without affecting the overall membrane dimensions.[60] A quick simplification for Fig. 4 is made assuming the polymers have equivalent densities and volumes, and therefore masses. According to Eq. (6), in the matrix droplet model, the total volume occupied by uniform P2 polymer spheres with radius r dispersed in polymer P1 is

$$\sum(V_{P2}) = N4\pi r^3/3, \tag{6}$$

where N is the total number of P2 spheres. In the layered model, the total volume of the P2 polymer is represented by $L_x L_y L_z$, where each dimension of the cube is represented by L (Fig. 4). Setting the volumes equal to each other and rearranging results in Eq. (7),

$$N = 3(L_X L_Y L_Z)/4\pi r^3, \tag{7}$$

while the total interfacial surface area for the matrix droplet model is represented by Eq. (8).

$$\sum(ISA_{MD}) = N4\pi r^2. \tag{8}$$

Combining Eqs. (7) and (8) and simplifying results in Eqs. (9)–(11).

$$ISA_{MD} = 3(L_X L_Y L_Z)/r, \quad (9)$$

$$ISA_{LBL} = L_X L_Y, \quad (10)$$

$$ISA_{MD}/ISA_{LBL} = 3L_Z/r. \quad (11)$$

The ratio $\mathrm{ISA}_{MD}/\mathrm{ISA}_{LBL}$ is plotted in Fig. 4 and demonstrates the dramatic increase in interfacial surface area as the radius of the dispersed droplets decreases.

In this simplified example, a one-micron thick membrane with a matrix droplet morphology and a 75 nm droplet radius has an interfacial surface area 40 times that of the layered morphology, and this interfacial surface area only increases with decreasing droplet size. Thus, the droplet morphology is desired for gas-separation membranes. Not all blend morphologies lead to improvements in performance, however. The cocontinuous morphology is undesirable for gas separations because it is likely that the more permeable component can extend uninterrupted across the entire thickness of the membrane, giving gas molecules a nonselective path to traverse the membrane. This morphology would result in increased permeability at the cost of decreased selectivity. With multiple possible blend morphologies, it is important for those testing blend membranes for gas separations to correlate their performance with the internal microstructure of the blend.

2.1. *Blends*

Single polymer membranes have a limited number of applications due to their finite properties. As mentioned before, a rudimentary way to increase gas flux in a pure polymer membrane is to decrease the thickness of the membrane while increasing the surface area accessible to the feed gas. The success of this approach is limited by the mechanical properties of the polymer, since defects and the possibility of mechanical failure increase with decreasing thickness.[2] One way to overcome this limitation is by blending a highly selective, mechanically robust polymer with a highly permeable polymer.[61] The major advantage of utilizing polymer blend membranes as compared to developing new polymers is the synergistic combination of desired properties. In addition, developing new polymer systems for gas separations with properties similar to those obtained from blends is both costly and time consuming,[12] effectively removing the possibility of commercializing completely novel systems. Because of the availability and accessibility of starting materials, polymer membrane research largely focuses on finding useful combinations with enhanced properties that can be easily scaled up. Polymer blend membranes are prepared in a variety of ways, most easily through solution and melt mixing as well as more involved molding techniques such as extrusion.[12,62] As noted earlier, polymer blends can be miscible or immiscible, represented by single and separated phases, respectively.

2.2. *Miscible Blends*

The properties of a polymer blend membrane are largely determined by the miscibility of its components. Miscibility is defined as the degree of mixing that results in properties that are expected of a pure material, which indicates approaching homogeneity at the molecular level.[61] Miscibility is easily determined by measuring the glass transition temperature of the blended membrane.[61] A miscible blend exhibits a single T_g like a homogeneous polymer, whereas an immiscible blend has multiple, discrete glass transitions. Other methods, such as light scattering and visualization through microscopy, can be utilized to determine the level of miscibility.[63] The gas-separation properties in a blend are controlled mainly by its phase behavior. Miscible blends have a homogenous single-phase structure, derived from the specific intermolecular interactions between the polymers. In the case of gas separations, performance is largely dependent on the membrane morphology resulting from the level of miscibility. In miscible blends, the permeability coefficient can be predicted by Eq. (12),

$$\ln P_b = \varphi_1 \ln P_1 + \varphi_2 \ln P_2, \tag{12}$$

where P_b, P_1, and P_2 are the permeability coefficients for the blend (P_b), polymer 1(P_1), and polymer 2(P_2), respectively, while φ_1 and φ_2

represent the volume fractions of the polymers. The theoretical derivation of Eq. (12), reported by Paul, predicts deviations from this relationship and experimental results for strongly interacting polymers.[64] Equation 12 predicts that miscible blends comprised of high-performance polymers with permselective properties at the upper bound will produce membranes with performance properties on the upper bound for all compositions.[13] The use of miscible blends for gas separations is attractive since their properties are usually intermediate between those of the individual components. Blending can compensate for the limitations of one or both of the polymers such as poor mechanical stability, low selectivity, low permeability, high cost, and susceptibility to plasticization. Miscible polymer blends tested in the last 10 years include those composed of Matrimid®, PIM-1, PVA, PSF, PES, PBI (Table 1), and other high-performance polymers. Since this chapter will focus on immiscible polymer blends, only a few of the more interesting examples of miscible blends will be discussed, even though numerous examples can be found elsewhere in the literature.[58,65–87]

Cross-linked poly(vinyl alcohol) (PVA) and polyvinylpyrrolidone (PVP) miscible blends were used by Mondal *et al.*[79,88–90] to prepare thin-film membranes. Different amine carriers, such as polyethylenimine (PEI), poly(allylamine) (PAA), 2-amino-2-hydroxymethyl-1,3-propanediol (AHPD), tetraethylenehexamine (TEPA), and hexaethylenehexamine (HEPA), were used to facilitate CO_2 transport through the membrane. As discussed before, polymeric membranes are governed by the solution-diffusion mechanism. To overcome the limitations of solution and diffusion, a reactive carrier that enhances the transport of a single species across a membrane is embedded in the membrane, usually by blending.[12,91] These carriers react reversibly with the target compound. In this case, the amines react with CO_2 to enhance the selectivity properties of the membrane. Cross-linking and blending with PVP increased the thermal stability of the PVA membrane, while the introduction of carriers greatly increased the permeability and selectivity. The most outstanding membrane performance resulted in a permeability increase from 7.8 Barrer for pure PVA to 1890 Barrer for the

PVA–PVP–PAA–AHPD blend, along with an increase in selectivity from 40 to 437. Improvements in both permeability and selectivity have been achieved through the use of blended amine carriers, providing a resourceful way to achieve facilitated CO_2 transport.

Deng *et al.*[92] prepared blend membranes with different ratios of polyvinylamine and poly(vinyl alcohol) (PVAm/PVA) and tested the CO_2–CH_4 separation performance. This particular polymer blend was tested because PVAm provides high CO_2 selectivity through a facilitated transport mechanism,[8] while PVA provides mechanical strength. The blend was cross-linked by heat treatment to prevent the overswelling of the membrane and to enhance the strength of the polymer network. A polysulfone flat sheet was used as a support for the blend membrane. An increase in permeability with an increase in PVAm content was observed, especially with loading values above 60 wt.%. The membrane with a PVAm content of 67% reached the highest CO_2–CH_4 selectivity (~45) of all the blend compositions, with a 30% improvement over the pure PVAm. Additionally, the 67–33 blend exhibited good stability over a 5-month test in humid conditions but rapidly aged under ambient conditions without any humidity treatment.

Poly(amide-12-b-ethylene oxide) (Pebax® 1074) was used in two different blend membranes reported by Feng *et al.*[93,94] Pebax® is a mechanically stable, commercially available polymer with inherent CO_2 selectivity.[95] Glycerol triacetate (GTA) and poly(ethylene glycol) (PEG1500) were blended with Pebax® in order to enhance CO_2 permeability. The first membrane was prepared by blending with GTA. An increase in GTA mass content from 0 to 80 wt.% resulted in an increase in CO_2 permeability from ~135 to ~1800 Barrers, while the CO_2–H_2 ideal selectivity increased from 10 to 19. In general, with a GTA content above 40 wt.%, all gas permeabilities improved. The second blend was prepared using PEG. Gas permeabilities improved with increasing content of PEG, achieving a 1000% increase of CO_2 permeability versus unblended Pebax®. These increases were observed only at operating conditions higher than 55°C, i.e., above the melting point of the PEG phase of Pebax®. Operating at lower temperatures resulted in the opposite effect.

Polybenzimidazole (PBI)–Matrimid® miscible blends have been investigated by Hosseini *et al.*[22,66,78] PBI and Matrimid® are both high-performance polymers with high thermal and chemical stabilities. Since each polymer has a glass transition temperature higher than 300°C, blending highly permeable Matrimid® with mechanically robust PBI results in a membrane capable of withstanding the high temperatures that are attained in industrial processes. By increasing the PBI content in the blend, the gas permeabilities decreased while the selectivities increased. Permeability values were in between those of the pure polymers, but H_2–N_2 and H_2–CO_2 selectivities improved over the unblended polymer values. Further modification of the blends has been done including cross-linking (decreasing permeability while improving selectivity) and carbonization (increasing both permeability and selectivity) with membranes surpassing the upper-bound limit.

The effect of blending Matrimid® with either poly(ethyleneglycol) (PEG) or a poly(ethylene oxide)–polydimethylsiloxane copolymer (PEO–PDMS) was investigated by Hu *et al.* PEG has been shown to improve CO_2 separation,[96] while PDMS has high permeability.[97] Utilizing a copolymer with each component was chosen in order to enhance CO_2 permeability without reducing selectivity. Hollow fibers of the blend were prepared. PEG concentration of 12 wt.% increased the CO_2 permeance to 16.1, an increase of 70%, with an 18% increase in CO_2–N_2 selectivity to 6.7, against pure Matrimid®. For a blend with a 4 wt.% loading of PEO–PDMS, both CO_2 permeance and CO_2–N_2 selectivity showed an $\sim$ 50% increase. Increasing the PEO-PDMS loading further led to a decrease in permeance with a further increase in selectivity. This was attributed to chain interactions between the copolymer and Matrimid® which resulted in decreased porosity of the support layer.

In the work by Yong *et al.*,[98] tuning of the permeability of Matrimid® was achieved by blending with PIM-1. These polymers were chosen in order to easily formulate a highly permeable membrane. Complete and partially miscible behaviors were observed in the membranes. The addition of 5 and 10 wt.% PIM-1 into Matrimid® lead to increase of 25% and 77% CO_2 permeability, to 12 and 17 Barrer, respectively. No significant alteration to the CO_2–N_2 and CO_2–CH_4 selectivities was observed with just 2.8% and 5.6% decreases, respectively. With the inversion of polymer compositions (5 wt.% Matrimid® addition into PIM-1), the O_2–N_2 selectivity has a slight increase, 3%, with a 24% increment in O_2 permeability (3.4 Barrer).

2.3. *Immiscible Blends*

Entropy of mixing does not favor homogenous blends since polymers are so much larger than solvent molecules that they tend to aggregate in solution.[99,100] Most polymer blends are immiscible by nature, but miscibility can be achieved by increasing the enthalpy of mixing through intermolecular interactions.[101] Immiscible blends present a heterogeneous phase-separated structure, where an interface between the polymers is observable (Fig. 3). Calculating the permeability and separation performance for an immiscible blend is inherently more complex due to the separation of phases. The most extreme separation results in compositions ranging from a continuous P_1 phase with P_2 dispersed throughout to a continuous P_2 phase with P_1 dispersed throughout. Typically, the parallel and series models are utilized for blends with these compositions, which represent the upper and lower bounds for membrane performance.[13]

$$\text{Parallel model: } P_b = \varphi_1 P_1 + \varphi_2 P_2 \tag{13}$$

$$\text{Series model: } P_b = \frac{P_1 P_2}{(\varphi_1 P_1 + \varphi_2 P_2)} \tag{14}$$

The lower bound can be represented by the parallel model, which predicts that phase-separated blends comprised of "upper bound" achieving polymers will have permselective properties below the upper bound for all compositions. The series model predicts that laminated films of polymers with differing permeability values will overcome the upper bound if the individual components approach or pass the upper bound on their own.[13] This prediction holds true for hollow fiber membranes, as well as some surface modified membranes, where a thin selective shell on top of a permeable support can enhance the permselective properties to surpass the upper bound.[2] Another model typically used to calculate the permeability coefficient for immiscible polymer blends is the Maxwell model, represented by the matrix droplet morphology in Fig. 3, where spheres of one polymer are dispersed in the matrix of the other polymer. In the Maxwell model,

$$P_b = P_m \left[\frac{P_d + 2P_m - \varphi_d(P_m - P_d)}{P_d + 2P_m + \varphi_d(P_m - P_d)} \right], \tag{15}$$

the subscripts b, m, and d represent the blend, matrix phase, and dispersed phase respectively. The Maxwell, parallel, and series models only predict reasonable values for blends where one phase is continuous with the other polymer dispersed throughout.[13] Blends that are of intermediate morphology where each component contributes to the continuous and dispersed phase can be modeled using the equivalent box model.[102] First proposed by Kolarik,[103] the equivalent box model (EBM) predicts reasonable values for the permeability coefficient by utilizing DeGennes percolation theory[104] as well as a combination of the series and parallel models. The derivations of Eq. (16) can be found elsewhere.[13] In the following equation,

$$P_b = P_1\varphi_{1p} + P_2\varphi_{2p} + \varphi_s^2 \Big/ \left[\frac{\varphi_1}{P_1} + \frac{\varphi_2}{P_2} \right], \tag{16}$$

φ_{1p}, φ_{2p}, φ_1, φ_2, and φ_s^2 are critical threshold percolation components for components 1 and 2, respectively, and can be considered adjustable parameters if the percolation theory is applied to fit data.[13] As an example, Fig. 5 illustrates O_2–N_2 separation as predicted by each model. The upper bound in Fig. 5, represented by the miscible blend model, is only overcome at low permeabilities when modeled using the EBM. The EBM predicts that only blends with the less permeable polymer as the continuous phase can surpass the upper bound.[13] As discussed by Robeson, phase-separated polymer blends will only achieve success as gas-separation membranes if this morphology is achieved, i.e., where the less permeable polymer makes up the continuous phase. The ultimate properties, such as

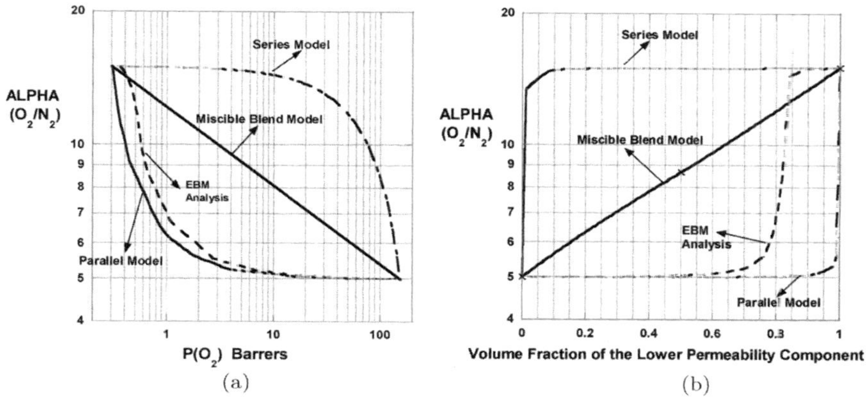

Fig. 5. Model predictions for the O_2–N_2 gas pair (a) as superimposed on an upper bound plot; and (b) selectivity versus volume fraction of the less permeable polymer of the blend. (Reprinted with permission from Ref. 13 Copyright 2010 American Chemical Society.)

flux and permeability of a polymeric membrane, are entirely dependent on the internal morphology of the blend. The importance of understanding and correctly modeling the internal morphology of polymeric membranes cannot be overstated. The challenge and most promising avenue for progress is the *control of blend microstructures.*

Immiscible polymer blends are often characterized by uncontrolled phase separation leading to inconsistent properties and poor mechanical stability. Nevertheless, polymeric membranes have been prepared from immiscible blends of several of the polymers listed in Table 1.[105–111] The more interesting immiscible blend membranes will be discussed in the following section.

Feng *et al.*[112] prepared blend membranes from polyvinyl acetate (PVAc) and block copolymer poly(amide-12-b-ethylene oxide) (Pebax® 1074). PVAc has been shown to improve CO_2 separation due to its polar acetate groups facilitating transport.[113] However, PVAc is not mechanically strong enough to act as a membrane on its own, so it is typically blended with a more robust polymer. Partial miscibility was observed for the blends, but no effort was made to define the morphology, such as utilizing solvent extraction to selectively remove one of the phases for better visualization. With the increase of Pebax® 1074 content in the blend, gas permeabilities of CO_2, H_2, N_2, and CH_4 all increase greatly. In particular, blend membranes with 20 wt.% Pebax® 1074 content enhanced CO_2–CH_4 selectivity, while those with high Pebax® 1074 content (50 wt.%) improved CO_2–H_2 selectivity. Plasticization was observed when the feed pressure was increased.

A poly(urethane) (PU)/poly(vinylacetate) (PVAc) immiscible blend was tested for gas-separation membranes by Ghalei *et al.*[21] No morphological trends were observed with increased PVAc content, but it was found that the membrane with 20 wt.% PVAc content displayed a more uniform dispersion of PVAc domains; for this reason this blend composition was selected for testing. Compared to the pure PU membrane, the blend had a slightly higher CO_2 permeability, 5% increase, and higher CO_2–N_2 and CO_2–CH_4 selectivities, both with ~70% increase.

Low molecular weight poly(ethylene glycol)s (LMWPEGs) with methyl end groups having a molecular weight of 500 g/mol, DM500 and A500, were dispersed in a poly ether-block-amide (Pebax® 1657) matrix to prepare dense membranes. Lilleparg *et al.*[114] then tested the gas-separation properties of these membranes. LMWPEGs formed stable blends with Pebax® with a mass fraction around 30 wt.%. The blends have higher gas permeabilities (three to Five times higher) for H_2, CH_4, N_2, O_2, and CO_2 while loss in selectivity if any was minimal for CO_2–N_2, CO_2–H_2, and O_2–N_2 separations, compared to pure Pebax® membranes. DM500-containing blends have superior permeabilities compared to blends with A500 which can be attributed to the presence of methyl end groups in DM500.

3. Morphology Control

3.1. *Copolymers*

Morphology control in a membrane can be achieved by synthesizing block copolymers composed of polymers with differing gas-separation properties.[42,115–117] Depending on the volume fraction of the components of the copolymer, different structures can be formed.

Single phase, spherical or cylindrical dispersion and cocontinuous morphology are just some examples of morphologies achieved with copolymers.[118] The biggest drawback to using copolymers for the production of a membrane is the complexity of their synthesis, so other methods of morphology control are desirable.

3.2. *Compatibilization*

Polymers that phase separate in a blend are said to be immiscible and typically are mechanically weaker materials than miscible blends due to poor interactions at the interface between the two polymers. This immiscibility is due to unfavorable thermodynamics of mixing, which can be alleviated to some extent through compatibilization methods such as utilizing copolymers or nanomaterials to obtain uniform microstructures.[119] Copolymers offer excellent control over the blend morphology since the interactions between the polymers can be tailored for specific polymer families.[119] Generally, copolymers used for compatibilization contain subunits of both polymers used in the blend so that the copolymer can localize at the interface and decrease interfacial tension. Unfortunately, the synthesis of copolymers adds extra time and complexity to membrane production. This reduces throughput since these compatibilizers can only be tested in blends composed of the copolymer's components.

3.3. *Immiscible Blends Compatibilized with Copolymers*

The blend of polymethylmethacrylate (PMMA) with polyvinylmethylether (PVME) was tested for gas separations by Moon *et al.*[120] The immiscible blend was compatibilized with a diblock copolymer of styrene and methylmethacrylate (PS-b-PMMA). This copolymer was chosen as an interfacial control agent that could interact with each component to produce enhanced structures. The blend presented a droplet morphology structure with the PVME being the dispersed phase while PMMA was the continuous one. Figure 6 depicts the morphology control achieved with different concentrations of the compatibilizer. The average diameter of the dispersed particles decreased from 9 μm to 2 μm when the addition of PS-b-PMMA was increased

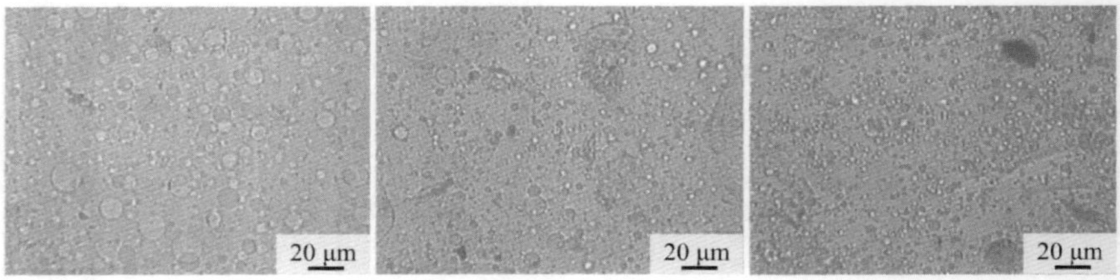

Fig. 6. SEM images of PMMA–PVME blend membranes (a) without PS-b-PMMA; and compatibilized (b) with 2 pph PS-b-PMMA; (c) with 5 pph PS-b-PMMA. (Adapted with permission from Ref. 120 Copyright 2002 Elsevier.)

from 0 parts per hundred (phr) to 5 phr. Incorporation of up to 20 wt.% of PVME into PMMA increased both N_2 and O_2 permeabilities but a decrease in selectivity was observed. The addition of the copolymer further exacerbated these tendencies: increasing permeability at the expense of selectivity as the PS-b-PMMA content increased.

Semsarzadeh *et al.*[121] prepared blend membranes of polyurethane (PU) and polyvinyl acetate (PVAc) in the presence of various polyethylene oxide–polypropylene oxide–polyethylene oxide triblock copolymer (Pluronic®) contents. The blends with 5 wt.% PVAc showed higher CO_2 permeability compared to the PU membrane, but in general, increasing content of PVAc decreased gas permeabilities (CO_2, CH_4, O_2, and N_2) while improving O_2–N_2, CO_2–O_2 and CO_2–N_2 selectivities. While the authors mention that PVAc domains were randomly dispersed in a PU continuous phase, no effort was made to confirm the identity of the phases. Since PVAc content is so much lower than PU, it is reasonable to assume that PVAc would be the dispersed phase. The PVAc domains have various sizes depending on the content of PVAc. For a 5 wt.% blend the shape of dispersed phase was spherical with an average domain size between 1 and 6 μm. By increasing the PVAc content to 10 wt.% the domains also presented a spherical shape but with an average domain size about 5–10 μm. Further increase of PVAc content to 15 wt.% led to an ellipsoid morphology and an average particle size of 10–15 μm. By adding the Pluronic® copolymer from 0.5 to 4 parts per hundred (phr), the PVAc domains became much narrower with a fibril-like morphology, decreasing from around 20 μm in width to nearly 10 μm. Compatibilization in this case reduced the permeabilities of the membrane with a slight increase in the ideal selectivities.

3.4. *Immiscible Blends Compatibilized with Nanoparticles and Small Molecules*

Another way to control microstructure is by using nanoparticles as compatibilization agents. Nanoparticle synthesis is less complex than copolymer synthesis and can be utilized in a variety of blends. The biggest challenge nanoparticles face is localization; if they do not localize at the interface, then compatibilization will most likely not occur. The control of morphology in nanoparticles is a result of decreased interfacial tension from localization at the interface between the polymers on the blend.[101,119]

Membranes from an acrylonitrile-butadiene-styrene (ABS)–poly(vinyl acetate) (PVAc) immiscible blend were prepared and tested for He–CH_4 separation by Bandehali *et al.*[122] ABS is a low cost, commercially available polymer with high flux and mechanical stability, making it a suitable polymer for blending with selective PVAc, which is too mechanically weak to fabricate into a membrane. The resultant morphology of the blend resembled an uncontrolled matrix droplet morphology, but the magnification of the SEM images provided were too low to be certain. Unfortunately, the microstructure of the blend is not discussed, so the morphology of the blend cannot be described with any confidence. Higher contents of PVAc (up to 40 wt.%) increased the He permeability ~20% with a minor increase in selectivity compared to pure ABS. Addition of SiO_2 nanoparticles to the blend with 20 wt.% PVAc content increased both He and CH_4 permeabilies while greatly decreasing the selectivities. Especially, the addition of 1 wt.% of nanoparticles increased permeabilities by more than 3000% of the unblended ABS membrane. In order to improve the adhesion between the polymers and additives, styrene–butadiene rubber (SBR) was added to the blend. The introduction of SBR (15 wt.%) to the membrane significantly improved selectivity (seven times) compared to the original blend, which is attributed to a large decrease in CH_4 permeability.

Membranes fabricated from Pebax® 2533 were blended with novel soft polymeric nanoparticles by Halim *et al.*[123] The nanoparticles were synthesized using grafted star polymers based on poly(ethylene glycol) (PEG) and the block copolymer poly(ethylene glycol)-b-poly(dimethylsiloxane) (PEO-b-PDMS). The size of the nanoparticles was controlled through radical polymerization, typically resulting in diameters near 20 nm, but larger aggregate particles (30–70 nm) were also observed. The soft nanoparticles act as CO_2-permeable domains, which are dispersed in the more selective matrix. This thin selective layer was deposited onto a substrate using spin coating. Increasing the loading of nanoparticles in the blend increased CO_2 permeability ~1000% against pure Pebax®, while decreasing selectivity. Increasing PDMS fraction in the copolymer also led to a CO_2 permeability increase.

An immiscible blend of polypropylene (PP) and ethylene vinyl acetate copolymer (EVA) prepared by Goodarzi *et al.*[124] was used to prepare gas-separation membranes. Lamellar clay nanoparticles (OMMT) and polypropylene-grafted maleic anhydride (PP-g-MAH) were also incorporated into the blend to investigate the influence they had on blend morphology and gas-separation properties. At 25 wt.% loading of EVA the blend had a droplet-type morphology, seen in Fig. 7, while at 50 wt.% the morphology changed to co-continuous,

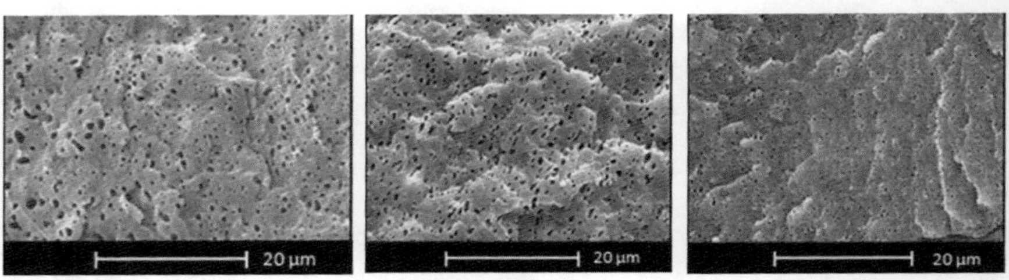

(a) (b) (c)

Fig. 7. SEM images of 25:75 EVA–PP blend membranes (a) without OMMT or PP-g-MAH and compatibilized with (b) 5 wt.% OMMT; and (c) 5 wt.% OMMT and 5 wt.% PP-g-MAH. (Adapted with permission Ref. 124 Copyright 2013 Elsevier.)

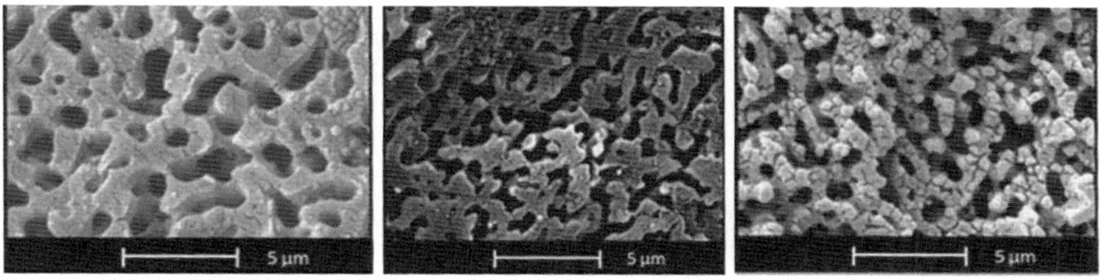

Fig. 8. SEM images of 50:50 EVA–PP blend membranes (a) without OMMT or PP-g-MAH and compatibilized with (b) 5 wt.% OMMT; and (c) 5 wt.% OMMT and 5 wt.% PP-g-MAH. (Adapted with permission from Ref. 124 Copyright 2013 Elsevier.)

shown in Fig. 8. This is a clear example of the morphological effect of adjusting polymer ratios in the blend. Increase in OMMT loading to 5 wt.% decreased the EVA droplet size seen in Fig. 7b while in the cocontinuous blend, there was an increase in the continuity index of EVA (Fig. 8b). PP-g-MAH was then added to the blend, which further intensified the effects of OMMT, as shown in Figs. 7c and 8c. Permeability in the blend was higher than in pure PP but lower than EVA, with the droplet size morphology having higher permeabilities than the co-continuous. Utilizing the clay nanoparticles in conjunction with the copolymer lowered the permeabilities of the blend but greatly increased the CO_2–N_2 and O_2–N_2 selectivities. The authors attribute this outcome to the polar compatibilizer improving OMMT dispersion in the blend, but the increased selectivity could also be due to the resultant morphology providing fewer permeable gaps for gas transport.

Despite many efforts to create or discover polymeric membranes with enhanced gas-separation properties, few systems have exceeded the nearly decade-old upper bound correlated by Robeson.[8] Since the synthesis of novel materials is both costly and time-consuming, alternative approaches such as using blends of known high-performance polymers with defined microstructures could be an effective way to improve membrane performance while remaining both time-and cost-effective.[12,13] The potential for improving gas-separation membranes using immiscible polymer blends is huge, especially if the resulting morphologies can be controlled. Since the ultimate gas-separation properties are dependent on the blend microstructure, the major hurdle is to control the phase separation typically seen in immiscible blends. Recent work by Panapitiya *et al.*[125] has shown that it may be beneficial to utilize metal organic frameworks (MOFs) to simultaneously enhance the performance and compatibility of the polymers in the blend. MOFs inherently possess unique pore apertures, high chemical and thermal stabilities, and have large surface areas which make them desirable additives for gas-separation membranes.[126] The work described utilized loadings of the zeolitic imidazolate framework-8 (ZIF-8), illustrated in Fig. 9, to compatibilize high-performance PBI and 6FDA-DAM:DABA (6FDD) polymers. ZIF-8 has a 0.34 nm pore aperture,[127] a cage size of 0.82 nm^3 (sphere in Fig. 9),[128] and a high surface area above 1600 m^2/g,[34] making it a suitable additive for the creation of a mixed matrix membrane. The advantage of utilizing porous additives for MMMs is increased selectivity due to the pores approaching sizes small enough to allow for molecular sieving and selective adsorption.[129]

The morphology of a 50:50 PBI–6FDD blend membrane resembles the matrix droplet morphology with an irregularly shaped dispersed phase and broad size distribution, as seen in Fig. 10. In this work, membranes were subjected to Sohxlet extraction of the soluble

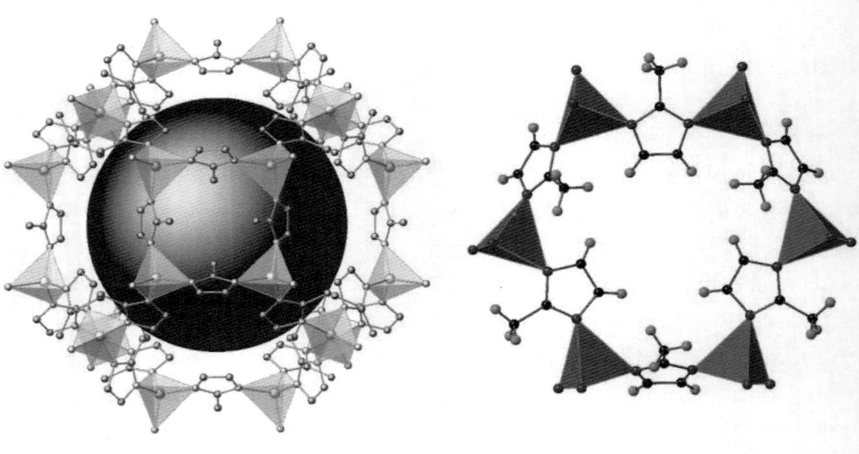

Fig. 9. Structure of ZIF-8.

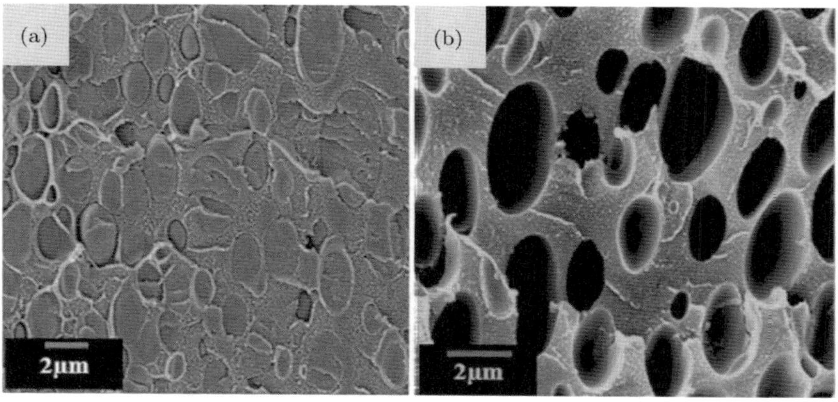

Fig. 10. SEM images of PBI–6FDD (50:50) membrane cross-section (a) before and (b) after Sohxlet extraction. (Reprinted with permission from Ref. 125 Copyright 2014 Elsevier.)

6FDD using tetrahydrofuran (THF) as the solvent so that identification of each phase would be possible using SEM. As seen in Fig. 10b, the irregular dispersed 6FDD domains were cleanly removed during Sohxlet extraction.

Upon the addition of 5 wt.% ZIF-8, both the size and polydispersity of the 6FDD domains decreased. Further increasing the ZIF-8 loading to 10 wt.% resulted in an additional reduction of domain size while uniformity was improved. As illustrated by the histograms in Fig. 11, the average domain size of the dispersed 6FDD phase decreased from $1.46 \pm 0.60\mu m$ in the uncompatibilized blend to $0.201 \pm 0.03\mu m$ upon the incorporation of 10 wt.% ZIF-8 in to the membrane.

It is clear that ZIF-8 has a compatibilization effect on this blend due to the more uniform dispersion of phases with increased loading. The authors attribute this to ZIF-8's ability to position itself at the interface between the two polymers, as depicted in Fig. 12.

One possible distribution of the ZIF-8 is randomly throughout either polymer phase, which is unlikely due to the increase in dispersion and uniformity with increased loadings. If ZIF-8 were distributed unequally in either polymer, the resulting morphology would be similar to the precursor blend seen in Fig. 11A, where there is a large

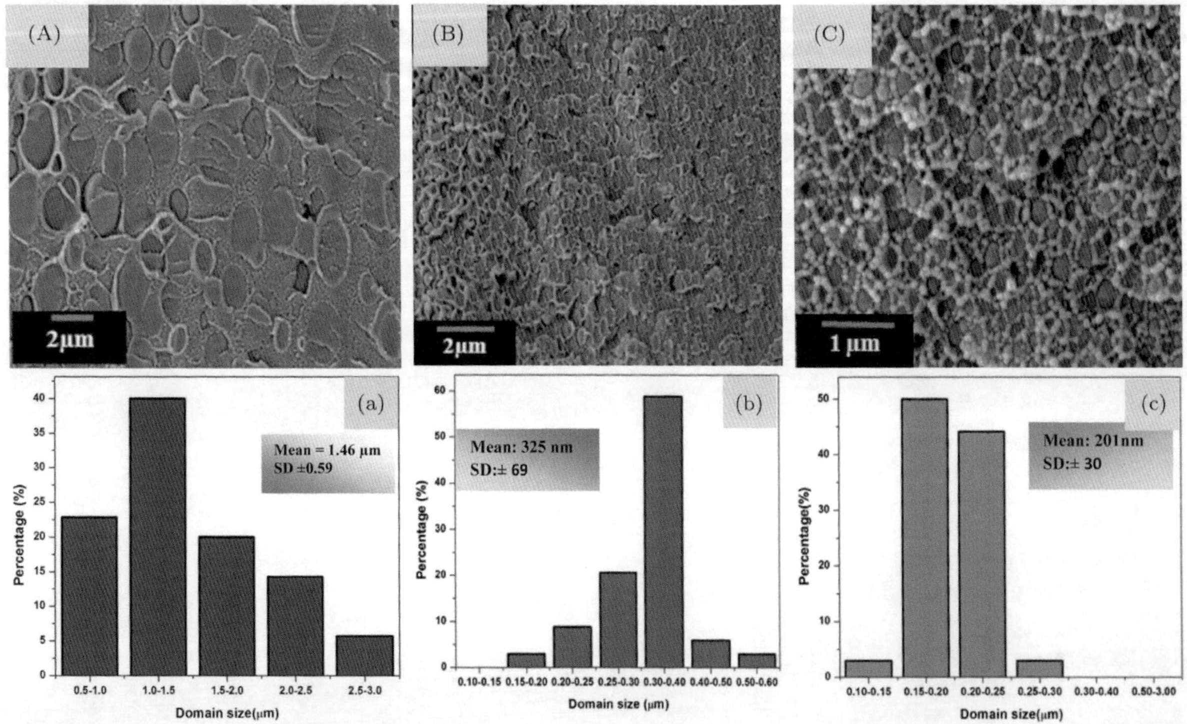

Fig. 11. SEM images and size histograms for (A, a) PBI–6FDD (50:50); (B, b) 5 wt.% ZIF-8 PBI/6FDD (50:50); (C, c) 10 wt.% ZIF-8 PBI–6FDD (50:50). (Adapted with permission from Ref. 125. Copyright 2014 Elsevier.)

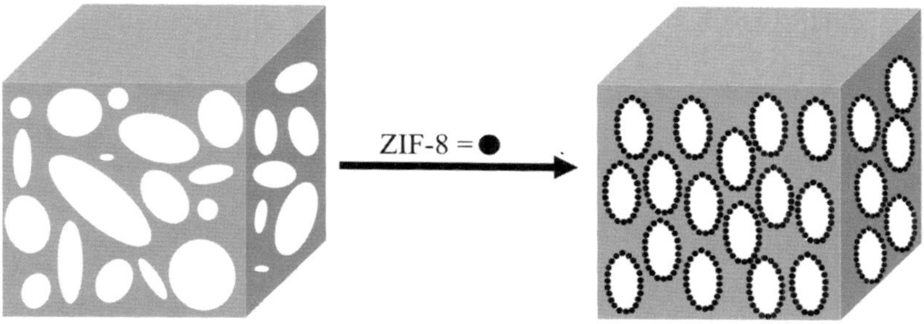

Fig. 12. Illustration of ZIF-8 localized at the interface of the two polymers.

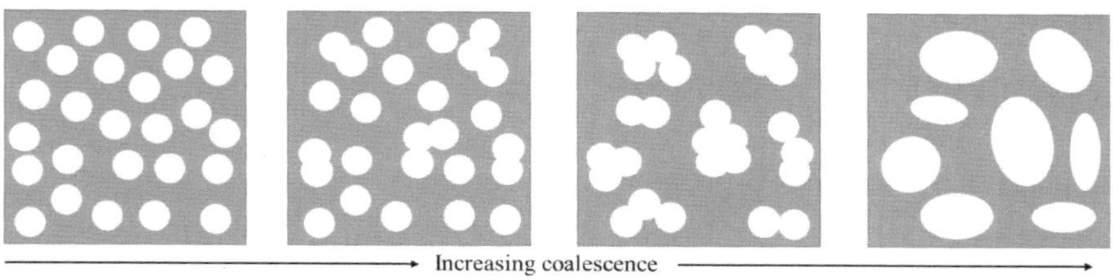

Fig. 13. Illustration of coalescence of dispersed (colorless) phases.

variance in the dispersed phase's size and dispersion due to a lack of restriction for coalescence (Fig. 13).

In order to further resolve ZIF-8's location in the blend, a goniometer was utilized to measure the water contact angles for ZIF-8, 6FDD, and PBI. The surface tension values of each component were then calculated from these measurements, which were substituted into the Girifalco–Good equation,[130] $\sigma_{xy} = \sigma_x + \sigma_y - 2\Phi(\sigma_x\sigma_y)^{1/2}$, to calculate the ratio of interfacial tension values $\sigma_{x/y}$. In the previous equation, Φ is considered a constant that defines the free energy of adhesion and cohesion for the phases. Both $\sigma_{ZIF-8/PBI}$ and $\sigma_{ZIF-8/6FDD}$ were found to be lower than $\sigma_{6FDD/PBI}$, which indicates a better compatibility between ZIF-8 and both polymers. Young's equation,[131] $\sigma_{lv}\cos\theta = \sigma_{sv} - \sigma_{sl}$, allows for a direct correlation between the measured contact angle and each interface, represented by liquid–vapor (lv), solid–vapor (sv), and solid-liquid (sl).

Substituting the interfacial tension values into Young's equation predicts a ZIF-8 contact angle in 6FDD approaching 90°, which corresponds to ZIF-8 localizing at the interface between each polymer. The improved compatibility in the 6FDD–PBI immiscible blend due to increased loadings of ZIF-8 is apparent and is evidence of a lowered interfacial energy due to interfacial localization.

In an effort to determine the specific interactions ZIF-8 has with 6FDD and PBI, the components of the MOF were blended into polymeric membranes in order to monitor their effects. Although no effect was observed for the incorporation of Zn^{+2} ions, Panapitiya *et al.*[24] reported that the incorporation of the MOF linker 2-methylimidazole (MI) was successful in compatibilizing otherwise immiscible blends of high performance PBI and 6FDD polymers, resulting in membranes with uniform microstructures. The resultant microstructure of these polymer blend membranes also contain the more permeable 6FDD dispersed in the continuous selective PBI in a matrix droplet morphology. As seen in Fig. 14, increasing the MI loading in the membrane reduced the size and ovoid shape of the 6FDD domains. The dispersed 6FDD phase was selectively removed by Sohxlet extraction using THF prior to imaging using SEM.

Infrared spectroscopy studies performed on membranes with and without MI loadings indicated that MI can potentially hydrogen bond with 6FDD, as well as form hydrophobic and π–π interactions with PBI. Tests were also performed to rule out phase coarsening or coalescence restriction by varying which polymer received the MI loading before blending. The development of the uniform matrix droplet morphology, regardless of where MI is loaded into the blend, showed the potential for small molecules to localize at the interface of immiscible polymer blends. Membranes containing higher loadings of MI surpassed the Robeson's upper bound for the H_2–CO_2 gas pair, displaying increased selectivity at the expense of somewhat decreased permeability. The success of incorporating this small molecule into an immiscible polymer blend as a convenient and cost-effective compatibilization agent opens up new opportunities for improving separation performances for polymeric gas-separation membranes.

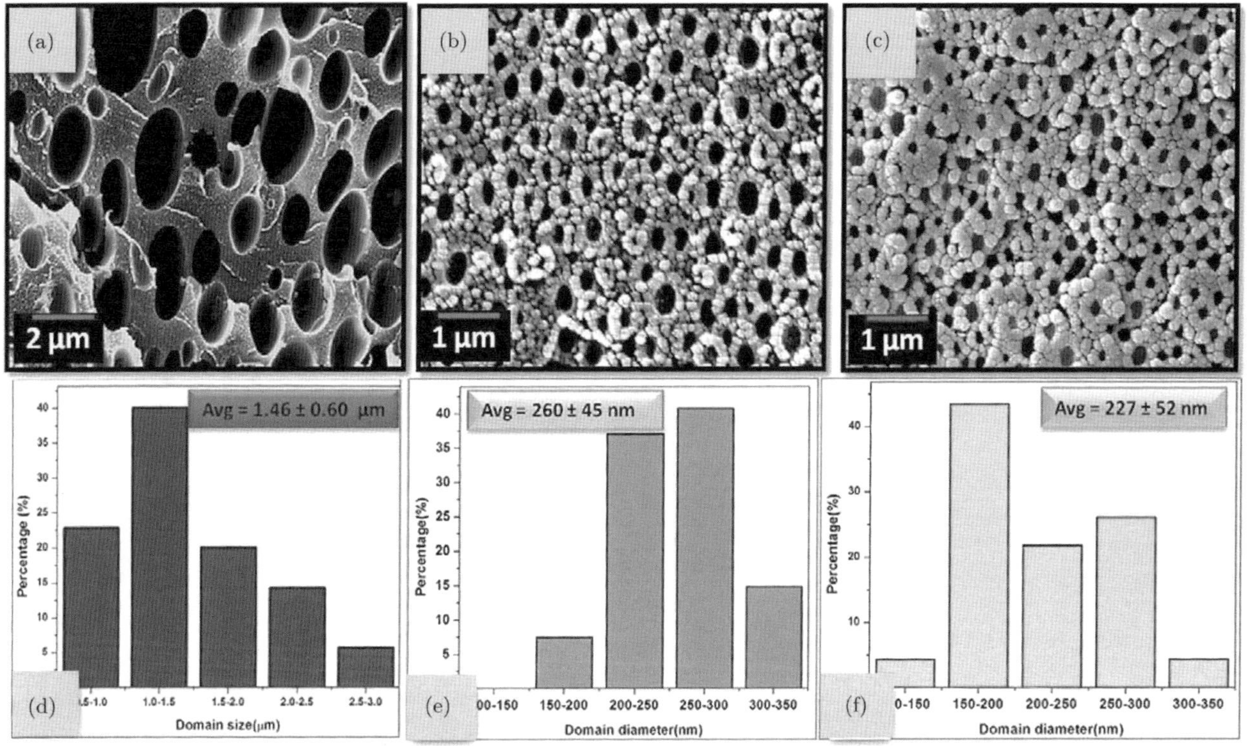

Fig. 14. SEM images and size histograms after Sohxlet extraction of 6FDD–PBI (50:50) blend membranes containing (a, d) 0 wt.%; (b, e) 5 wt.%; (c, f) 9 wt.% MI loadings. (Adapted from Ref. 24 Copyright 2015 American Chemical Society.)

4. Summary and Outlook

There is a promising future for blends in membrane research. Blending polymers into a single membrane is the most effective way to obtain materials with enhanced properties since designing a novel polymer is time consuming and can be expensive. Pure polymers without modifications have finite properties and fall below the Robeson upper bound of performance. While not heavily discussed in this chapter, inorganic membranes are also utilized for gas separations. Typically, inorganic membranes display better separation performance and stability than pure or blended polymeric membranes but are hindered by their associated higher cost. Controlling the morphology of phase-separated blends is a major factor in performance. The theoretical maximum limit of performance could be achieved in the case of the matrix droplet morphology, where the dispersed spheres are separated by a thin layer of selective polymer. Control of this morphology was achieved with MOFs and small molecules. The thickness of the selective polymer that separates each droplet was around 100 nm. Decreasing the thickness of the selective separating matrix will likely result in further increases in performance. A number of different inorganic and organic additives have been used to compatibilize polymers, but control of microstructure is still a challenge. Metal organic frameworks and the MI small molecule have shown promise as structure directing additives. The basic structure of MOFs is a combination of organic ligands and inorganic clusters, which allows MOFs to be easily tuned and readily utilized as additives for MMM fabrication.[126,132] Metal and linker choice could be adjusted to enhance interactions with polymers in an immiscible blend, possibly influencing the location of interfaces. The biggest advantage of utilizing MOFs as additives for MMM fabrication is the potential for enhanced selectivity due to molecular-sieving effects since the pore sizes of MOFs approach the size of gas molecules.[133] Further research should be undertaken to better understand the structure-directing effects of small molecules. The potential for small molecules to direct the phase behavior of immiscible blends as a cost-effective compatibilizer has been shown for MI in a 6FDD–PBI blend. This molecule was easily removed through thermal annealing of the membrane under vacuum, which potentially created small gas transport pathways that enhanced H_2–CO_2 selectivity.[24] Studies utilizing small molecules as compatibilizers for polymer blend membranes are current and ongoing, but MI's ability to direct immiscible blend structure in the 6FDD–PBI blend has been shown. Methylimidazole was most likely effective as a compatibilizer because of its similar structure to the polymers in the blend, but the mechanism of morphology control is still not known. The success of this small molecule as a compatabilizer in the 6FDD–PBI is not indicative of universal effectiveness since it was specifically chosen for this polymer pair. As mentioned previously, facilitated transport of CO_2 can be achieved by incorporating certain functional groups, such as amines, into the membrane. Other small aliphatic amines may prove useful as compatibilization agents for different polymer blends, but further study is required. The MI small molecule was removed from the membrane during thermal annealing under vacuum, and while it is not certain how this affects membrane performance, it may prove a useful method to produce enhanced microstructures without drastically changing membrane composition. Additives have proven their effectiveness at compatibilizing immiscible blends. Tuning the ratio of polymer components in the blend is another possible optimization towards enhanced morphologies. A change from matrix droplet to cocontinuous morphology seen (Figs. 7 and 8) above, but it is likely that adjusting polymer ratios can also result in beneficial morphologies. Considering the numerous adjustable parameters for polymer blend membranes, it is likely that the high-performance membranes of tomorrow will largely focus on parameter optimization to control internal morphology.

Immiscible blends are largely fabricated as dense membranes and are not typically seen in hollow fiber geometries. This is most likely due to phase-separation issues that can happen during membrane preparation. Several applications of miscible blend hollow fiber membranes can be found in the literature, including gas separation.[134–136] and water-purification applications.[137] The stability provided by compatibilization may allow for the fabrication of immiscible blends in geometries more similar to those used commercially.

As discussed throughout the chapter, the morphology of blends plays a key role in the final gas-separation properties of the membrane. While the chapter focused on polymer–polymer blends, it is worth noting that the morphology of immiscible ternary blends has been studied.[138,139] These blends can present interesting core–shell morphologies[140] inaccessible to two polymer systems, and the control of phase separation through compatibilization has been achieved.[141] Ultimately, immiscible ternary blends have not been applied to gas separations, but the morphologies presented may lead to new developments in the field of polymer blend gas-separation membranes.

Currently high-performance polymers are blended to produce high-performance polymer blend membranes, but it may be beneficial to test more cost-effective materials. Efforts should be made to test membranes in conditions more similar to the industrial applications they are targeting, such as testing gas mixtures at higher temperatures and pressures. It is of utmost importance to record as many details as possible for each step of the membrane fabrication process, as there are several factors such as concentration of casting solution, casting and drying conditions, and reagent consistency that can prevent successful reproduction. A systematic and controlled method of fabrication must be utilized in order to eliminate any variables, which is necessary for consistent results. Research projects that are closer to deployment than development must also investigate scalability in order to factor cost.

Acknowledgments

The authors acknowledge support from the National Science Foundation (CBET-1403950) and CONACYT.

References

1. Goettler, L. A., Scobbo, J. J., Applications of Polymer Blends. In *Polymer Blends Handbook*, Utracki, A. L., Wilkie, A. C. (Eds). Springer Netherlands: Dordrecht, 2014, pp. 1433–1458.
2. Sanders, D. F., Smith, Z. P., Guo, R., Robeson, L. M., McGrath, J. E., Paul, D. R., Freeman, B. D. Energy-efficient polymeric gas separation membranes for a sustainable future: A review. *Polymer.* 2013, 54(18), 4729–4761.

3. Goh, P. S., Ismail, A. F., Sanip, S. M., Ng, B. C., Aziz, M., Recent advances of inorganic fillers in mixed matrix membrane for gas separation. *Sep. Purif. Technol.* 2011, 81(3), 243–264.
4. Spillman, R. W. Economics of gas separation membranes. *Chem. Eng. Prog.* 1989, 85(1), 41–62.
5. Kesting, R. E., Fritzsche, A. K. *Polymeric Gas Separation Membranes.* Wiley: New York, 1993.
6. Puri, P. S. Commercial applications of membranes in gas separations, Chapter 8. In *Membrane Engineering for the Treatment of Gases: Volume 1: Gas-separation Problems with Membranes*, The Royal Society of Chemistry, 2011, pp. 215–244.
7. EERE, Multi-Year Research, Development, and Demonstration Plan. 2007.
8. Robeson, L. M. The upper bound revisited. *J. Memb. Sci.* 2008, 320(1–2), 390–400.
9. Robeson, L. M. Correlation of separation factor versus permeability for polymeric membranes. *J. Memb. Sci.* 1991, 62(2), 165–85.
10. Stern, S. A. The "barrer" permeability unit. *J. Polym. Sci. Poly. Phys.* 1968, 6(11), 1933–1934.
11. Koros, W. J., Fleming, G. K. Membrane-based gas separation. *J. Memb. Sci.* 1993, 83(1), 1–80.
12. Mannan, H. A., Mukhtar, H., Murugesan, T., Nasir, R., Mohshim, D. F., Mushtaq, A. Recent applications of polymer blends in gas separation membranes. *Chem. Eng. Technol.* 2013, 36(11), 1838–1846.
13. Robeson, L. M. Polymer blends in membrane transport processes. *Ind. Eng. Chem. Res.* 2010, 49(23), 11859–11865.
14. Li, J., Wang, S., Nagai, K., Nakagawa, T., Mau, A. W. H. Effect of polyethyleneglycol (PEG) on gas permeabilities and permselectivities in its cellulose acetate (CA) blend membranes. *J. Memb. Sci.* 1998, 138(2), 143–152.
15. Laghaei, M., Sadeghi, M., Ghalei, B., Shahrooz, M. The role of compatibility between polymeric matrix and silane coupling agents on the performance of mixed matrix membranes: Polyethersulfone/MCM-41. *J. Memb. Sci.* 2016, 513, 20–32.
16. Reid, B. D., Ruiz-Trevino, F. A., Musselman, I. H., Balkus, K. J., Jr., Ferraris, J. P. Gas Permeability properties of polysulfone membranes containing the mesoporous molecular sieve MCM-41. *Chem. Mater.* 2001, 13(7), 2366–2373.
17. Kim, S., Pechar, T. W., Marand, E. Poly(imide siloxane) and carbon nanotube mixed matrix membranes for gas separation. *Desalination* 2006, 192(1–3), 330–339.
18. Yoshimune, M., Fujiwara, I., Suda, H., Haraya, K. Novel carbon molecular sieve membranes derived from poly(phenylene oxide) and its derivatives for gas separation. *Chem. Lett.* 2005, 34(7), 958–959.
19. Min, K. E., Paul, D. R. Effect of tacticity on permeation properties of poly(methyl methacrylate). *J. Poly. Sci. Poly. Phy.* 1988, 26(5), 1021–1033.

20. Yong, W. F., Li, F. Y., Chung, T.-S., Tong, Y. W. Highly permeable chemically modified PIM-1/Matrimid membranes for green hydrogen purification. *J. Mater. Chem. A* 2013, 1(44), 13914–13925.
21. Ghalei, B., Semsarzadeh, M.-A. A Novel nano structured blend membrane for gas separation. *Macromol. Symp.* 2007, 249–250(1), 330–335.
22. Hosseini, S. S., Chung, T. S. Carbon membranes from blends of PBI and polyimides for N2/CH4 and CO2/CH4 separation and hydrogen purification. *J. Memb. Sci.* 2009, 328(1–2), 174–185.
23. Hsieh, J. O., Balkus, K. J., Jr., Ferraris, J. P., Musselman, I. H. MIL-53 frameworks in mixed-matrix membranes. *Microporous Mesoporous Mater.* 2014, 196, 165–174.
24. Panapitiya, N. P., Wijenayake, S. N., Nguyen, D. D., Huang, Y., Musselman, I. H., Balkus, K. J., Ferraris, J. P. Gas separation membranes derived from high-performance immiscible polymer blends compatibilized with small molecules. *ACS Appl. Mater. Interfaces* 2015, 7(33), 18618–18627.
25. Wijenayake, S. N., Panapitiya, N. P., Versteeg, S. H., Nguyen, C. N., Goel, S., Balkus, K. J., Musselman, I. H., Ferraris, J. P. Surface cross-linking of ZIF-8/polyimide mixed matrix membranes (MMMs) for gas separation. *Ind. Eng. Chem. Res.* 2013, 52(21), 6991–7001.
26. Baker, R. W., Lokhandwala, K., Natural gas processing with membranes: An overview. *Ind. Eng. Chem. Res.* 2008, 47(7), 2109–2121.
27. Yasuda, H. Units of gas permeability constants. *J. Appl. Polym. Sci.* 1975, 19(9), 2529–2536.
28. Baker, R. W. Membranes and modules. In *Membrane Technology and Applications*, John Wiley & Sons, Ltd NewYork, 2004, pp. 89–160.
29. Feng, C. Y., Khulbe, K. C., Matsuura, T., Ismail, A. F. Recent progresses in polymeric hollow fiber membrane preparation, characterization and applications. *Sep. Purif. Technol.* 2013, 111, 43–71.
30. Liu, J. Gas Permselectivity Properties of PBI-based Asymmetric Flat and Hollow Fiber Membranes for H2/CO2 Separation. The University of Texas at Dallas, 2015.
31. Hunger, K., Schmeling, N., Jeazet, H. B. T., Janiak, C., Staudt, C., Kleinermanns, K. Investigation of cross-linked and additive containing polymer materials for membranes with improved performance in pervaporation and gas separation. *Membranes.* 2012, 2(4), 727.
32. Zhao, C., Xue, J., Ran, F., Sun, S. Modification of polyethersulfone membranes — A review of methods. *Prog. Mater. Sci.* 2013, 58(1), 76–150.
33. Kim, S., Lee, Y. M., Rigid and microporous polymers for gas separation membranes. *Prog. Polym. Sci.* 2015, 43, 1–32.
34. Wijenayake, S. N., Panapitiya, N. P., Nguyen, C. N., Huang, Y., Balkus, K. J., Musselman, I. H., Ferraris, J. P. Composite membranes with a highly selective polymer skin for hydrogen separation. *Sep. Purif. Technol.* 2014, 135, 190–198.
35. Buonomenna, M. G., Yave, W., Golemme, G. Some approaches for high performance polymer based membranes for gas separation: Block copolymers,

carbon molecular sieves and mixed matrix membranes. *RSC Advances* 2012, 2(29), 10745–10773.

36. Ordonez, M. J. C., Balkus, K. J., Jr., Ferraris, J. P., Musselman, I. H. Molecular sieving realized with ZIF-8/Matrimid mixed-matrix membranes. *J. Memb. Sci.* 2010, 361(1–2), 28–37.
37. Basu, S., Khan, A. L., Cano-Odena, A., Liu, C., Vankelecom, I. F. J. Membrane-based technologies for biogas separations. *Chem. Soc. Rev.* 2010, 39(2), 750–768.
38. Wright, C. T., Paul, D. R. Gas sorption and transport in UV-irradiated poly(2,6-dimethyl-1,4-phenylene oxide) films. *J. Appl. Polym. Sci.* 1998, 67(5), 875–883.
39. Fang, D. Z., Striemer, C. C., Gaborski, T. R., McGrath, J. L., Fauchet, P. M. Pore size control of ultrathin silicon membranes by rapid thermal carbonization. *Nano Letters.* 2010, 10(10), 3904–3908.
40. Tillet, G., Boutevin, B., Ameduri, B., Chemical reactions of polymer crosslinking and post-crosslinking at room and medium temperature. *Prog. Polym. Sci.* 2011, 36(2), 191–217.
41. Tin, P. S., Chung, T. S., Liu, Y., Wang, R., Liu, S. L., Pramoda, K. P. Effects of cross-linking modification on gas separation performance of matrimid membranes. *J. Memb. Sci.* 2003, 225(1–2), 77–90.
42. Querelle, S. E., Chen, L., Hillmyer, M. A., Cussler, E. L., Nijmeijer, K., Wessling, M. Block copolymer derived membranes for sustained carbon dioxide-methane separations. *Ind. Eng. Chem. Res.* 2010, 49(23), 12051–12059.
43. Benhamou, M., El Fazni, A., Bettachy, A. Derouiche, A. Theory of microphase separation in crosslinked polymer blends immersed in a theta-solvent. *Eur. Phys. J. E Soft Matter* 2010, 32(4), 391–8.
44. Gennes, P. G. *Scaling Concept in Polymer Physics.* Cornell University Press: Ithaca, NY, 1979.
45. Mahajan, R., Koros, W. J. Factors controlling successful formation of mixed-matrix gas separation materials. *Ind. Eng. Chem. Res.* 2000, 39(8), 2692–2696.
46. Perez, E. V., Balkus, K. J., Jr., Ferraris, J. P., Musselman, I. H. Metal-organic polyhedra 18 mixed-matrix membranes for gas separation. *J. Memb. Sci.* 2014, 463, 82–93.
47. Miller, S. J., Koros, W. J., Vu, D. Q. Mixed matrix membrane technology: Enhancing gas separations with polymer/molecular sieve composites. In *Studies in Surface Science and Catalysis*, Vol. 170, Ruren Xu, Z. G. J. C., Wenfu, Y. Eds. Elsevier: Amsterdam, 2007, pp. 1590–1596.
48. Pascu, M., Vasile, C., Gheorghiu, M. Modification of polymer blend properties by argon plasma/electron beam treatment: Surface properties. *Mater. Chem. Phys.* 2003, 80(2), 548–554.
49. Durigon, R. P. E., Petri, S. D. F., Drings, H., Schimmel, T., Bruns, M. Characterization and modification of polymer blend films. *Colloid Polym. Sci.* 2001, 279(10), 1013–1019.

50. Schaub, T. F., Kellogg, G. J., Mayes, A. M., Kulasekere, R., Ankner, J. F., Kaiser, H. Surface Modification via Chain End Segregation in Polymer Blends. *Macromolecules.* 1996, 29(11), 3982–3990.
51. Rajagopalan, P., McCarthy, T. J. Two-step surface modification of Chemically resistant polymers:? Blend formation and subsequent chemistry. *Macromolecules* 1998, 31(15), 4791–4797.
52. Binder, K., Frisch, H. L. Phase stability of weakly crosslinked interpenetrating polymer networks. *J. Chem. Phys.* 1984, 81(4), 2126–2136.
53. Inoue, T. Selective crosslinking in polymer blends. II. Its effect on impact strength and other mechanical properties of polypropylene/unsaturated elastomer blends. *J. Appl. Polym. Sci.* 1994, 54(6), 723–733.
54. Mamun, C. K. Free energy of mixing of cross-linked polymer blends. *Langmuir* 2005, 21(1), 240–250.
55. Spenadel, L. Radiation crosslinking of polymer blends. *Radiat. Phys. Chem. (1977)* 1979, 14(3), 683–697.
56. Kusworo, T. D., Ismail, A. F., Mustafa, A., Matsuura, T. Dependence of membrane morphology and performance on preparation conditions: The shear rate effect in membrane casting. *Sep. Purif. Technol.* 2008, 61(3), 249–257.
57. Dorosti, F., Omidkhah, M. R., Pedram, M. Z., Moghadam, F. Fabrication and characterization of polysulfone/polyimide–zeolite mixed matrix membrane for gas separation. *Chem. Eng. J.* 2011, 171(3), 1469–1476.
58. Ismail, A. F., Rahim, R. A., Rahman, W. A. W. A. Characterization of polyethersulfone/Matrimid® 5218 miscible blend mixed matrix membranes for O2/N2 gas separation. *Sep. Purif. Technol.* 2008, 63(1), 200–206.
59. Moghadassi, A. R., Rajabi, Z., Hosseini, S. M., Mohammadi, M. Fabrication and modification of cellulose acetate based mixed matrix membrane: Gas separation and physical properties. *J. Ind. Eng. Chem.* 2014, 20(3), 1050–1060.
60. Panapitiya, N. P. Novel compatibilized immiscible polymer blend based membranes for gas separations. The University of Texas at Dallas, 2014.
61. Robeson, L. M. *Polymer Blends: A Comprehensive Review.* Hanser Gardener publications, Inc.: cinannadi, 2007.
62. Utracki, L. A. History of commercial polymer alloys and blends (from a perspective of the patent literature). *Polym. Eng. Sci.* 1995, 35(1), 2–17.
63. Yong, W. F., Li, F. Y., Xiao, Y. C., Li, P., Pramoda, K. P., Tong, Y. W., Chung, T. S. Molecular engineering of PIM-1/Matrimid blend membranes for gas separation. *J. Memb. Sci.* 2012, 407–408, 47–57.
64. Paul, D. R. Gas transport in homogeneous multicomponent polymers. *J. Memb. Sci.* 1984, 18, 75–86.
65. Acharya, N., Kulshrestha, V., Awasthi, K., Jain, A., Singh, M., Vijay, Y. Hydrogen separation in doped and blend polymer membranes. *Int. J. Hydrogen Energy.* 2008, 33(1), 327–331.
66. Hosseini, S. S., Teoh, M. M., Chung, T. S. Hydrogen separation and purification in membranes of miscible polymer blends with interpenetration networks. *Polymer.* 2008, 49(6), 1594–1603.

67. Basu, S., Cano-Odena, A., Vankelecom, I. F. J. Asymmetric membrane based on Matrimid® and polysulphone blends for enhanced permeance and stability in binary gas (CO2/CH4) mixture separations. *Sep. Purif. Technol.* 2010, 75(1), 15–21.
68. Han, J., Lee, W., Choi, J. M., Patel, R., Min, B.-R. Characterization of polyethersulfone/polyimide blend membranes prepared by a dry/wet phase inversion: Precipitation kinetics, morphology and gas separation. *J. Memb. Sci.* 2010, 351(1–2), 141–148.
69. Hosseini, S. S., Peng, N., Chung, T. S. Gas separation membranes developed through integration of polymer blending and dual-layer hollow fiber spinning process for hydrogen and natural gas enrichments. *J. Memb. Sci.* 2010, 349(1–2), 156–166.
70. Khan, A. L., Li, X., Vankelecom, I. F. J. SPEEK/Matrimid blend membranes for CO2 separation. *J. Memb. Sci.* 2011, 380(1–2), 55–62.
71. Suzuki, T., Miki, M., Yamada, Y. Gas transport properties of hyperbranched polyimide/hydroxy polyimide blend membranes. *Euro. Polym. J.* 2012, 48(8), 1504–1512.
72. Mondal, A., Mandal, B. Synthesis and characterization of crosslinked poly(vinyl alcohol)/poly(allylamine)/2-amino-2-hydroxymethyl-1,3-propanediol/polysulfone composite membrane for CO2/N2 separation. *J. Memb. Sci.* 2013, 446, 383–394.
73. Moradihamedani, P., Ibrahim, N. A., Yunus, W. M. Z. W., Yusof, N. A. Separation of CO2 from CH4 by pure PSF and PSF/PVP blend membranes: Effects of type of nonsolvent, solvent, and PVP concentration. *J. Appl. Polym. Sci.* 2013, 130(2), 1139–1147.
74. Shen, J. N., Yu, C. C., Zeng, G. N., van der Bruggen, B. Preparation of a facilitated transport membrane composed of carboxymethyl chitosan and polyethylenimine for CO2/N2 separation. *Int J Mol Sci.* 2013, 14(2), 3621–38.
75. Wolinska-Grabczyk, A., Kubica, P., Jankowski, A. Effect of the acetate group content on gas permeation through membranes based on poly(ethylene-co-vinyl acetate) and its blends. *J. Memb. Sci.* 2013, 443, 227–236.
76. Yong, W. F., Li, F. Y., Chung, T.-S., Tong, Y. W. Highly permeable chemically modified PIM-1/Matrimid membranes for green hydrogen purification. *J. Mate. Chem. A* 2013, 1(44), 13914.
77. Yong, W. F., Li, F. Y., Xiao, Y. C., Chung, T. S., Tong, Y. W. High performance PIM-1/Matrimid hollow fiber membranes for CO2/CH4, O2/N2 and CO2/N2 separation. *J. Memb. Sci.* 2013, 443, 156–169.
78. Hosseini, S. S., Omidkhah, M. R., Zarringhalam Moghaddam, A., Pirouzfar, V., Krantz, W. B., Tan, N. R. Enhancing the properties and gas separation performance of PBI–polyimides blend carbon molecular sieve membranes via optimization of the pyrolysis process. *Sep. Purif. Technol.* 2014, 122, 278–289.
79. Mondal, A., Mandal, B. CO2 separation using thermally stable crosslinked poly(vinyl alcohol) membrane blended with polyvinylpyrrolidone/

polyethyleneimine/tetraethylenepentamine. *J. Memb. Sci.* 2014, 460, 126–138.

80. Saedi, S., Madaeni, S. S., Hassanzadeh, K., Shamsabadi, A. A., Laki, S. The effect of polyurethane on the structure and performance of PES membrane for separation of carbon dioxide from methane. *J. Ind. Eng. Chem.* 2014, 20(4), 1916–1929.
81. Scholes, C. A., Ribeiro, C. P., Kentish, S. E., Freeman, B. D. Thermal rearranged poly(benzoxazole)/polyimide blended membranes for CO2 separation. *Sep. Purif. Technol.* 2014, 124, 134–140.
82. Yong, W. F., Li, F. Y., Chung, T. S., Tong, Y. W. Molecular interaction, gas transport properties and plasticization behavior of cPIM-1/Torlon blend membranes. *J. Memb. Sci.* 2014, 462, 119–130.
83. Rabiee, H., Ghadimi, A., Abbasi, S., mohammadi, T. CO2 separation performance of poly(ether-b-amide6)/PTMEG blended membranes: Permeation and sorption properties. *Chem. Eng. Res. Des.* 2015, 98, 96–106.
84. Yong, W. F., Chung, T.-S. Miscible blends of carboxylated polymers of intrinsic microporosity (cPIM-1) and Matrimid. *Polymer.* 2015, 59, 290–297.
85. Abdul Mannan, H., Mukhtar, H., Shima Shaharun, M., Roslee Othman, M., Murugesan, T. Polysulfone/poly(ether sulfone) blended membranes for CO2separation. *J. Appl. Polym. Sci.* 2016, 133(5).
86. Giel, V., Kredatusová, J., Trchová, M., Brus, J. Žitka, J., Peter, J., Polyaniline/polybenzimidazole blends: Characterisation of its physicochemical properties and gas separation behaviour. *Euro. Polym. J.* 2016, 77, 98–113.
87. Qiu, Y., Ren, J., Zhao, D., Li, H., Deng, M. Poly(amide-6-b-ethylene oxide)/[Bmim][Tf2N] blend membranes for carbon dioxide separation. *J. Energ. Chem.* 2016, 25(1), 122–130.
88. Mondal, A., Mandal, B., Synthesis and characterization of crosslinked poly(vinyl alcohol)/poly(allylamine)/2-amino-2-hydroxymethyl-1,3-propanediol/polysulfone composite membrane for CO2/N2 separation. *J. Memb. Sci.* 2013, 446, 383–394.
89. Mondal, A., Barooah, M., Mandal, B., Effect of single and blended amine carriers on CO2 separation from CO2/N2 mixtures using crosslinked thin-film poly(vinyl alcohol) composite membrane. *Int. J. Greenhouse Gas Control* 2015, 39, 27–38.
90. Mondal, A., Mandal, B. CO2 separation using thermally stable crosslinked poly(vinyl alcohol) membrane blended with polyvinylpyrrolidone/polyethyleneimine/tetraethylenepentamine. *J. Memb. Sci.* 2014, 460, 126–138.
91. Mondal, A., Mandal, B. Novel CO2-selective cross-linked poly(vinyl alcohol)/polyvinylpyrrolidone blend membrane containing amine carrier for CO2-N2 separation: Synthesis, characterization, and gas permeation study. *Ind. Eng. Chem. Res.* 2014, 53(51), 19736–19746.

92. Deng, L., Hagg, M.-B. Fabrication and evaluation of a blend facilitated transport membrane for CO2/CH4 separation. *Ind. Eng. Chem. Res.* 2015, 54(44), 11139–11150.
93. Feng, S., Ren, J., Hua, K., Li, H., Ren, X., Deng, M. Poly(amide-12-b-ethylene oxide)/polyethylene glycol blend membranes for carbon dioxide separation. *Sep. Purif. Technol.* 2013, 116, 25–34.
94. Feng, S., Ren, J., Li, Z., Li, H., Hua, K., Li, X., Deng, M. Poly(amide-12-b-ethylene oxide)/glycerol triacetate blend membranes for CO2 separation. *Inte. J. Greenhouse Gas Control.* 2013, 19, 41–48.
95. Sampranpiboon, P., Jiraratananon, R., Uttapap, D., Feng, X., Huang, R. Y. M. Pervaporation separation of ethyl butyrate and isopropanol with polyether block amide (PEBA) membranes. *J. Memb. Sci.* 2000, 173(1), 53–59.
96. Reijerkerk, S. R., Knoef, M. H., Nijmeijer, K., Wessling, M. Poly(ethylene glycol) and poly(dimethyl siloxane): Combining their advantages into efficient CO2 gas separation membranes. *J. Memb. Sci.* 2010, 352(1–2), 126–135.
97. Merkel, T. C., Bondar, V. I., Nagai, K., Freeman, B. D., Pinnau, I. Gas sorption, diffusion, and permeation in poly(dimethylsiloxane). *J. Polym. Sci. Polym. Phys.* 2000, 38(3), 415–434.
98. Yong, W. F., Li, F. Y., Xiao, Y. C., Li, P., Pramoda, K. P., Tong, Y. W., Chung, T. S. Molecular engineering of PIM-1/Matrimid blend membranes for gas separation. *J. Memb. Sci.* 2012, 407–408, 47–57.
99. Wang, D., Yuan, Y., Mardiyati, Y., Bubeck, C., Koynov, K. From single chains to aggregates, how conjugated polymers behave in dilute solutions. *Macromolecules* 2013, 46(15), 6217–6224.
100. Teraoka, I. *Polymer Solutions: An Introduction to Physical Properties.* Wiley: New York, 2002.
101. Isayev, A. I. *Encyclopedia of Polymer Blends. Vol. I: Fundamentals.* WILEY-VCH: New York, 2010.
102. Kolařík, J., Fambri, L., Pegoretti, A., Penati, A. Prediction of the gas permeability of heterogeneous polymer blends. *Polym. Eng. Sci.* 2000, 40(1), 127–131.
103. Kolařk, J. Simultaneous prediction of the modulus, tensile strength and gas permeability of binary polymer blends. *Euro.Polym. J.* 1998, 34(5–6), 585–590.
104. DeGennes, P. G. *J. Phys. Lett. (Paris).* 1976, 37(L1).
105. Li, X.-G., Kresse, I., Springer, J., Nissen, J., Yang, Y.-L. Morphology and gas permselectivity of blend membranes of polyvinylpyridine with ethylcellulose. *Polymer.* 2001, 42(16), 6859–6869.
106. de Sales, J. A., Patrício, P. S. O., Machado, J. C., Silva, G. G., Windmöller, D. Systematic investigation of the effects of temperature and pressure on gas transport through polyurethane/poly(methylmethacrylate) phase-separated blends. *J. Memb. Sci.* 2008, 310(1–2), 129–140.

107. Weng, T.-H., Tseng, H.-H., Wey, M.-Y. Preparation and characterization of PPSU/PBNPI blend membrane for hydrogen separation. *Int. J. Hydrogen Energy.* 2008, 33(15), 4178–4182.
108. Ben Hamouda, S., Nguyen, Q. T., Langevin, D., Roudesli, S. Poly(vinylalcohol)/poly(ethyleneglycol)/poly(ethyleneimine) blend membranes — structure and CO2 facilitated transport. *Comptes Rendus Chimie.* 2010, 13(3), 372–379.
109. Sanaeepur, H., Amooghin, A. E., Moghadassi, A., Kargari, A. Preparation and characterization of acrylonitrile–butadiene–styrene/poly(vinyl acetate) membrane for CO2 removal. *Sep. Purif. Technol.* 2011, 80(3), 499–508.
110. Liu, Y., Yu, S., Wu, H., Li, Y., Wang, S., Tian, Z., Jiang, Z. High permeability hydrogel membranes of chitosan/poly ether-block-amide blends for CO2 separation. *J. Memb. Sci.* 2014, 469, 198–208.
111. Mei Wu, X., Gen Zhang, Q., Ju Lin, P., Qu, Y., Mei Zhu, A., Lin Liu, Q. Towards enhanced CO2 selectivity of the PIM-1 membrane by blending with polyethylene glycol. *J. Memb. Sci.* 2015, 493, 147–155.
112. Feng, S., Ren, J., Li, H., Hua, K., Li, X., Deng, M. Polyvinyl acetate/ poly(amide-12-b-ethylene oxide) blend membranes for carbon dioxide separation. *J. Energ. Chem.* 2013, 22(6), 837–844.
113. Shen, Z., McHugh, M. A., Xu, J., Belardi, J., Kilic, S., Mesiano, A., Bane, S., Karnikas, C., Beckman, E., Enick, R. CO2-solubility of oligomers and Polymers that contain the carbonyl group. *Polymer.* 2003, 44(5), 1491–1498.
114. Lillepärg, J., Georgopanos, P., Shishatskiy, S. Stability of blended Polymeric materials for CO2 separation. *J. Memb. Sci.* 2014, 467, 269–278.
115. Car, A., Stropnik, C., Yave, W., Peinemann, K.-V. Tailor-made polymeric membranes based on segmented block coPolymers for CO2 separation. *Adv. Funct. Mater.* 2008, 18(18), 2815–2823.
116. Phillip, W. A., Martono, E., Chen, L., Hillmyer, M. A., Cussler, E. L. Seeking an ammonia selective membrane based on nanostructured sulfonated block copolymers. *J. Memb. Sci.* 2009, 337(1–2), 39–46.
117. Song, K. W., Ka, K. R., Kim, C. K. Changes in gas-transport properties with the phase structure of blends containing styrene-butadiene-styrene triblock coPolymer and poly(2,6-dimethyl-1,4-phenylene oxide). *Ind. Eng. Chem. Res.* 2010, 49(14), 6587–6592.
118. Lape, N. K., Mao, H., Camper, D., Hillmyer, M. A., Cussler, E. L. Barrier membranes of self-assembled lamellar poly(lactide-isoprene-lactide) triblock copolymers. *J. Memb. Sci.* 2005, 259(1–2), 1–9.
119. Utracki, L. A., Mukhopadhyay, P., Gupta, R. K. Polymer Blends: Introduction. 2014, 3–170.
120. Moon, E. J., Yoo, J. E., Choi, H. W., Kim, C. K. Gas transport and thermodynamic properties of PMMA/PVME blends containing PS-b-PMMA as a compatibilizer. *J. Memb. Sci.* 2002, 204(1–2), 283–294.
121. Semsarzadeh, M. A., Ghalei, B. Characterization and gas permeability of polyurethane and polyvinyl acetate blend membranes with polyethylene

oxide–polypropylene oxide block coPolymer. *J. Memb. Sci.* 2012, 401–402, 97–108.

122. Bandehali, S., Kargari, A., Moghadassi, A., Saneepur, H., Ghanbari, D. Acrylonitrile-butadiene-styrene/poly(vinyl acetate)/nanosilica mixed matrix membrane for He/CH4separation. *Asia Pac. J. Chem. Eng.* 2014, 9(5), 638–644.
123. Halim, A., Fu, Q., Yong, Q., Gurr, P. A., Kentish, S. E., Qiao, G. G. Soft Polymeric nanoparticle additives for next generation gas separation membranes. *J. Mater. Chem. A* 2014, 2(14), 4999.
124. Goodarzi, V., Hassan Jafari, S., Ali Khonakdar, H., Ghalei, B., Mortazavi, M. Assessment of role of morphology in gas permselectivity of membranes based on polypropylene/ethylene vinyl acetate/clay nanocomposite. *J. Memb. Sci.* 2013, 445, 76–87.
125. Panapitiya, N. P., Wijenayake, S. N., Huang, Y., Bushdiecker, D., Nguyen, D., Ratanawanate, C., Kalaw, G. J., Gilpin, C. J., Musselman, I. H., Balkus, K. J., Ferraris, J. P. Stabilization of immiscible Polymer blends using structure directing metal organic frameworks (MOFs). *Polymer.* 2014, 55(8), 2028–2034.
126. Furukawa, H., Cordova, K. E., O'Keeffe, M., Yaghi, O. M. The Chemistry and Applications of metal-organic frameworks. *Science.* 2013, 341(6149).
127. Eddaoudi, M., Moler, D. B., Li, H., Chen, B., Reineke, T. M., O'Keeffe, M., Yaghi, O. M. Modular chemistry: Secondary building units as a basis for the design of highly porous and robust metal-organic carboxylate frameworks. *Acc. Chem. Res.* 2001, 34(4), 319–330.
128. Park, K. S., Ni, Z., Côté, A. P., Choi, J. Y., Huang, R., Uribe-Romo, F. J., Chae, H. K., O'Keeffe, M., Yaghi, O. M. Exceptional chemical and thermal stability of zeolitic imidazolate frameworks. *Proc. Nat. Acad. Sci.* 2006, 103(27), 10186–10191.
129. Ordonez, M. J. C., Balkus, K. J., Jr., Ferraris, J. P., Musselman, I. H. Molecular sieving of ZIF-8/Matrimid mixed-matrix membranes. *Polym. Prepr. (Am. Chem. Soc., Div. Polym. Chem.)* 2009, 50(2), 804–805.
130. Girifalco, L. A., Good, R. J. A Theory for the estimation of surface and interfacial energies. I. derivation and application to interfacial tension. *J. Phys. Chem.* 1957, 61(7), 904–909.
131. Young, T. An Essay on the Cohesion of Fluids. *Phil. Trans. Royal Soc. London.* 1805, 95, 65–87.
132. Furukawa, H., Müller, U., Yaghi, O. M. "Heterogeneity within Order" in metal–organic frameworks. *Angew. Chemie Int. Ed.* 2015, 54(11), 3417–3430.
133. Rowsell, J. L. C., Yaghi, O. M. Metal–organic frameworks: A new class of porous materials. *Micropor. Mesopor. Mate.* 2004, 73(1–2), 3–14.
134. Yong, W. F., Li, F. Y., Xiao, Y. C., Chung, T. S., Tong, Y. W. High performance PIM-1/Matrimid hollow fiber membranes for CO2/CH4, O2/N2 and CO2/N2 separation. *J. Memb. Sci.* 2013, 443, 156–169.

135. Kapantaidakis, G. C., Koops, G. H. High flux polyethersulfone–polyimide blend hollow fiber membranes for gas separation. *J. Memb. Sci.* 2002, 204(1–2), 153–171.
136. Chung, T.-S., Xu, Z.-L. Asymmetric hollow fiber membranes prepared from miscible polybenzimidazole and polyetherimide blends. *J. Memb. Sci.* 1998, 147(1), 35–47.
137. Zhu, W.-P., Sun, S.-P., Gao, J., Fu, F.-J., Chung, T.-S. Dual-layer polybenzimidazole/polyethersulfone (PBI/PES) nanofiltration (NF) hollow fiber membranes for heavy metals removal from wastewater. *J. Memb. Sci.* 2014, 456, 117–127.
138. Reignier, J., Favis, B. D. Control of the subinclusion microstructure in HDPE/PS/PMMA ternary blends. *Macromolecules* 2000, 33(19), 6998–7008.
139. Ravati, S., Favis, B. D. Morphological states for a ternary Polymer blend demonstrating complete wetting. *Polymer.* 2010, 51(20), 4547–4561.
140. Reignier, J., Favis, B. D., Core–shell structure and segregation effects in composite droplet Polymer blends. *AIChE J.* 2003, 49(4), 1014–1023.
141. Wang, D., Li, Y., Xie, X.-M., Guo, B.-H. Compatibilization and morphology development of immiscible ternary Polymer. blends. *Polymer.* 2011, 52(1), 191–200.

Chapter 6

Manipulating Polyimide Nanostructures via Cross-linking for Membrane Gas Separation

Lingxiang Zhu, Maryam Omidvar and Haiqing Lin*

Department of Chemical and Biological Engineering,
University at Buffalo, The State University of New York,
Buffalo, NY 14260, USA
**haiqingl@buffalo.edu*

Cross-linking has been widely utilized to modify polyimide nanostructures for membrane gas separations, such as increasing size-sieving ability and diffusivity selectivity for H_2–CO_2 and CO_2–CH_4 separation and improving resistance to plasticization derived from CO_2 and heavy hydrocarbons for CO_2–CH_4 and C_3H_6–C_3H_8 separations. However, there is a lack of fundamental understanding of the relationship between cross-linked structure and membrane gas-separation properties. This chapter critically reviews the effect of cross-linking on polymer physical properties (such as glass transition temperature, T_g) and current strategies adopted to cross-link polyimides for membrane gas separation. The information is synthesized to elucidate the effect of cross-linking on T_g and cross-linking density in polyimides, which is then used to interpret the changes of gas permeability and selectivity. The benefits of cross-linking in improving gas-separation properties are also illustrated in Robeson's upper bound plots for H_2–CO_2, CO_2–CH_4, and C_3H_6–C_3H_8 separations.

1. Introduction

Membrane gas separation has enjoyed significant growth in the last three decades for industrial gas separation, due to its high energy-efficiency, small footprint, simplicity in operation, and easiness in scale-up.[1–3] The key to advanced membrane technology is a membrane with high productivity (or gas permeance) and high separation efficiency (or selectivity). However, there is a trade-off between membrane productivity and selectivity, i.e., higher permeability results in lower selectivity.[4–7] There have been intensive efforts in design and synthesis of novel polymers for membrane gas separation, such as polyimides, perfluoropolymers, poly(ethylene oxide) containing materials, poly(ionic liquids), polymers of intrinsic microporosity, and thermally rearranged polymers.[1,2] These efforts lead to significant understanding of structure–property relationship for design of new materials, and move the trade-off lines towards higher permeability and higher selectivity.

Cross-linking has been explored as an effective way to modify polymer nanostructures, such as increasing chain rigidity and even disrupting chain crystallization.[8–10] As shown in Fig. 1, cross-linking may improve polymer chain packing efficiency, decrease polymer free volume, and thus increase size-sieving ability.[11,12] This approach has been particularly attractive for important separations such as CO_2–CH_4 and olefin–paraffin separations, when gas mixtures contain highly plasticizing components such as CO_2 and heavy hydrocarbons. The CO_2 and hydrocarbons can be significantly absorbed in the polymer, swell the polymer matrix and increase polymer interchain spacing and chain mobility (i.e., plasticization), leading to weaker size-sieving ability and thus reduced gas selectivity.[7,13–15] On the other hand, cross-linking can restrict the swelling, preventing the significant loss of size-sieving ability or mixed-gas selectivity. In addition, cross-linking modification may be

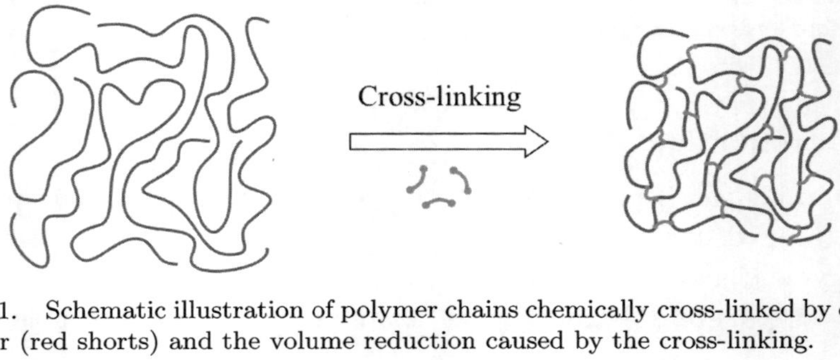

Fig. 1. Schematic illustration of polymer chains chemically cross-linked by cross-linker (red shorts) and the volume reduction caused by the cross-linking.

facilely incorporated into the post treatment step for current industrial membrane fabrication.[16]

Recently, glassy polyimides are considered as a promising platform for cross-linking modification because of their good intrinsic size-sieving ability, versatile structures prepared from a variety of diamines and dianhydrides, and good processability.[17] However, the effect of cross-linking has not been clearly determined, in major part due to the complexity of the polymer systems. For example, the reaction conversion may not be well characterized, and the addition of cross-linkers in polyimides changes the chemical composition and thus influences the gas transport properties.[8,18,19]

This chapter reviews the effect of cross-linking on polymer properties, such as glass transition temperature (T_g). The various chemical routes for polyimide cross-linking are reviewed, and the effect of cross-linking on gas separation properties is summarized for important applications such as H_2–CO_2, CO_2–CH_4, and C_3H_6–C_3H_8 separations. Finally, the relationship between cross-linking density and gas transport properties is elucidated. This study may shed some light on the fundamental mechanism of polymer cross-linking and how the modified polymer nanostructures can be harnessed to improve gas separation properties for practical applications.

2. Background on Membrane Gas Separation

The permeation process of penetrants through dense polymeric films can be described by the solution-diffusion model.[20] Penetrants dissolve into the film surface on the upstream side, then diffuse across the film due to the concentration gradient, and finally desorb from the film on the downstream side. The permeability of a polymer film to gas component A, P_A, is defined as[20]:

$$P_A = \frac{N_A l}{p_2 - p_1}, \tag{1}$$

where N_A [cm^3 (STP) cm^{-2} s^{-1}] is steady-state gas flux across the film, l (cm) is the film thickness, and p_1 (cmHg) and p_2 (cmHg) are the downstream and upstream pressure of component A, respectively.

Gas permeability is usually expressed in Barrers, where 1 Barrer $= 1 \times 10^{-10}$ cm^3 (STP) cm cm^{-2} s^{-1} cmHg^{-1}.

Based on the solution-diffusion model, gas permeability can also be expressed as:

$$P_A = S_A \times D_A, \tag{2}$$

where S_A [cm^3 (STP) cm^{-3} polymer atm^{-1}] is solubility coefficient of gas component A in the polymer, and D_A (cm^2 s^{-1}) is a concentration averaged effective diffusivity coefficient. Gas selectivity, characterizing separation efficiency of membranes, is a combination of solubility selectivity (S_A/S_B) and diffusivity selectivity (D_A/D_B):

$$\alpha_{A/B} = \frac{P_A}{P_B} = \left(\frac{S_A}{S_B}\right) \times \left(\frac{D_A}{D_B}\right). \tag{3}$$

Gas solubility in polymers is primarily determined by its condensability (indicated by critical temperature, as shown in Table 1), and influenced by its interaction with polymers and free volume in polymers. Gas diffusivity is governed by penetrant size (indicated by kinetic diameter or critical volume, as shown in Table 1), polymer free volume, and polymer chain flexibility. Gas diffusivity can be described using a simplified free-volume model[7,21]:

$$D_A = A \exp\left(-\frac{B}{\text{FFV}}\right), \tag{4}$$

Table 1. Physical properties of gases and vapors governing their transport properties in polymers.[22,23]

Gas	Kinetic diameter (Å)	Critical volume (cm^3 mol^{-1})	Critical temperature (K)
H_2	2.89	65.1	33.2
CO_2	3.30	93.9	304
N_2	3.64	89.8	126
CH_4	3.80	99.2	191
C_3H_6	4.68*	181.0	365
C_3H_8	5.06*	203.0	370

*Lennard–Jones diameter.[24]

where A is a front factor depending weakly on temperature, B is a constant related to size of gas penetrant, and FFV is the polymer fractional free volume.

The free volume indicates the space generated by inefficient chain packing or transient gaps created by thermally stimulated segmental motion of polymer chains. The FFV is defined as follows[21]:

$$\mathrm{FFV} = \frac{V - V_0}{V}, \tag{5}$$

where V is polymer specific volume ($cm^3\,g^{-1}$), and V_0 is the volume occupied by polymer chains ($cm^3\,g^{-1}$).

The separation of H_2–CO_2 and C_3H_6–C_3H_8 reflects a combination of favorable diffusivity selectivity and unfavorable solubility selectivity, while the separation of CO_2–CH_4 benefits from both favorable solubility selectivity and diffusivity selectivity. Cross-linking often decreases polymer free volume, leading to lower diffusivity and stronger size-sieving ability (as indicated by Eqs. (3) and (4)), which is expected to increase the selectivity of all three separations.

3. Effect of Cross-linking on Polymer Physical Properties

The degree of cross-linking is often characterized using the molecular weight of polymer chains between cross-links (M_c).[8,25] The M_c value can be estimated from the known compositions of the polymers and cross-linking agents or from elastic modulus at rubbery state using the following expression[8,25,26]:

$$M_c = \frac{3\varphi\rho RT}{E_e}, \tag{6}$$

where φ is a front factor usually assumed to be 1 for highly cross-linked polymers, ρ is polymer density ($g\,cm^{-3}$), R is the gas constant ($8.314\,cm^3$ MPa K^{-1} mol^{-1}), T is absolute temperature (K) and E_e is equilibrium tensile modulus (MPa) at approximately 40°C above T_g. Lower M_c value denotes higher cross-linking density, and highly cross-linked polymers typically have M_c less than $1000\,g\,mol^{-1}$.[1,25]

Nielsen proposed an empirical equation to correlate the decreased M_c values (or increased cross-linking density) with the increased T_g values in cross-linked polymers, as shown below[27]:

$$T_g - T_{g,0} = \frac{K}{M_c}, \tag{7}$$

where $T_{g,0}$ is the glass transition temperature of an uncross-linked polymer, and K is a constant related to chemical compositions of cross-linked polymers (K g mol^{-1}).

The empirical correlation as described in Eq. (7) has been validated. For example, Fig. 2 shows the linear relationship between $(T_g - T_{g,0})$ and $1/M_c$ for six polymer networks. These polymers are based on well-studied epoxide resins (as summarized in Table 2), including phenyl glycidyl ether (PGE), Bisphenol A diglycidyl ether (DGEBA), elastomer-modified Bisphenol A diglycidyl ether (e-DGEBA), N,N-bis-(2,3-epoxypropyl)-N,N-dimethyl-4,4′-diamino-diphenylene-methane (G2A) and N,N-bis-(2,3-epoxypropyl)-N,

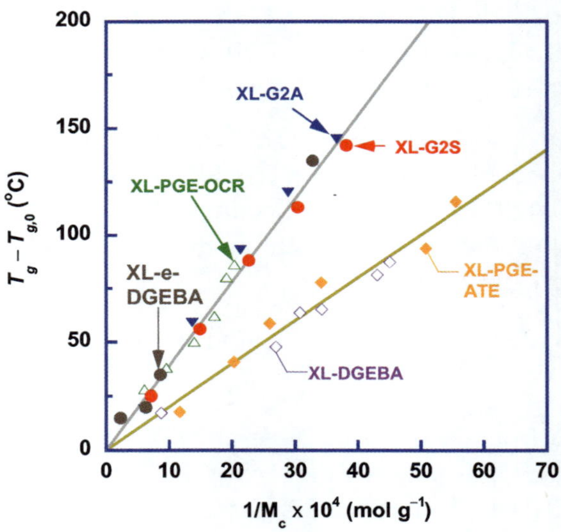

Fig. 2. Effect of cross-linking density on glass transition temperature for 6 series of polymer networks based on well-known epoxide resins.[28–32] The details of polymer structures are shown in Table 2.

Table 2. Chemical structure of resins and cross-linkers and physical properties of the resulted cross-linked networks.

Resin	Cross-linker	T_g(°C)	M_c(g mol^{-1})	E_e (MPa)	Ref.
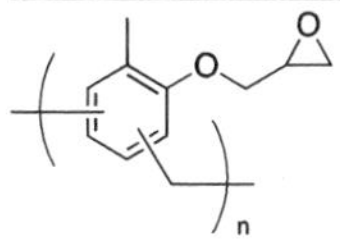	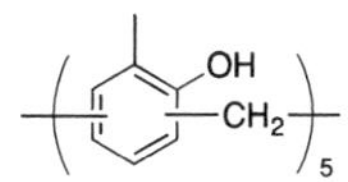	57	1653[a]		28
		67	1047[a]		
		79	714[a]		
		91	581[a]		
(PGE)	(OCR)	109	523[a]		
		115	491[a]		
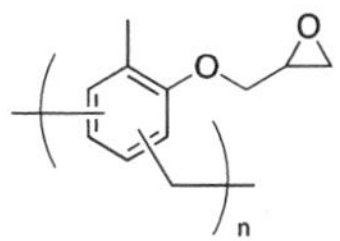	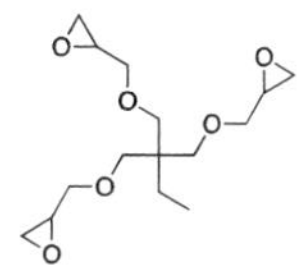	41	856[b]	12.4	29
		64	492[b]	23.4	
		82	385[b]	31.0	
		101	292[b]	42.6	
		117	197[b]	65.0	
(PGE)	(ATE)	139	180[b]	75.0	
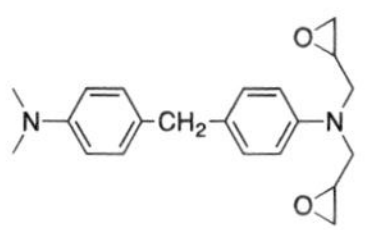	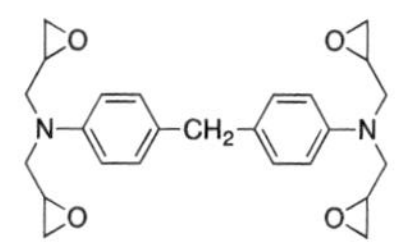	112	1650[c]	5.5	30
		152	731[c]	13.2	
		186	469[c]	28.2	
		213	346[c]	26.0	
G2A		238	273		

(Continued)

Table 2. (*Continued*)

Resin	Cross-linker	T_g(°C)	M_c(g mol^{-1})	E_e(MPa)	Ref.
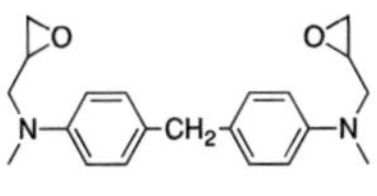 G2S	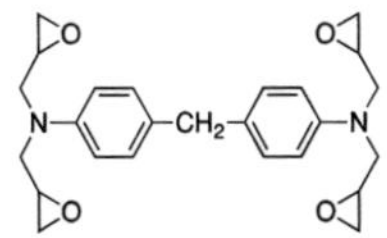	124	1402^c	7.4	30
		155	671^c	15.3	
		187	441^c	30.5	
		212	329^c	28.0	
		241	262		
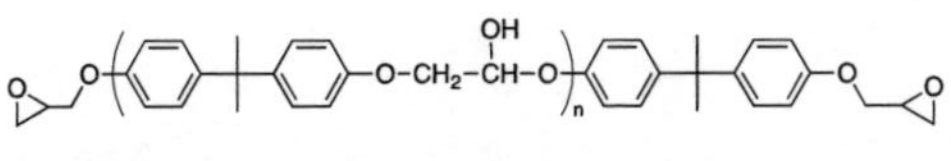 e-DGEBA	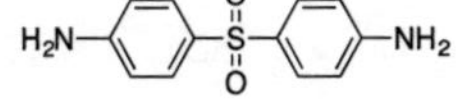	100	4350^d	2.9	31
		105	1600^d	8.0	
		120	1160^d	11.4	
		220	305^d	53.3	
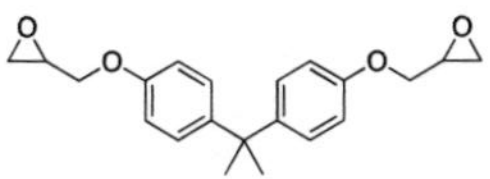 DGEBA	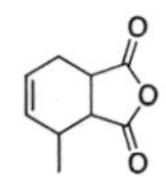	68	1147^b	7.9	32
		99	371^b	24.4	
		115	325^b	24.3	
		116	292^b	31.0	
		132	232^b	38.0	
		138	222^b	40.1	

[a] M_c values are calculated from compositions of epoxy resins and cross-linkers.
[b] M_c values are estimated based on the E_e values at 40°C above T_g.
[c] M_c values are estimated based on the E_e values at 30°C above T_g.
[d] M_c values are estimated based on the E_e values at 50°C above T_g.

N-dimethyl-4,4′-diamino-diphenylene-methane (G2S), and their corresponding cross-linkers (or curing agents).[28–32] The detailed structures are shown in Table 2. The M_c values are estimated based on either the compositions of epoxide and curing agent or the E_e values using Eq. (6).

As shown in Fig. 2, there are two linear lines for the 6 polymer networks with two K values, 2.0×10^4 and 3.9×10^4 K g mol^{-1}. The K constant is often influenced by polymer chemical compositions.[25,27] Interestingly, the value of 3.9×10^4 K g mol^{-1} is the same as that estimated by Nielsen.

Nielsen's empirical equation provides good estimates of T_g in cross-linked polymers, despite the simplicity of Eq. (7) and the complexity of the polymer systems. This equation will be used to interpret the effect of cross-linking on the T_g of polyimides, which will then be related to gas permeability.

4. Cross-linking of Polyimides for Membrane Gas Separation

4.1. *Aromatic Polyimides for Gas Separation*

Aromatic polyimides have been extensively investigated for membrane gas separation, due to their superior gas separation properties, flexibility in polymer structure design and good processability.[33–35] Figure 3 shows a typical structure of aromatic polyimides comprising rigid five-member imide rings in polymer backbones. These polyimides are often prepared by polycondensation of aromatic dianhydrides (containing Ar groups) and aromatic diamines (containing R groups) in a two-step process. First, polyamic acids are obtained by nucleophilic substitution of amino groups to carbonyl carbons of anhydride groups. Second, polyimides are produced by thermal imidization of polyamic acids at temperatures up to 300°C.[36] Both Ar and R groups can be varied separately to obtain a large collection of polyimides with high free volume and good solution processability, as shown in Fig. 3. For example, introduction of 6FDA and durene groups are well known to increase polymer free volume and their solubility in solvents.[17,37,38]

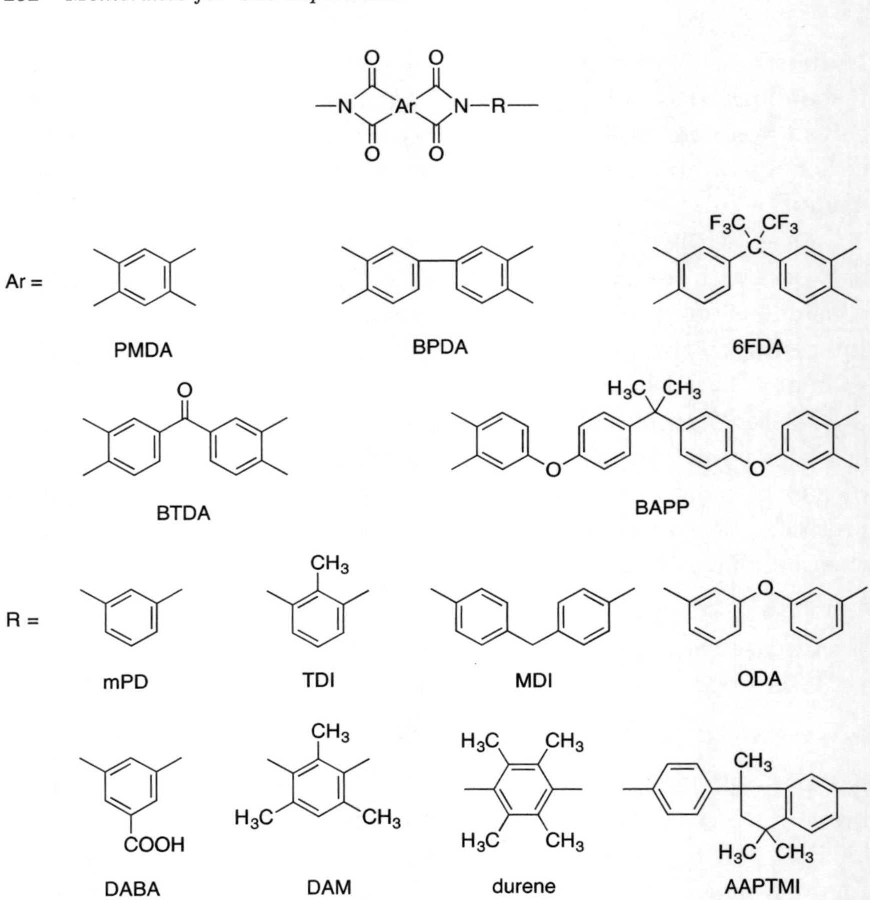

Fig. 3. Chemical structures of representative aromatic polyimides used for membrane gas separations. Adapted from Refs. 33 and 39.

Table 3 summarizes the physical properties and gas separation performance of some well-studied polyimides, which are compared with other commercial membrane polymers such as polysulfone (PSf) and cellulose triacetate (CTA). Generally, polyimides have higher T_g than PSf and CTA, and they have a wide range of gas-separation properties to meet different separation targets. For example, 6FDA-DABA has CO_2 permeability of 13 Barrers and CO_2–CH_4 selectivity of 62, while 6FDA-DAM shows CO_2 permeability of 842 Barrers and CO_2–CH_4 selectivity of only 18. A copolymer of these

Table 3. Physical properties and gas-separation performance of well-known aromatic polyimides and some other commercialized membrane polymers at 35°C.

		Permeability (Barrer)		Gas selectivity			
Polymer	Trade name	CO_2	N_2	H_2/CO_2	CO_2/CH_4	T_g(°C)	FFV
BTDA-AAPTMI[41–43]	Matrimid®	9.5	0.31	3.1	40	313	0.170
BTDA-TDA/MDI[44,45]	P84®	1.4	0.05	6.5	49	300	0.14
BAPP-mPD[42,46]	Ultem®	1.5	0.05	—	38	209	0.180
6FDA-durene[47,48]	—	600	39	1.0	19	424	0.180
6FDA-DABA[40]	—	13	0.51	—	62	363	—
6FDA-DAM[40,49]	—	842	55	—	18	395	0.190
Polysulfone[50,51]	Udel®	5.5	0.22	2.5	23	186	0.156
Cellulose triacetate[52]	—	8.1	0.27	2.4[53]	30	198	0.215

two polyimides such as 6FDA-DAM:DABA (3:2) has a good balance of high CO_2 permeability (144 Barrers) and high CO_2–CH_4 selectivity of 34.[40]

4.2. *Chemical Cross-Linking of Polyimides*

Polyimides can be chemically cross-linked via two routes based on the location of cross-links on polyimide backbones, i.e., the five-member imide ring and the functional R groups, as shown in Fig. 3. Both cross-linking approaches are described in detail below.

Figure 4 shows the schematic of the reaction between imide rings and diamines to form an amide-based intermolecular cross-link. For example, a 6FDA-durene film was immersed in methanol solution containing *p*-xylenediamine.[13] The *p*-xylenediamine diffuses inside the polyimide swollen by methanol and reacts with imide rings.[13] Figure 4 also lists other diamines used to cross-link polyimides, such as ethylenediamine (EDA),[54,55] propane-1,3-diamine (PDA),[54,55]

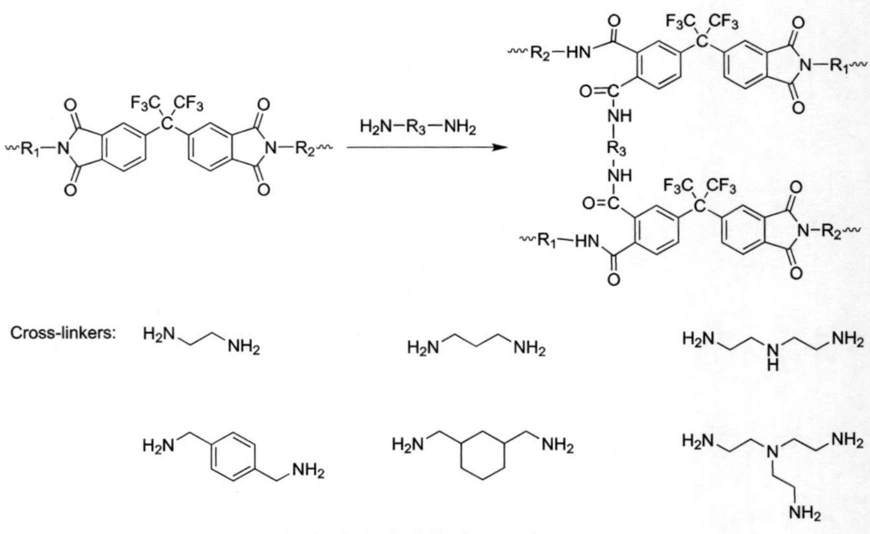

Fig. 4. Schematic of chemical cross-linking of 6FDA based polyimides using diamines, and examples of diamine cross-linkers. Adapted from Ref. 13.

xylenediamine,[56–58] 1,3-cyclohexanebis-(methylamine),[59] polypropylenimine dendrimers,[60] and polyamidoamine (PAMAM) dendrimers.[61,62]

The reaction of imide rings and diamines can also occur in vapor phase using volatile diamines such as EDA,[63] diethylenetriamine,[64] and tris(2-aminoethyl) amine.[65] For example, 6FDA-durene was cross-linked using EDA vapor, which yields extremely high size-sieving ability. The H_2-CO_2 selectivity increases from 1.0 in the virgin polymer to 100 in the cross-linked one. This increase in selectivity is consistent with the decrease in *d*-spacing from 6.18 Å to 6.02 Å after cross-linking.[63]

The degree of cross-linking depends on the reactivity and diffusion rate of diamines in the swollen polyimides.[66] The obtained cross-linked polyimides films from this immersion/cross-linking method are usually not homogenous, with higher cross-linking density on or near surface region than the center.[54,55] Due to this non-homogenous structure, the effect of cross-linking on physical properties such as T_g and gas transport properties is not well understood.

The second approach of cross-linking is to functionalize R groups with cross-linkable groups. For example, copolyimides with DABA blocks containing carboxylic acid (as shown in Fig. 3) can be cross-linked by diols (or glycidol) following a two-step procedure, as shown in Fig. 5.[17] First, the carboxylic acid groups react with excessive

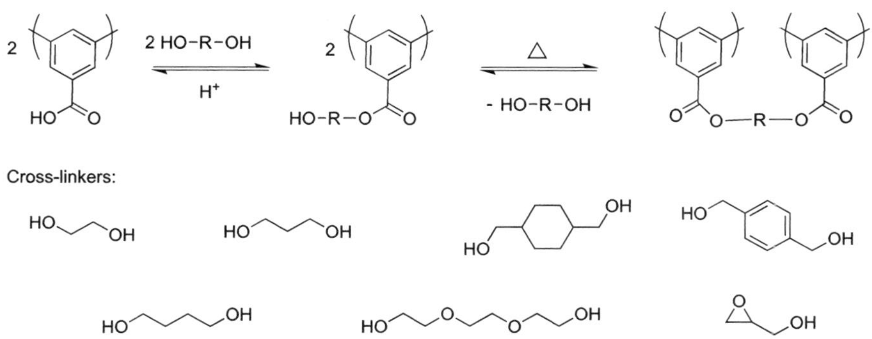

Fig. 5. Two-step cross-linking scheme for carboxylic acid-containing polyimides and cross-linking agents including diols and glycidol. Adapted from reference.[14]

diols to form a mono-esterified polymer. Second, the polymer chains are cross-linked via a transesterification reaction at temperatures of about 220°C in vacuum.[14,18,67] The cross-linking density can be controlled by varying the content of cross-linkable DABA blocks in the copolymers. Unlike the first approach of immersion/cross-linking, the second approach based on DABA provides homogeneous films, since it is not diffusion-limited. As a result, cross-linked polyimide did not plasticize even at a CO_2 feed pressure of 35 atm while the pristine polymer can be plasticized by CO_2 at pressures as low as 14 atm.[14] Such cross-linked systems are also ideal to investigate the relationship between physical properties, such as T_g and cross-linking density, and gas separation properties. The details will be described in Section 5.4.

The gas-separation properties of the cross-linked polyimides are determined by the chemical structure of base polyimides, cross-linking agents, and thermal treatment conditions. Figure 5 lists a variety of diols (or glycidol) investigated as cross-linkers for polyimides containing DABA block.[18,67–76] 6FDA-DAM:DABA (3:2) copolyimide has been extensively investigated, due to its balanced CO_2 permeability and CO_2–CH_4 selectivity.[73–77] In addition, the DABA containing polyimides can be thermally cross-linked in the absence of cross-linkers at sub-T_g temperatures (such as 300–370°C) due to decarboxylation of the DABA moieties.[78]

5. Gas Separation Properties of Cross-linked Polyimides

Cross-linking has been an effective route to modify polyimides to increase size-sieving ability, and resistance to plasticization by high pressure CO_2 and heavy hydrocarbons.[13,69,73,76] In this section, we will review the gas-separation properties of cross-linked polyimides for the separation of H_2–CO_2, CO_2–CH_4, and C_3H_6–C_3H_8, aiming to uncover how cross-linking modifies the polymer structure and thus gas separation properties.

5.1. H_2–CO_2 *Separation*

Membranes with high H_2–CO_2 separation properties can be used for CO_2 capture from syngas (i.e., precombustion processes).[79] To achieve high H_2–CO_2 selectivity, polymers should have very strong size-sieving ability to compensate for the unfavorable solubility selectivity (cf. Table 1). Conventional polyimides do not show the promise for this application, due to the low selectivity, as shown in Table 3.

Cross-linking of polyimides using diamines is found to be a convenient way to dramatically enhance H_2–CO_2 separation performance, as shown in Fig. 6.[54,59,60,66,80] The upper bound indicates the highest selectivity possible for any H_2 permeability. In general, cross-linking decreases H_2 permeability and increases H_2–CO_2 selectivity, in a large part due to the decrease in free volume and increase in size-sieving ability. For example, 6FDA-durene shows a H_2–CO_2 selectivity of only 1.3 at 35°C. The cross-linking of 6FDA-durene using

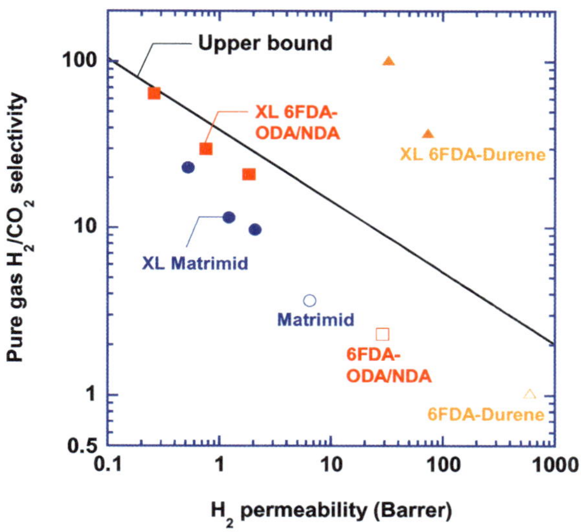

Fig. 6. Advantages of cross-linking polyimides for H_2–CO_2 separation in Robeson's upper bound plot, including Matrimid (○), XL Matrimid (•), 6FDA-ODA/NDA 25:75 (□), XL 6FDA-ODA/NDA 25:75 (■), 6FDA-durene (△), and XL 6FDA-durene (▲). The label of "XL" indicates cross-linked polymers.[63,81]

BuDA increases the H_2–CO_2 selectivity to 102.[63] Together with a H_2 permeability of 32 Barrers, this cross-linked polymer demonstrates the H_2–CO_2 separation properties well above the upper bound and a great potential for CO_2 capture from the precombustion processes.[63]

Other cross-linking methods have been explored. For example, 6FDA-durene was also cross-linked using different linear diamines such as EDA and PDA in the solution or vapor phase. Matrimid was also cross-linked using PDA at 50°C for 1 h, and the H_2–CO_2 selectivity increased from 5.4 to 49.[81]

5.2. CO_2–CH_4 *Separation*

Removal of CO_2 from natural gas is one of the most important industrial gas separation applications.[82] Raw natural gas may contain high content of CO_2 and heavy hydrocarbons (i.e., C_2H_6, C_3H_8, and higher aromatic and aliphatic hydrocarbons), which are highly condensable and strongly sorbed by polymers.[67,83,84] Figure 7a presents

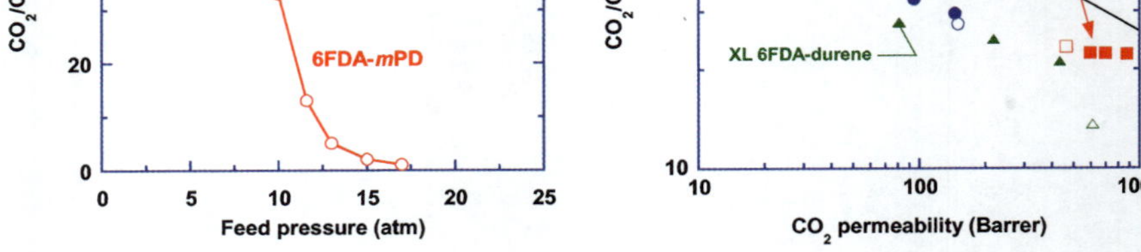

(a) (b)

Fig. 7. (a) Advantage of cross-linked polyimides with strong resistance to plasticization for CO_2–CH_4 separation. This graph is adapted from Ref. 14. (b) Cross-linked polyimides for CO_2–CH_4 separation in Robeson's upper bound plot, including 6FDA-DAM:DABA (○), XL 6FDA-DAM:DABA (•), 6FDA-durene (△), XL 6FDA-durene (▲), 6FDA-durene:DABA (□), and XL 6FDA-durene:DABA (■).[66,73,85] the effect of feed pressure on mixed gas CO_2–CH_4 selectivity in 6FDA-*m*PD at 35°C with a feed gas containing 50:50 CO_2:CH_4. At low CO_2 pressures, the polyimide exhibits high CO_2–CH_4 selectivity due to high diffusivity selectivity. However, the selectivity dramatically decreases as the feed pressure increases to above 10 atm. On the other hand, the cross-linked 6FDA-DABA using ethylene glycol maintains the high selectivity at pressures up to 20 atm, presumably due to the strong resistance to the CO_2 plasticization. Interestingly, the cross-linking even increases the CO_2 permeability from 6.5 Barrers to 9.5 Barrers.[14,18,19]

Cross-linking of other polyimides has also been reported. For example, co-polyimides of 6FDA-durene:DABA and 6FDA-DAM:DABA were cross-linked using glycols. 6FDA based polyimides were also cross-linked using glycidol. The 41% glycidol-modified cross-linked membrane has excellent selectivity, and permeability, as well as good plasticization resistance.[73]

Figure 7b shows the effect of cross-linking on CO_2–CH_4 separation in Robeson's plot. Generally, the cross-linking decreases CO_2 permeability and increases CO_2–CH_4 selectivity. The separation properties of cross-linked polyimides approach the upper bound.

Cross-linked polyimides have demonstrated their potentials for practical CO_2–CH_4 separations. Chinn *et al.* reported that cross-linked DABA-based polyimides have been fabricated into bench-scale hollow fiber membranes and field tests of these membranes have been performed at natural gas plants.[89] These membranes exhibited high CO_2 permeance and CO_2–CH_4 selectivity with good long-term stability in real natural gas environment.

5.3. C_3H_6–C_3H_8 Separation

Olefin–paraffin separation such as C_3H_6–C_3H_8 separation is one of the most energy-intensive gas separations.[1,24] A variety of polymers have been explored for this application. However, both C_3H_6 and C_3H_8 are highly soluble in polymers and thus plasticize the polymers, leading to a loss of size-sieving ability.[24] For example, BPDA–TmPD shows a pure-gas C_3H_6–C_3H_8 selectivity of 50. However, with a feed

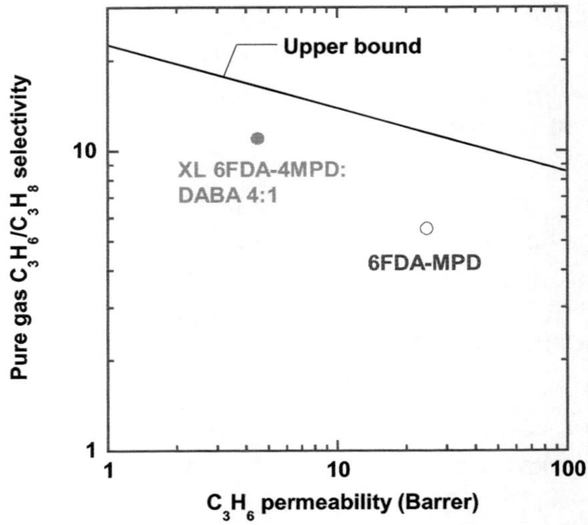

Fig. 8. Effect of cross-linking on C_3H_6–C_3H_8 separation properties at 35°C with a mixture containing 50:50 C_3H_6: C_3H_8 at 4 atm for 6FDA-MPD (○), and cross-linked 6FDA-4MPD:DABA 4:1 using ethylene glycol (•).[87]

gas mixture of 50:50 at 10 atm, the mixed-gas C_3H_6–C_3H_8 selectivity decreases to below 2.[86]

Cross-linking has been explored as a way to improve the membrane performance by decreasing plasticization effect and improving selectivity.[87] As shown in Fig. 8, the cross-linked 6FDA-4MPD:DABA 4:1 copolyimide still exhibits C_3H_6–C_3H_8 selectivity of 11 at the feed gas mixture of 4 atm, while the uncross-linked 6FDA-4MPD shows a selectivity of less than 6. The cross-linked polyimides demonstrate the C_3H_6–C_3H_8 separation properties closer to the upper bound than the uncross-linked one.

5.4. *Correlation of Gas Permeability and Cross-Linking Density*

In this section, we aim to interpret the effect of cross-linking on polymer physical properties including gas permeability. Table 4 summarizes chemical compositions and physical properties in selected cross-linked polyimides. Cross-linking modification slightly increases density except 6FDA-durene:DABA (3:2).[85] Interestingly,

Table 4. Effect of chemical cross-linking on the physical properties of polyimides, including cross-linking density (M_c), T_g, density, and CO_2–CH_4 separation properties at 35°C.

Polyimides	Cross-linker	M_c ($g\,mol^{-1}$)	T_g (°C)	Density ($g\,cm^{-3}$)	$P(CO_2)$ (Barrer)	CO_2–CH_4 selectivity	Refs.
6FDA-DAM:DABA (3:2)	None		386	1.400	151	28	73
	glycidol	14800	383	1.399	146	30	
		6300	375	1.403	95	33	
	None		385	1.402	85	20	76
	dietheylene glycol	2800[a]	374	1.397	55	23	
	trietheylene glycol	1500[a]	384	1.399	48	30	
	tetraetheylene glycol	1800[a]	365	1.398	40	31	
6FDA-DAM:DABA (2:1)	None		364	1.388	115	29	19
	butylene glycol	3100[b]	363	1.394	44	34	
6FDA-6FpDA:DABA (2:1)	None		319	1.482	45	44	88
	butylene glycol	3800[b,c]	328	1.532	43	47	
	None		313	1.483	28	45	18
	ethylene glycol	14000[b,c]	323	1.485	35	42	
6FDA-durene:DABA (3:2)	None		375	1.375	465	23	85
	Fe(III) acetylacetonate		377	1.350	599	22	
			380	1.339	700	22	
			390	1.296	877	22	

[a]Conversion rate of transesterification reaction is estimated based on the gel content.
[b]Conversion rate of transesterification reaction is from Ref. 73.
[c]Conversion rate of monoesterification is from Ref. 19.

cross-linking does not necessarily increase polymer T_g. The cross-linking may increase the T_g due to enhanced polymer chain rigidity; on the other hand, the addition of cross-linking agents such as butylene glycol may lead to lower T_g because of the copolymer effect.[27,73,85] As a result of these two competing effects, the T_g of most cross-linked polyimides changes within ±10°C from that of the uncross-linked ones, which is much less than that of highly cross-linked epoxies (up to 100°C, as shown in Fig. 2).

Unlike the well-studied epoxy networks as shown in Table 2, there is often no quantitative analysis of cross-linking density (or M_c values) for cross-linked polyimides. Herein, M_c values are estimated using the conversion rate of cross-linking agents. Figure 9a shows the correlation between the measured $(T_g - T_{g,0})$ values and $1/M_c$ values. However, there is no clear correlation between $(T_g - T_{g,0})$ and $1/M_c$ as described in Eq. (7), presumably due to the low cross-linking density (or large M_c values) in polyimides or the copolymer effect derived from the cross-linkers.[27,73,85] Equation (7) generally works for highly cross-linked networks with M_c values less than 1000 g mol^{-1};[25] whereas the M_c values for cross-linked polyimides are often much greater than 1000 g mol^{-1}.

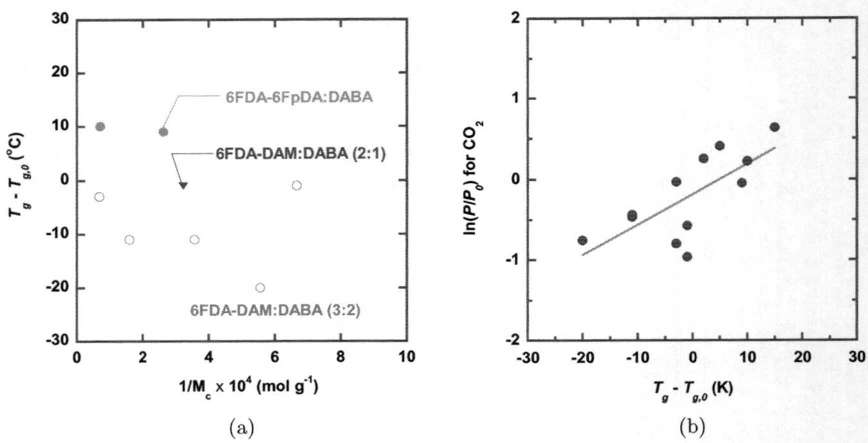

Fig. 9. (a) Effect of cross-linking density on the deviation of T_g from that of uncross-linked polyimide ($T_{g,0}$). (b) Relative CO_2 permeability as a function of T_g deviation for cross-linked polyimides. The line in (b) is to guide the eye.

Figure 9b demonstrates the effect of T_g on the relative CO_2 permeability (defined as the CO_2 permeability in cross-linked polyimides to that of uncross-linked ones). Generally, increasing T_g increases CO_2 permeability. However, there is a lack of FFV values in these systems to directly correlate with gas permeability values.

6. Concluding Remarks

6.1. *Summary*

This study aims to elucidate the effect of cross-linking on membrane gas separation properties in polyimides. Though cross-linking has been extensively utilized to improve separation properties for important industrial gas separations such as H_2–CO_2, CO_2–CH_4, and C_3H_6–C_3H_8 separations, it remains unexplored how the cross-linking influences physical properties of the polyimides such as T_g, FFV and gas permeability. This chapter critically reviews the strategies utilized to chemically cross-link polyimides (including the reaction via imide rings or DABA functional groups) and the gas separation properties in the resulted polymer networks.

Cross-linking of polyimides may slightly increase or decrease T_g due to the competitive effect of cross-linking and the change of chemical compositions due to the addition of cross-linkers with low T_g. There is no clear correlation between T_g and cross-linking density (as indicated by M_c), as described by Nielsen's empirical model. This behavior is very different from that observed for well-studied epoxides networks. In cross-linked polyimides, gas permeability often increases with increasing T_g. However, due to the lack of the FFV information, the change of permeability cannot be examined using the free volume model. Nevertheless, hollow fiber membranes based on cross-linked polyimides have been successfully demonstrated for CO_2 removal from natural gas with good long-term stability at natural gas processing plants.

6.2. *Outlook*

Systematic understanding of the effect of cross-linking on polymer nanostructures is the key to design advanced polymers for those important separations in practical conditions. There is an urgent need in the design and synthesis of cross-linked polymers that can be accurately characterized for their nanostructures, in order to study structure–property relationship.

A critical step to move these cross-linked polymers with high performance to industrial membranes is to seamlessly incorporate the cross-linking step in the membrane production processes. The ability to form defect-free membranes with ~100 nm thick selective layers would be required. The successful commercialization of membranes based on cross-linked polymers would also need to address the membrane production scale-up (i.e., reproducibility in membrane production) and the long-term stability of these membranes in the real gas streams.

Acknowledgment and Disclaimer

This material is based upon work supported by the U.S. National Science Foundation under Award Number 1554236 and U.S. Department of Energy under Award Number DE-FE0026463. This report was prepared as an account of work sponsored by an agency of the United States Government. Neither the United States Government nor any agency thereof, nor any of their employees, makes any warranty, express or implied, or assumes any legal liability or responsibility for the accuracy, completeness, or usefulness of any information, apparatus, product, or process disclosed, or represents that its use would not infringe privately owned rights. Reference herein to any specific commercial product, process, or service by trade name, trademark, manufacturer, or otherwise does not necessarily constitute or imply its endorsement, recommendation, or favoring by the United States Government or any agency thereof. The views and opinions of authors expressed herein do not necessarily state or reflect those of the United States Government or any agency thereof.

References

1. Sanders, D. F., Smith, Z. P., Guo, R., Robeson, L. M., McGrath, J. E., Paul, D. R., Freeman, B. D. Energy-efficient polymeric gas separation membranes for a sustainable future: A review. *Polymer*. 2013, 54(18), 4729–4761.

2. Baker, R. W., Low, B. Gas separation membrane materials: a perspective. *Macromolecules.* 2014, 47(20), 6999–7013.
3. Lin, H. Integrated membrane material and process development for gas separation. *Curr. Opin. Chem. Eng.* 2014, 4, 54–61.
4. Robeson, L. M. Correlation of separation factor versus permeability for polymeric membrance. *J. Memb. Sci.* 1991, 62(2), 165–185.
5. Robeson, L. M. The upper bound revisited. *J. Memb. Sci.* 2008, 320(1–2), 390–400.
6. Freeman, B. D. Basis of permeability/selectivity tradeoff relations in polymeric gas separation membranes. *Macromolecules.* 1999, 32(2), 375–380.
7. Lin, H., Yavari, M. Upper bound of polymeric membranes for mixed-gas CO_2/CH_4 separation. *J. Memb. Sci.* 2015, 475, 101–109.
8. Lin, H., Kai, T., Freeman, B. D., Kalakkunnath, S., Kalika, D. S. The effect of cross-linking on gas permeability in crosslinked poly(ethylene glycol diacrylate). *Macromolecules.* 2005, 38, 8381–8393.
9. Kalakkunnath, S., Kalika, D. S., Lin, H., Raharjo, R. D., Freeman, B. D. Molecular dynamics of poly(ethylene glycol) and poly(propylene glycol) copolymer networks by broadband dielectric spectroscopy. *Macromolecules.* 2007, 40(8), 2773–2781.
10. Kalakkunnath, S., Kalika, D. S., Lin, H., Raharjo, R. D., Freeman, B. D. Molecular relaxation in cross-linked poly(ethylene glycol) and poly(propylene glycol) diacrylate networks by dielectric spectroscopy. *Polymer.* 2007, 48(2), 579–589.
11. Song, Q. L., Cao, S., Pritchard, R. H., Ghalei, B., Al-Muhtaseb, S. A., Terentjev, E. M., Cheetham, A. K., Sivaniah, E. Controlled thermal oxidative crosslinking of polymers of intrinsic microporosity towards tunable molecular sieve membranes. *Nat. Commun.* 2014, 5, 3705.
12. Qiao, Z. A., Chai, S. H., Nelson, K., Bi, Z. G., Chen, J. H., Mahurin, S. M., Zhu, X., Dai, S. Polymeric molecular sieve membranes via in situ cross-linking of non-porous polymer membrane templates. *Nat. Commun.* 2014, 5, 4813.
13. Liu, Y., Wang, R., Chung, T. S. Chemical cross-linking modification of polyimide membranes for gas separation. *J. Memb. Sci.* 2001, 189(2), 231–239.
14. Staudt-Bickel, C., Koros, W. J. Improvement of CO_2/CH_4 separation characteristics of polyimides by chemical crosslinking. *J. Memb. Sci.* 1999, 155(1), 145–154.
15. Askari, M., Chung, T. S. Natural gas purification and olefin/paraffin separation using thermal cross-linkable co-polyimide/ZIF-8 mixed matrix membranes. *J. Memb. Sci.* 2013, 444, 173–183.
16. Koros, W. J., Mahajan, R. Pushing the limits on possibilities for large scale gas separation: Which strategies?. *J. Membr. Sci.* 2000, 175(2), 181–196.
17. Vanherck, K., Koeckelberghs, G., Vankelecom, I. F. J. Crosslinking polyimides for membrane applications: A review. *Prog. Polym. Sci.* 2013, 38(6), 874–896.

18. Wind, J. D., Staudt-Bickel, C., Paul, D. R., Koros, W. J. The effects of crosslinking chemistry on CO_2 plasticization of polyimide gas separation membranes. *Ind. Eng. Chem. Res.* 2002, 41(24), 6139–6148.
19. Wind, J. D., Staudt-Bickel, C., Paul, D. R., Koros, W. J. Solid-state covalent cross-linking of a polyimide membranes for carbon dioxide plasticization reduction. *Macromolecules.* 2003, 36(6), 1882–1888.
20. Wijmans, J. G., Baker, R. W. The solution-diffusion model: a review. *J. Memb. Sci.* 1995, 107(1–2), 1–21.
21. Park, J. Y., Paul, D. R. Correlation and prediction of gas permeability in glassy polymer membrane materials via a modified free volume based group contribution method. *J. Memb. Sci.* 1997, 125(1), 23–39.
22. Lin, H., Freeman, B. D. Gas solubility, diffusivity and permeability in poly(ethylene oxide). *J. Memb. Sci.* 2004, 239(1), 105–117.
23. Poling, B. E., Prausnitz, J. M., O'Connell, J. P. *The Properties of Gases and Liquids.* 5th edn. McGraw-Hill: New York, 2000.
24. Burns, R. L., Koros, W. J. Defining the challenges for C_3H_6/C_3H_8 separation using polymeric membranes. *J. Memb. Sci.* 2003, 211(2), 299–309.
25. Kaiser, T. Highly crosslinked polymers. *Prog. Polym. Sci.* 1989, 14(3), 373–450.
26. Flory, P. J. *Principles of Polymer Chemistry.* Cornell University Press: Ithaca and London, 1953.
27. Nielsen, L. E. Cross-linking — effect on physical properties of polymers. *J. Macromol. Sci. Polym. Rev.* 1969, 3(1), 69–103.
28. Hale, A., Macosko, C. W., Bair, H. E. Glass transition temperature as a function of conversion thermosetting polymers. *Macromolecules.* 1991, 24(9), 2610–2621.
29. Schroeder, J. A., Madsen, P. A., Foister, R. T. Structure/property relationships for a series of crosslinked aromatic/aliphatic epoxy mixtures. *Polymer.* 1987, 28(6), 929–940.
30. Lau, C. H., Hodd, K. A., Wright, W. W. Structure and properties relationships of epoxy resins Part 1: Crosslink density of cured resin:(II) Model networks properties. *Brit. polym. J.* 1986, 18(5), 316–322.
31. Pearson, R. A., Yee, A. F. Toughening mechanisms in elastomer-modified epoxies. *J. Mater. Sci.* 1989, 24(7), 2571–2580.
32. Rahul, R., Kitey, R. Effect of cross-linking on dynamic mechanical and fracture behavior of epoxy variants. *Composites Part B.* 2016, 85, 336–342.
33. Hirayama, Y., Yoshinaga, T., Kusuki, Y., Ninomiya, K., Sakakibara, T., Tamari, T. Relation of gas permeability with structure of aromatic polyimides I. *J. Memb. Sci.* 1996, 111(2), 169–182.
34. Hirayama, Y., Yoshinaga, T., Kusuki, Y., Ninomiya, K., Sakakibara, T., Tamari, T. Relation of gas permeability with structure of aromatic polyimides II. *J. Memb. Sci.* 1996, 111(2), 183–192.
35. Kim, T. H., Koros, W. J., Husk, G. R., O'Brien, K. C. Relationship between gas separation properties and chemical structure in a series of aromatic polyimides. *J. Memb. Sci.* 1988, 37(1), 45–62.

36. Kirby, A. J. *Polyimides: Materials, Processing and Applications*, Vol. 59. Pergamon Press: New York, 1992.
37. Shao, L., Low, B. T., Chung, T. S., Greenberg, A. R. Polymeric membranes for the hydrogen economy: contemporary approaches and prospects for the future. *J. Memb. Sci.* 2009, 327, 18–31.
38. Xiao, Y., Low, B. T., Hosseini, S. S., Chung, T. S., Paul, D. R. The strategies of molecular architecture and modification of polyimide-based membranes for CO_2 removal from natural Gas — A review. *Prog. Polym. Sci.* 2009, 34, 561–580.
39. Kanehashi, S., Sato, S., Nagai, K. Synthesis and gas permeability of hyper-branched and cross-linked polyimide membranes. In *Membrane Gas Separation*, John Wiley & Sons, Ltd: New Jersey, 2010, pp. 1–27.
40. Qiu, W., Xu, L., Chen, C. C., Paul, D. R., Koros, W. J. Gas separation performance of 6FDA-based polyimides with different chemical structures. *Polymer.* 2013, 54(22), 6226–6235.
41. Ordoñez, M. J. C., Balkus Jr, K. J., Ferraris, J. P., Musselman, I. H. Molecular sieving realized with ZIF-8/Matrimid® mixed-matrix membranes. *J. Memb. Sci.* 2010, 361(1–2), 28–37.
42. Bos, A., Pünt, I. G. M., Wessling, M., Strathmann, H. CO_2-induced plasticization phenomena in glassy polymers. *J. Memb. Sci.* 1999, 155(1), 67–78.
43. Huang, Y., Wang, X., Paul, D. R. Physical aging of thin glassy polymer films: Free volume interpretation. *J. Memb. Sci.* 2006, 277(1–2), 219–229.
44. Tin, P. S., Chung, T. S., Liu, Y., Wang, R. Separation of CO_2/CH_4 through carbon molecular sieve membranes derived from P84 polyimide. *Carbon.* 2004, 42(15), 3123–3131.
45. Hosseini, S. S., Chung, T. S. Carbon membranes from blends of PBI and polyimides for N_2/CH_4 and CO_2/CH_4 separation and hydrogen purification. *J. Memb. Sci.* 2009, 328(1–2), 174–185.
46. Vu, D. Q., Koros, W. J., Miller, S. J. Mixed matrix membranes using carbon molecular sieves: I. Preparation and experimental results. *J. Memb. Sci.* 2003, 211(2), 311–334.
47. Lin, W. H., Chung, T. S. Gas permeability, diffusivity, solubility, and aging characteristics of 6FDA-durene polyimide membranes. *J. Memb. Sci.* 2001, 186(2), 183–193.
48. Lin, W. H., Vora, R. H., Chung, T. S. Gas transport properties of 6FDA-durene/1,4-phenylenediamine (pPDA) copolyimides. *J. Polym. Sci., Part B: Polym. Phys.* 2000, 38(21), 2703–2713.
49. Kim, J. H., Koros, W. J., Paul, D. R. Physical aging of thin 6FDA-based polyimide membranes containing carboxyl acid groups. Part I. Transport properties. *Polymer.* 2006, 47(9), 3094–3103.
50. Ghosal, K., Chern, R. T., Freeman, B. D., Daly, W. H., Negulescu, II. Effect of basic substituents gas sorption and permeation in polysulfone. *Macromolecules.* 1996, 29(12), 4360–4369.
51. Aitken, C. L., Koros, W. J., Paul, D. R. Effect of structural symmetry on gas transport properties of polysulfones. *Macromolecules.* 1992, 25(13), 3424–3434.

52. Lam, B., Wei, M., Zhu, L., Luo, S., Guo, R., Morisato, A., Alexandridis, P., Lin, H. Cellulose triacetate doped with ionic liquids for membrane gas separation. *Polymer.* 2016, 1–11.
53. Puleo, A. C., Paul, D. R., Kelley, S. S. The effect of degree of acetylation on gas sorption and transport behavior in cellulose acetate. *J. Memb. Sci.* 1989, 47(3), 301–332.
54. Chung, T. S., Shao, L., Tin, P. S. Surface modification of polyimide membranes by diamines for H_2 and CO_2 separation. *Macromol. Rapid Commun.* 2006, 27(13), 998–1003.
55. Low, B. T., Xiao, Y., Chung, T. S., Liu, Y. Simultaneous Occurrence of Chemical Grafting, Cross-Linking, and Etching on the Surface of Polyimide Membranes and Their Impact on H_2/CO_2 Separation. *Macromolecules.* 2008, 41(4), 1297–1309.
56. Cao, C., Chung, T. S., Liu, Y., Wang, R., Pramoda, K. P. Chemical cross-linking modification of 6FDA-2,6-DAT hollow fiber membranes for natural gas separation. *J. Memb. Sci.* 2003, 216(1–2), 257–268.
57. Liu, Y., Chung, T. S., Wang, R., Li, D. F., Chng, M. L. Chemical Cross-Linking Modification of Polyimide/Poly(ether sulfone) Dual-Layer Hollow-Fiber Membranes for Gas Separation. *Ind. Eng. Chem. Res.* 2003, 42(6), 1190–1195.
58. Tin, P. S., Chung, T. S., Liu, Y., Wang, R., Liu, S. L., Pramoda, K. P. Effects of cross-linking modification on gas separation performance of Matrimid membranes. *J. Memb. Sci.* 2003, 225(1–2), 77–90.
59. Shao, L., Chung, T. S., Goh, S. H., Pramoda, K. P. The effects of 1, 3-cyclohexanebis (methylamine) modification on gas transport and plasticization resistance of polyimide membranes. *J. Memb. Sci.* 2005, 267(1), 78–89.
60. Shao, L., Chung, T. S., Goh, S. H., Pramoda, K. P. Transport properties of cross-linked polyimide membranes induced by different generations of diaminobutane (DAB) dendrimers. *J. Memb. Sci.* 2004, 238(1–2), 153–163.
61. Chung, T. S., Chng, M. L., Pramoda, K. P., Xiao, Y. PAMAM dendrimer-induced cross-linking modification of polyimide membranes. *Langmuir.* 2004, 20(7), 2966–2969.
62. Xiao, Y., Chung, T. S., Chung, M. L. Surface characterization, modification chemistry, and separation performance of polyimide and polyamidoamine dendrimer composite film. *Langmuir.* 2004, 20(19), 8230–8238.
63. Shao, L., Lau, C.-H., Chung, T. S. A novel strategy for surface modification of polyimide membranes by vapor-phase ethylenediamine (EDA) for hydrogen purification. *Int. J. Hydrogen Energy.* 2009, 34(20), 8716–8722.
64. Wang, H., Paul, D. R., Chung, T. S. Surface modification of polyimide membranes by diethylenetriamine (DETA) vapor for H_2 purification and moisture effect on gas permeation. *J. Memb. Sci.* 2013, 430, 223–233.
65. Japip, S., Liao, K.-S., Xiao, Y., Chung, T. S. Enhancement of molecular-sieving properties by constructing surface nano-metric layer via vapor cross-linking. *J. Memb. Sci.* 2016, 497, 248–258.

66. Shao, L., Liu, L., Cheng, S. X., Huang, Y. D., Ma, J. Comparison of diamino cross-linking in different polyimide solutions and membranes by precipitation observation and gas transport. *J. Memb. Sci.* 2008, 312(1–2), 174–185.
67. Wind, J. D., Paul, D. R., Koros, W. J. Natural gas permeation in polyimide membranes. *J. Memb. Sci.* 2004, 228(2), 227–236.
68. Wind, J. D., Sirard, S. M., Paul, D. R., Green, P. F., Johnston, K. P., Koros, W. J. Carbon dioxide-induced plasticization of polyimide membranes: pseudo-equilibrium relationships of diffusion, sorption, and swelling. *Macromolecules.* 2003, 36(17), 6433–6441.
69. Hillock, A. M. W., Koros, W. J. Cross-linkable polyimide membrane for natural gas purification and carbon dioxide plasticization reduction. *Macromolecules.* 2007, 40(3), 583–587.
70. Hillock, A. M. W., Miller, S. J., Koros, W. J. Crosslinked mixed matrix membranes for the purification of natural gas: Effects of sieve surface modification. *J. Memb. Sci.* 2008, 314(1–2), 193–199.
71. Kratochvil, A. M., Damle-Mogri, S., Koros, W. J. Effects of supercritical CO_2 conditioning on un-cross-linked polyimide membranes for natural gas purification. *Macromolecules.* 2009, 42(15), 5670–5675.
72. Kratochvil, A. M., Koros, W. J. Effects of supercritical CO_2 conditioning on cross-linked polyimide membranes. *Macromolecules.* 2010, 43(10), 4679–4687.
73. Eguchi, H., Kim, D. J., Koros, W. J. Chemically cross-linkable polyimide membranes for improved transport plasticization resistance for natural gas separation. *Polymer.* 2015, 58, 121–129.
74. Ma, C., Zhang, C., Labreche, Y., Fu, S., Liu, L., Koros, W. J. Thin-skinned intrinsically defect-free asymmetric mono-esterified hollow fiber precursors for crosslinkable polyimide gas separation. *J. Memb. Sci.* 2015, 493, 252–262.
75. Ma, C., Koros, W. J. High-performance ester-crosslinked hollow fiber membranes for natural gas separations. *J. Memb. Sci.* 2013, 428, 251–259.
76. Kraftschik, B., Koros, W. J. Cross-linkable polyimide membranes for improved plasticization resistance and permselectivity in sour gas separations. *Macromolecules.* 2013, 46(17), 6908–6921.
77. Ma, C., Koros, W. J. Ester-cross-linkable composite hollow fiber membranes for CO_2 removal from natural gas. *Ind. Eng. Chem. Res.* 2013, 52(31), 10495–10505.
78. Qiu, W., Chen, C. C., Xu, L., Cui, L., Paul, D. R., Koros, W. J. Sub-T_g cross-linking of a polyimide membrane for enhanced CO_2 plasticization resistance for natural gas separation. *Macromolecules.* 2011, 44(15), 6046–6056.
79. Merkel, T. C., Zhou, M., Baker, R. W. Carbon dioxide capture with membranes at an IGCC power plant. *J. Memb. Sci.* 2012, 389, 441–450.
80. Low, B. T., Xiao, Y., Chung, T. S. Amplifying the molecular sieving capability of polyimide membranes via coupling of diamine networking and molecular architecture. *Polymer.* 2009, 50(14), 3250–3258.
81. Aberg, C. M., Ozcam, A. E., Majikes, J. M., Seyam, M. A., Spontak, R. J. Extended chemical crosslinking of a thermoplastic polyimide: macroscopic

and microscopic property development. *Macromol. Rapid Commun.* 2008, 29(17), 1461–1466.

82. Lin, H., White, L. S., Lokhandwala, K. A., Baker, R. W. Natural gas purification. In *Encyclopedia of Membrane Science and Technology*, Hoek, E. M. V., Tarabara, V. V. (Eds.) Wiley: New Jersey, 2013, pp. 1644–1667.
83. Bos, A., Pünt, I. G. M., Wessling, M., Strathmann, H. Plasticization-resistant glassy polyimide membranes for CO_2/CH_4 separations. *Sep. Purif. Technol.* 1998, 14(1–3), 27–39.
84. Ismail, A. F., Lorna, W. Penetrant-induced plasticization phenomenon in glassy polymers for gas separation membrane. *Sep. Purif. Technol.* 2002, 27(3), 173–194.
85. Chua, M. L., Xiao, Y., Chung, T. S. Using iron (III) acetylacetonate as both a cross-linker and micropore former to develop polyimide membranes with enhanced gas separation performance. *Sep. Purif. Technol.* 2014, 133, 120–128.
86. Chinn, D., Bhuwania, N., Husain, S., Miller, S., Daniel, C., Bell, K. Crosslinked polyimide hollow fiber membranes for CO_2 removal from natural gas: From laboratory to field. In *North American Membrane Society 26th Annual Meeting*, Bellevue, WA USA, 2016.
87. Merkel, T. C., Pinnau, I., Prabhakar, R. S., Freeman, B. D. Gas and vapor transport properties of perfluoropolymers. In *Materials Science of Membranes for Gas and Vapor Separation*, Freeman, B. D., Yampolskii, Y. P., Pinnau, I. (Eds.) Wiley: New Jersey, 2006; pp. 251–270.
88. Hunger, K., Schmeling, N., Jeazet, H. B., Janiak, C., Staudt, C., Kleinermanns, K. Investigation of cross-linked and additive containing polymer materials for membranes with improved performance in pervaporation and gas separation. *Membranes.* 2012, 2(4), 727–763.
89. Kim, J. H., Koros, W. J., Paul, D. R. Effects of CO_2 exposure and physical aging on the gas permeability of thin 6FDA-based polyimide membranes: Part 2. with crosslinking. *J. Memb. Sci.* 2006, 282(1–2), 32–43.

Chapter 7

Dense Inorganic Membranes for Hydrogen Separation

Sean-Thomas B. Lundin*, Neil S. Patki*, Thomas F. Fuerst*, Sandrine Ricote†, Colin A. Wolden* and J. Douglas Way*,‡

**Department of Chemical and Biological Engineering, Colorado School of Mines, Golden, CO 80401, USA*

†Department of Mechanical Engineering, Colorado School of Mines, Golden, CO 80401, USA

‡dway@mines.edu

Hydrogen production has become an increasingly important field due to the rise of fuel cells and the search for a renewable source of energy. Yet, simply producing the hydrogen is not enough. It must also be separated from the various other effluent gases, whether the hydrogen comes from natural gas, a petroleum source, or is split from water. Various membrane technologies include porous silica membranes that rely on molecular sieving and dense technologies wherein the hydrogen is transported through a metal or ceramic lattice. Because of their perfect selectivity toward hydrogen, dense membranes using either certain metals or proton-conducting ceramics have remained a topic of great interest for decades and will be the subject of this chapter.

First, the mechanism of transport and various material-selection criteria will be discussed for both hydrogen-soluble metal membranes and proton-conducting ceramic membranes. The primary difference between the two transport mechanisms is that dense metal membranes rely solely on a thermochemical potential gradient to drive hydrogen transport, whereas proton-conducting ceramics more often use electrochemical potentials to increase the rate of hydrogen transport. In each case, the rate of transport is heavily influenced by careful material selection.

After reviewing the various materials available, an extended review of several membrane-reactor concepts will be detailed. Membrane reactors can be broken into two types: dehydrogenation reactors that generate and separate hydrogen and hydrogenation reactors that consume hydrogen. Dehydrogenation reactors have most often been the target for improvements to steam methane reforming or, more recently, in the decomposition of ammonia. Hydrogenation reactors, on the other hand, have focused on several different reactions and attempt to utilize the highly reactive atomic/protonic hydrogen that permeates the membrane. Finally, a brief overview of membrane applications in the separation of hydrogen isotopes at nuclear reactors will be discussed.

1. Introduction

Over 150 years ago, in 1863, the French group of Deville and Troost[1,2] first reported the diffusion of hydrogen through Group VIII metals such as iron and platinum. In 1866, while he was the curator of the British Royal Mint, Thomas Graham[3] reported the transport of hydrogen through heated palladium. Graham's initial diffusion research used palladium foils and, later, palladium tubes manufactured for him by George Matthey.[4] He also investigated the effects of alloying palladium with silver, noting that the hydrogen solubility was substantial — up to 50% Ag by mass. With this work as the basis, the field of dense metal membranes for hydrogen separation was born.

The study of proton-conducting ceramics, on the other hand, represents a relatively new field with only a few decades of history. According to a brief history written by Grover Coors,[5] the first mention of a proton-conducting ceramic came from the French group of Forrat *et al.*[6] in 1964. It later turned out that this apparent proton transport was most likely due to hydroxyl ion transport through the oxygen ion vacancies, as first proposed by Stotz and Wagner.[7] In fact, it would be nearly a decade after an extensive review on oxygen ion-conducting ceramics by Takahashi and Iwahara[8] until these two Japanese researchers would publish the first paper on verified proton-conducting ceramics. This first paper detailed the proton conductivities of both lanthanum-yttrium oxide and strontium zirconate in 1980,[9] and within another year Iwahara *et al.* published results on strontium cerate.[10] With the knowledge of proton conductivities in certain high-temperature ceramics now verified, the research into better structures with higher conductivities could begin.

One of the primary reasons for the continued interest in these dense membranes, both metallic and ceramic, is the perfectly selective nature of hydrogen transport. Unlike other methods of separation, both dense metallic and dense ceramic membranes have a transport mechanism for hydrogen that cannot be used by other gases such as nitrogen or even helium. To understand this, we must first review the methods of gas transport within various membrane technologies.

Transport mechanisms in hydrogen-separation membranes can be classified into five primary regimes (Fig. 1): (I) viscous flow, (II) Knudsen diffusion, (III) surface diffusion, (IV) molecular sieving, and (V) solution diffusion. Remember that within any one membrane, it may be possible to have contributions from multiple regimes, such as a surface-modified microporous membrane.

(I) **Viscous flow:** This regime is the resultant flow when a membrane has pores so large that gas molecules do not interact significantly with the membrane during transport — as such, no effective separation of differing gas molecules can be achieved.

(II) **Knudsen diffusion:** If the pore size is reduced to the point where its radius is smaller than the mean free path of the gas molecules, Knudsen diffusion begins to dominate. In this regime, transport is controlled by the reciprocal of the square

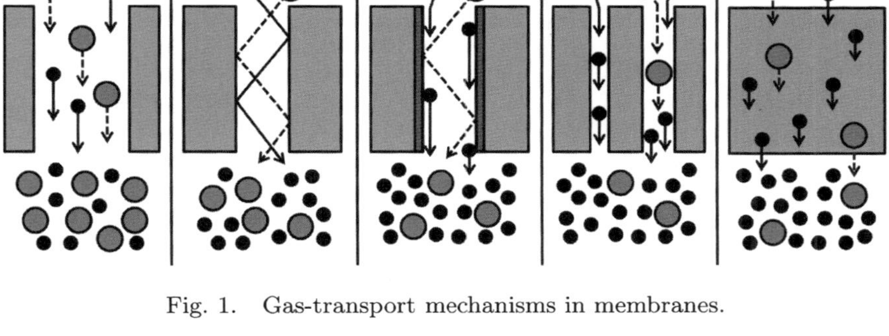

Fig. 1. Gas-transport mechanisms in membranes.

root of the molecule's mass, and a separation factor between two gases can be achieved based on the equation:

$$\alpha_{AB} = \left[\frac{M_B}{M_A}\right]^{1/2}. \tag{1}$$

(III) **Surface diffusion:** If the pores of a membrane are modified to adsorb hydrogen, a surface-diffusion mechanism can potentially greatly increase the separation factor over an unmodified support. For instance, Uemiya and coworkers coated various noble metals including Ru,[11] Rh,[12] Ir,[12] Pd,[11] and Pt[11,13] throughout an alumina support with 200 nm pores using chemical vapor deposition (CVD) and achieved enhanced hydrogen transport due to surface diffusion of hydrogen across a network of metal particles. Depending on the metal, this enhanced the H_2–N_2 selectivity from near Knudsen to anywhere from 80 to 240 at 773 K.

(IV) **Molecular sieving**: Once the pore size approaches that of the individual gas molecules, transport is highly biased by the size of the gas molecule and is known as molecular sieving. Although hydrogen is one of the smallest gas species, it remains impossible to be perfectly selective because helium and neon are smaller. Theoretical selectivity, however, is infinite for any gas molecule larger than hydrogen if the pores are perfectly formed. This is the primary mechanism of separation used in microporous silica membranes, but technical challenges of fabricating a perfect pore size distribution prevent high hydrogen flux and perfectly selective membranes from being formed.

(V) **Solution diffusion**: This method of transport is based entirely on the solubility and diffusivity of certain gases within a solid material. Both dense metals and polymers fall under this category, wherein hydrogen must absorb into the material and diffuse through the membrane before desorbing on the permeate side. Polymer membranes tend to suffer from low selectivity because they often allow for other gas species, particularly CO_2, to get transported as well. Due to the high solubility of CO_2, most polymers have CO_2–H_2 separation factors of order one, despite the significant size difference. The dense metallic membranes, on the other hand, have a transport mechanism based on atomic hydrogen that no other gas molecule or atom can partake. This is the reason dense metallic membranes are theoretically perfectly selective, even toward smaller molecules such as helium.

Essentially, mechanisms I–IV can never be perfectly selective toward hydrogen transport because pores that other gas molecules can penetrate will always exist (Table 1). The best one can achieve is perfect selectivity for gases larger than hydrogen, which is the goal of the work on microporous silica membranes. These membranes are typically formed by either CVD or sol–gel processes, creating a network of pores approximately 1 nm in diameter. It has been shown that these membranes can offer relatively high hydrogen selectivities for certain gases but suffer from much lower selectivity for gases close to the size of hydrogen. Tsuru *et al.*,[14] for example, obtained an H_2–N_2 ratio of 720, but the H_2–H_2O ratio was only 5–40. Other groups have also observed that hydrogen selectivity increased with increasing size of the comparison gas.[15–19]

The best selectivity reported in the literature for silica microporous membranes appears to be from the work of Gu and Oyama[20] wherein they obtained an H_2–CH_4 ratio of 5500 which would be very competitive for membrane work on steam methane reforming. Unfortunately, another issue with silica-based membranes is the densification of the pore network under steam atmospheres.[21] Because densification tends to decrease pore size, this densification is unlikely to affect the selectivity of these membranes, but the total hydrogen throughput may be greatly diminished. Although work has progressed regarding tolerance to steam, technical challenges remain in solving this issue at higher temperatures that have thus far prevented a stable high flux and high selectivity membrane from being realized.

The only mechanism of transport in membranes that has the potential to be perfectly selective for hydrogen transport over all other gases is the solution-diffusion regime. This regime is what dominates in polymer membranes as well as how hydrogen travels

Table 1. Various gases, their molecular diameters, and H_2 selectivity for specific membrane-sample studies.

Gas	Molecular diameter (nm)	Molecular mass (amu)	(II) Knudsen selectivity	(III) Surface diffusion (Pt on alumina)	(IV) Molecular sieving (microporous silica)		(V) Solution diffusion (dense polymer)
He	0.256 (a)	4.003	1.40	44	0.25	0.5	—
Ne	0.275 (a)	20.18	3.15	—	—	10	—
H_2	0.293 (a)	2.032	1.00	1	1	1	1
H_2O	0.296 (b)	18.02	2.98	—	22	—	—
NH_3	0.326 (b)	17.03	2.89	—	—	—	—
CO_2	0.330 (a)	44.01	4.65	—	—	1250	29
Ar	0.340 (a)	39.95	4.43	—	—	—	—
CO	0.362 (b)	28.01	3.71	—	—	5000	—
N_2	0.370 (a)	28.01	3.71	198	80	4700	341
CH_4	0.376 (a)	16.04	2.81	—	110	5500	278
				Ref. 13	Ref. 18	Ref. 20	Ref. 25

Note: Molecular diameters were calculated using either (a) Lennard–Jones potentials,[22,23] or (b) Stockmayer potentials.[24]

through dense metals. As will be discussed in Sec. 2, both dense metals and proton-conducting ceramics follow a dissociative transport mechanism for hydrogen in which the H_2 must first be broken into individual H-species for transport. The proton and electron of the H-species travel together through dense metal membranes whereas the electron is transported via an external circuit while the proton diffuses through the ceramic-based membranes by the Grotthus mechanism.[26] Dense polymer membranes, on the other hand, do not require dissociation of H_2 and can be viewed as allowing the bulk molecule to diffuse through nanoscale gaps in the polymer structure.

The issue with polymer membranes is that absorption into the polymer matrix is only partially selective to hydrogen because multiple gas species may get absorbed into the bulk, which can ultimately reduce selectivity. One of the best results for dense polymer membranes is from the work of Wijenayake *et al.*[25] who combined high permeability ZIF-8 adsorbent particles with a cross-linked 6FDA-durene polymer to increase selectivity while maintaining a hydrogen permeability of 500 Barrer ($1.67 \times 10^{-13}\,\mathrm{mol\,m\,m^{-2}\,s^{-1}\,Pa^{-1}}$). Yet, even with the good selectivity toward nitrogen, the selectivity of hydrogen toward carbon dioxide was low. Although this polymer may not allow for significant nitrogen absorption, like most polymers capable of hydrogen transport it has a high solubility for CO_2 that allows for its passage. To illustrate the difference in hydrogen-transport properties between membrane technologies, we can consider some recent results for several hydrogen-transport membrane technologies (Table 2). In addition, for a more detailed overview of competing hydrogen-separation technologies, the reader is referred to a recent review by Li *et al.*[27]

In the end, the only membranes that can be perfectly selective for hydrogen must involve a transport mechanism for either protons or atomic hydrogen. The dense metallic and dense ceramic membranes to be discussed in this chapter are meant to be an introduction into future possibilities for perfect hydrogen separations. Although it can certainly be argued that the benefits of the cheap material cost for polymers and silica may give rise to potential industrial applications using imperfect separations, this chapter will focus on the current research status of the perfectly selective technologies.

Table 2. Summary of performance characteristics for several membrane technologies.

	Porous carbon	Microporous ceramic	Dense polymer	Dense metallic	Dense ceramic
Transport regime	Knudsen to molecular sieving	molecular sieving	solution diffusion	solution diffusion	solution diffusion
Temperature (°C)	500–900	200–600	<100	300–600	600–900
H_2 selectivity	4–20	1–5000	4–400	10^2–10^4	10^2–10^4
H_2 flux ($\mathrm{mol\,m^{-2}\,s^{-1}}$)	0.01–0.2	0.06–0.3	<0.001	0.06–0.3	<0.01

Note: Dense metallic membrane data represents recent thin-film fabricated membranes, giving imperfect selectivity. Hydrogen flux given at a driving force of 100 kPa.
Source: Select data taken from Ref. 28.

2. Types of Dense Membranes

Although both dense metallic and proton-conducting ceramic membranes are theoretically perfectly selective to hydrogen transport, they operate under quite different transport mechanisms. Due to the high conductivity of most metals, dense metal membranes are never operated under an applied voltage. Various dense ceramics, on the other hand, have a range of electrical conductivities and can possibly be operated either under a galvanic (electrochemical gradient) or nongalvanic (thermochemical gradient) mode. In each of the primary sections below, the methods of operation will be detailed along with certain material-selection criteria and membrane-performance characteristics.

2.1. *Hydrogen Soluble Metal Membranes (Thermochemical Transport)*

In the introduction to this chapter, the concept of a solution-diffusion transport mechanism was briefly mentioned. Although this concept applies to dense polymer membranes as well, in the case of dense metal membranes this mechanism becomes perfectly selective to hydrogen transport. The reason for this high selectivity is due to the fact that hydrogen is not transported as a gas but as individual atoms through the metal lattice (Fig. 2). During transport, the feed hydrogen gas goes through several steps: (I) the gas adsorbs onto the metal surface, (II) the hydrogen molecule dissociates into two adsorbed atoms, (III) each atom then gets absorbed into the metal and diffuses to the far side, (IV) the atoms recombine on the surface into an adsorbed hydrogen molecule, and (V) the hydrogen molecule finally desorbs into the purified hydrogen permeate.

In general, the hydrogen permeability in metals can be described by Eq. (2), which states that the permeability coefficient, π_H, is the product of the solubility and diffusivity.

$$\pi_H = S \times D. \tag{2}$$

Assuming equilibrium at the gas–metal interfaces, the flux, J, of molecular hydrogen through Pd can be determined using Fick's law.

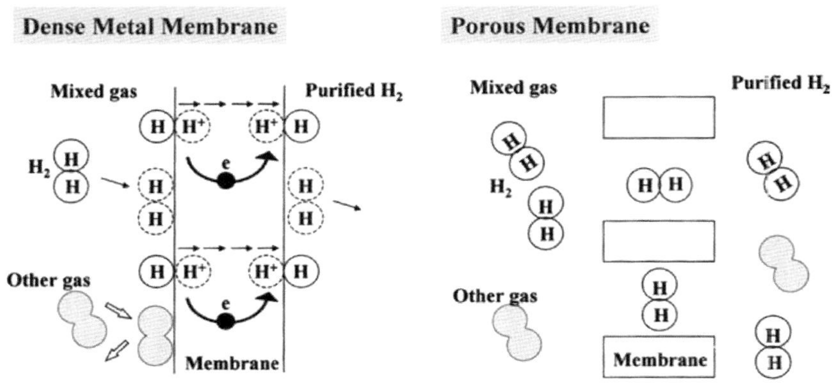

Fig. 2. Schematic of membrane operation for dense metal membranes compared to a porous membrane.

Source: Reprinted from Ref. 29 with permission from Springer.

The solution to this is Eq. (3), which is commonly known as Sieverts's law.

$$J = \frac{\pi_H}{L}\left(P^n_{H2,F} - P^n_{H2,P}\right). \tag{3}$$

In this equation, the flux, J, is determined by using the permeability of the metal, π_H, along with its thickness, L, and the pressure gradient of hydrogen between the feed, $P^n_{H2,F}$, and permeate, $P^n_{H2,P}$, sides. The pressure exponent, n, also known as the Sieverts's exponent, is known to vary between 0.5 and 1, depending on what portion of transport is rate-limiting. Bulk diffusion limitations, for instance, will result in a square root dependence, whereas surface adsorption/desorption limitations will result in a dependence of unity.

As early as the 1930s, the solubility of hydrogen in palladium and other metals was investigated by several groups, including those of Gillespie[30] and Sieverts[31–33] as reviewed by Lewis.[34] In their work, it was determined that the hydrogen solubility in Pd was proportional to the square root of the equilibrium hydrogen pressure. This can be rationalized by remembering that the transport mechanism of H_2 gas through these membranes occurs via dissociation and transport of two individual H-atoms in the bulk. Thus, solubility must account for twice the number of gas-phase molecules that enter the lattice.

However, deviations from the ideal n-value of 0.5 have been known to occur. Caravella and coworkers[35,36] discussed the value of the Sieverts's exponent, including circumstances where this exponent deviates from the theoretical value of 0.5. Meanwhile, Deveau *et al.*[37] developed a microkinetics model to deal with the non idealities of Sieverts's law as it pertains to very thin metal films at low temperatures. Alternatively, Vadrucci *et al.*[38] modified Sieverts's law to include surface effects, where non idealities have been observed even for thick films (84–200 μm).

When considering metals for hydrogen-transport membranes, only a select few appear as viable options. Figure 3 shows the measured permeability for several transition metals as well as the Pd–20Ag alloy, where it is apparent that the highest permeabilities

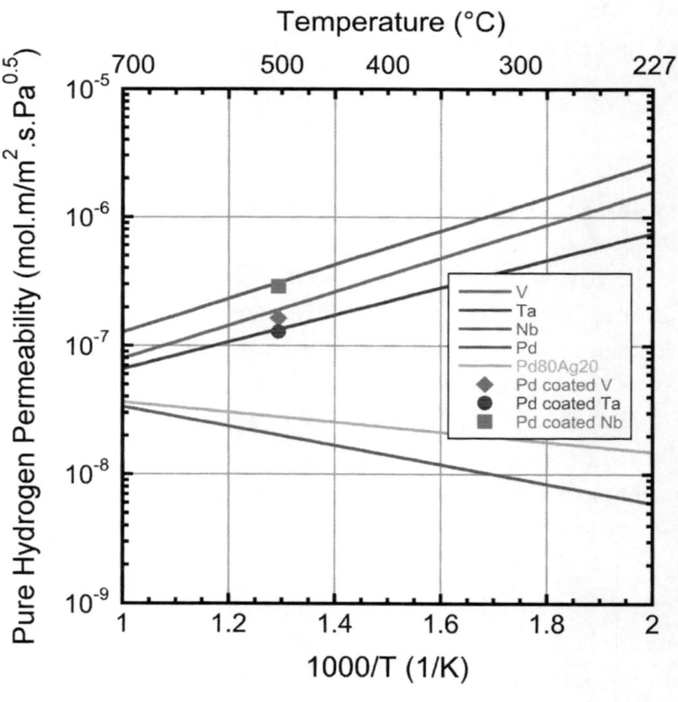

Fig. 3. Permeability of various metals with select experimental results. *Source*: Most data are from Ref. 39; Pd-coated BCC metal data are from Ref. 40, and Pd–Ag data are from Ref. 41.

are for those of V, Nb, Ta, and Pd. Metals such as Y and Zr, which are left of group Vb (V, Nb, and Ta), have such extreme solubilities that they suffer mechanical failure long before they can be used as hydrogen membranes; additionally, these metals also tend to suffer from a high propensity to oxidize, which substantially lowers their hydrogen solubility. On the other side of the table, Pd is the only metal to have a high permeability.

Outside of Pd and its alloys, group Vb metals (V, Nb, Ta) have previously been studied.[40,42–45] as potential hydrogen membranes due to their high hydrogen solubility and relatively cheap cost. These metals are about two orders of magnitude cheaper than Pd but have severe embrittlement issues at low temperatures.[46] In fact, even though their critical temperatures (T_c) for hydride formation are all below ca. 170°C, Owen[47] has argued that hydride formation does not fully explain embrittlement in the group Vb metals because they are known to embrittle well above these T_c values. It has instead been proposed[44,45] that membranes involving V, Nb, or Ta must be operated above 450°C to avoid embrittlement. The failure mechanism in these membranes is likely due to lattice expansion, which has been shown to increase 3–4% upon exposure to 700 kPa hydrogen, and transition from a ductile to brittle regime.[48]

Pd and its alloys can also suffer from hydride formation and embrittlement failure but to a much lesser extent than the group Vb metals. The drawback is lower permeability of the materials and thus lower hydrogen productivity. Table 3 gives the solubility, diffusivity, and permeability for each of the group Vb metals and Pd, showing

Table 3. Solubility, diffusivity, and permeability of dense metal membrane materials at specified temperatures.

Metal/ alloy	Temperature	Solubility ($mol\,m^{-3}\,Pa^{-0.5}$)	Diffusivity ($m^2\,s^{-1}$)	Permeability ($mol\,m\,m^{-2}\,s^{-1}\,Pa^{-0.5}$)
V[39]	500°C	12.6	1.5×10^{-8}	1.8×10^{-7}
Nb[39]	500°C	30.4	1.0×10^{-8}	3.0×10^{-7}
Ta[39]	500°C	24.8	5.4×10^{-9}	1.3×10^{-7}
Pd[49]	500°C	6.0	3.2×10^{-9}	1.9×10^{-8}

that the group Vb metals all have higher hydrogen solubilities and diffusivities. Yet, the benefit of Pd over the group Vb elements is its ability to dissociate hydrogen, whereas group Vb metals must be coated with a catalyst to dissociate and reassociate hydrogen during the transport process. In each of the following sections, we will discuss these metals and some of their alloys in detail, attempting to elucidate the costs and benefits associated with each.

2.1.1. *Palladium and Its Alloys*

As mentioned in the opening paragraph of this chapter, the work of Thomas Graham laid the very foundation for the development of dense metal membranes based on palladium. Nearly a century later, the interaction of hydrogen with palladium and palladium alloys was extensively investigated and reviewed by Lewis in 1967.[34] The potential use of pure palladium as a hydrogen-separation membrane, however, was severely hampered by the extremely high solubility of hydrogen in palladium below 300°C.[34] This extreme solubility leads to the formation of a separate palladium-hydride phase with a much larger lattice constant compared to pure Pd, causing subsequent distortion of the lattice and resulting in hydrogen embrittlement.

Hydrogen embrittlement is particularly problematic when cooling down a pure Pd membrane without removal of the dissolved hydrogen, causing membrane cracking/failure. To combat this issue, alloys of palladium with other metals, particularly silver, were investigated by Hunter and coworkers at the Atlantic Refining Company in collaboration with J. Bishop and Company which later became Johnson Matthey.[50,51] In his 1956 patent, Hunter disclosed[50] that alloying palladium with 27% by mass of Ag not only prevented hydrogen embrittlement but also improved the hydrogen permeability of the alloy by 70% compared to pure palladium at 450°C.

Several metals that can potentially lower the embrittlement temperature of Pd include Ag, Au, Ni, Pt, and Rh. Select properties of these alloys are summarized in Table 4. The embrittlement temperatures are approximations taken from Pd-H-X phase diagrams and simply indicate the temperature limit above which a single phase exists for all hydrogen concentrations. The hydrogen flux for

Table 4. Pd alloys for reducing the embrittlement temperature.

Comparison of phase diagrams		Comparison of hydrogen flux		Comparison of membrane cost	
Pd alloy (wt.%)	Embrittlement temperature (K)	Pd alloy (wt.%)	Hydrogen flux (alloy/pure Pd)	Alloy metal price (US$/oz)	Membrane cost (alloy/pure Pd)
Pd[54]	~570	Pd	1.0	540	1.00
Pd–25Ag[55]	~300	Pd–27Ag[52]	1.7	16	0.74
Pd–25Au[55]	~300	Pd–20Au[52]	1.0	1200	1.24
Pd–34Cu[55]	~300	Pd–39Cu[52]	1.3	0.13	0.60
Pd–9Ni[55]	~350	Pd–10Ni[52]	0.2	0.24	0.90
Pd–17Pt[55]	~340	Pd–20Pt[53]	0.6	980	1.16
Pd–10Rh[55]	~400	Pd–10Rh[53]	0.9	660	1.02

Note: Spot price of metals were taken on May 27, 2016.

each wt.% of alloy is then compared to pure Pd using compositions from the literature, where McKinley data[52] was taken at 450°C, and Gryaznov data,[53] at 500°C. Finally, a relative cost to Pd for the hydrogen flux alloys was calculated using the spot price of the metals and assuming a constant membrane thickness for each alloy.

As indicated by Table 4, a Pd–Ag membrane offers three primary benefits over the other alloys: an ability to operate as low as 300 K without embrittlement, a higher hydrogen flux, and a cheaper overall membrane cost. The ternary phase diagram for Pd-H-Ag system indicates that the two-phase region is lowered to 23°C (296 K) upon addition of a mass fraction of ca. 25% Ag to Pd. This result was verified by Paglieri *et al.*[45] who exposed a Pd–23Ag foil to 101 kPa hydrogen and did not observe any embrittlement at 298 K. Incidentally, this is also considered the optimum Ag addition to increase hydrogen flux in Pd membranes, with permeability increasing about 1.7 times for a Pd–23Ag over a pure Pd membrane. Finally, the addition of Ag lowers the price of the membrane, unlike Au, Pt, or Rh. Although Ni can also lower the cost, the permeability of Pd–Ni membranes is only expected to be 12% of the Pd–Ag membrane, which could be an issue for delivering hydrogen at lower temperatures.

Presence of Pt in the lattice improves the high temperature stability of the membrane as well as the performance in the presence of carbon species,[56] whereas Au greatly improves the sulfur resistance of the membrane.[57] Copper as an alloying element has also been studied in the literature, but the composition range in which it would be beneficial is much smaller, as seen from Fig. 4. Unfortunately, controlling the composition of the alloy can be quite difficult, making Cu an unfavorable alloying metal. Based on the data published by McKinley one can identify the alloy compositions for the Pd–Au and Pd–Ag systems which will have permeabilities higher than that of pure Pd.[52] These data are shown in Fig. 4.

A lot of work has been done to study the Pd–Au system over the last few years. The two metals have the same FCC crystal structure and are known to form solid solutions at the temperatures (673–873 K) and compositions (0–40 wt.% Au) of interest.[58] Pd–Au membranes are fabricated by using electroless plating to deposit alternating layers of Pd and Au followed by an annealing step.[59–61] In addition to understanding the effect of Au content on hydrogen solubility Gade *et al.*[62] demonstrated that a membrane with a mass

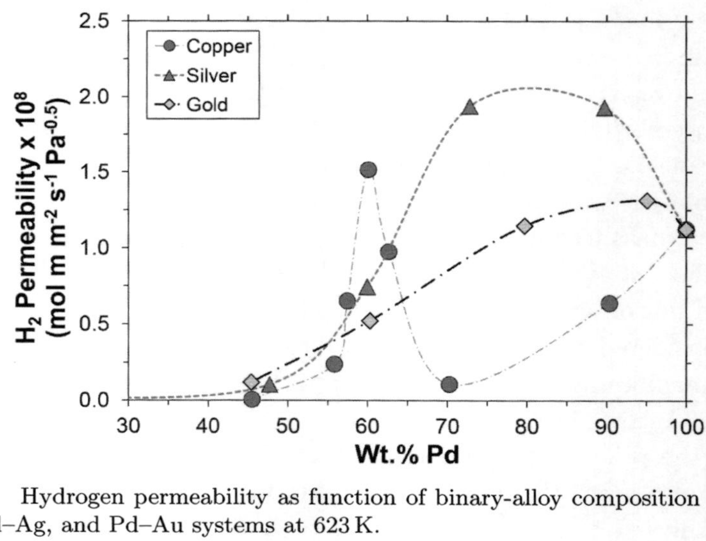

Fig. 4. Hydrogen permeability as function of binary-alloy composition for Pd–Cu, Pd–Ag, and Pd–Au systems at 623 K.

Source: Adapted from Ref. 52.

fraction of 40% Au had better hydrogen permeability in a stream containing 4 ppm of hydrogen sulfide in hydrogen than pure Pd as well as select high permeability Pd–Cu, and Pd–Ag alloys. They also observed[62] that increasing the Au content in membranes that were tested in a synthetic water–gas shift mixture containing 20 ppm H_2S resulted in better resistance to sulfur poisoning. Chen and Ma[63] observed that flux inhibition for a Pd–8Au membrane due to sulfur species decreased with increasing temperatures which points towards an equilibrium adsorption nature of the inhibition, unlike the formation of bulk sulfides for a pure Pd membrane.

Gade *et al.*[64] observed that fabrication technique (cold-rolled versus sputtered foils) and alloy content (7–20 wt.%) affected the performance of unsupported Pd–Au alloys in the presence of carbon- and sulfur-containing species. Sputtered foils did not perform as well as the cold-rolled foils in the presence of H_2S (0.03–0.06 kPa partial pressure or 20–50 ppm) and degraded quicker as well; both those results were attributed to brittle and porous structure resulting from the sputtering process.[64] The ternary alloy based on the addition of Ag to the Pd–Au system combines the sulfur resistance of the Pd–Au alloys with the high permeance of the Pd–Ag alloys.[65–68] Thus, the depression of the hydrogen permeance caused due to alloying of Au with Pd can be offset to some extent.

2.1.2. *BCC–Catalyst Combinations*

Body-centered cubic (BCC) metals are of keen interest as dense metallic membranes for hydrogen separation because, as mentioned previously, their bulk material cost is a mere fraction of Pd. The BCC lattice structure is common among many transition metals including W, V, Cr, Rb, Nb, Mo, and Ta; however, in particular, the Group Vb metals (V, Nb, and Ta) dominate research efforts due to their high hydrogen permeability. Therefore, in the dense metallic membrane literature, BCC metals refer to these Group Vb elements and their alloys.

The relationship between hydrogen and these metals was first studied in 1929 when Hagen and Sieverts[69] studied the solubility of hydrogen in Nb. The application of BCC metals for nuclear and high-temperature applications was quickly realized shortly thereafter[70] and is a continued area of research today. Jewet *et al.*[71] later patented the idea of using Group Vb elements coated with Pd as a membrane to produce high purity hydrogen in 1967. Although BCC metals maintain mechanical strength in harsh environments and provide the highest hydrogen permeability among all materials, they suffer from several deficiencies that limited their implementation. Not only do V, Nb, and Ta readily oxidize to form catalytically inactive and impermeable oxide layers on their surfaces, but these metals are also easily embrittled in hydrogen environments due to their extremely high solubility. This section will discuss an overview of the current knowledge of the material properties, alloys, and catalyst coatings for V, Nb, and Ta.

Group Vb metals are unique because their hydrogen permeability is an order of magnitude larger than pure Pd and its alloys. However, because permeability is a product of solubility and diffusivity (Eq. (2)), it is important to analyze both properties to understand why the permeability is so much higher. Solubility of hydrogen in V, Nb, and Ta is much larger than other metals because they exhibit large exothermic hydrogen binding energies in the lattice as modeled by DFT literature.[72] Materials with an exothermic binding energy (exothermic occluders), such as Pd, V, Nb, and Ta, observe solubility to decrease with an increase in temperature; however, the permeability of Pd increases with increasing temperature whereas that of BCC metals decreases. This is due to solubility dominating over diffusive terms in the hydrogen permeability for BCC metals.

The standard method of investigating solubility in dense metals is to use a Sieverts type apparatus to form pressure–composition isotherm (PCT) curves,[73] which show that the solubility of each metal follows the same trend as permeability for the BCC metals. Diffusivity, on the other hand, is a measure of H migration by hopping between interstitial sites in the lattice; this hopping increases as temperature increases with Arrhenius dependence.[74] Additionally, the activation energy for the hop between interstitial sites increases with distance between the sites.[75] The diffusion coefficients of V, Nb, and Ta at 500°C are 1.5×10^{-8}, 1.0×10^{-8}, and

5.4×10^{-9} ($m^2 s^{-1}$) respectively.[39] Although the diffusivities of the BCC metals are only slightly higher than Pd (Table 3), the solubilities of the BCC metals are also significantly higher. This combined increase results in the order of magnitude difference in permeability shown in Fig. 3. This provides BCC metal permeabilities at 500°C of 1.8×10^{-7}, 3.0×10^{-7}, and 1.3×10^{-7} ($mol\, m\, m^{-2} s^{-1} Pa^{-0.5}$) respectively, with Nb exhibiting the highest followed by V and then Ta (Table 3).

The high solubility of BCC metals is partially explained by comparing the BCC and FCC crystal lattices. The unit cell of a BCC structure consists of an atom at each corner of the cubic cell and one in the center. Similar to Pd's FCC lattice, the BCC lattice also contains tetrahedral (four nearest neighbors) and octahedral (eight nearest neighbors) interstitial sites. However, the number of sites per atom, site size, and packing density differ from those of FCC. For BCC the number of octahedral and tetrahedral sites is 3 and 6 per atom, respectively, with the size of the site relative to the metal-atom radius being 0.155 and 0.291 for octahedral and tetrahedral sites, respectively. The packing density of BCC is 68% whereas the packing density of FCC is 74%.[76] These features, shown in Table 5, lead to differences in solubility and transport properties between the two lattices. The lower-energy site has the smaller size, which shows that hydrogen is preferred in the octahedral site for BCC metals and the tetrahedral site for FCC metals. The increase in the number of sites for the BCC structure, with a relatively similar packing density, alludes to the higher solubility.

Although the large solubility of BCC metals is beneficial for achieving high hydrogen flux, it creates several issues with

Table 5. FCC and BCC lattices parameters.

	Octahedral		Tetrahedral		
	Sites	Relative size	Sites	Relative size	Packing density
FCC	1	0.414	2	0.225	74%
BCC	3	0.155	6	0.291	68%

mechanical stability. Group Vb metals remain ductile in hydrogen at near atmospheric pressures for a range of temperatures, but hydride-BCC metals become very brittle at moderate pressures that limit their applicability as hydrogen-separation membranes. Fortunately, the knowledge of the mechanical properties of BCC metals in hydrogen environments has improved drastically in the past decade, allowing for a much better determination of potential operating conditions.

In 1977–1979, Gahr and Birnbaum[77,78] published that Nb was ductile in hydrogen environments. Although these data are accurate, the study only tested pure Nb which has negligible hydrogen permeation when uncoated with a protective catalyst because the hydrogen cannot easily dissociate on the surface. Nambu *et al.*[44] have since shown that Nb becomes very brittle in hydrogen environments when Nb is coated with Pd and is thus readily able to dissociate and absorb hydrogen. Embrittlement of metals was tested with an *in situ* punch apparatus, and Nb was measured to have a ductile to brittle transition concentration (DBTC) at ca. 0.20–0.25 H/M (hydrogen atom to metal atom ratio).[44] Very little difference in the DBTC was measured between samples prepared from different manufacturing methods, despite the difference in grain sizes.[44] Similar DBTC for V has also been reported[79] which constrains these metals to an operational concentration of <0.2 H/M. The PCT curves for Nb and V with the DBTC are shown in Fig. 5, where the critical equilibrium hydrogen pressure corresponding to the DBTC at 400°C is observed to be 0.005 MPa for pure Nb and 0.04 MPa for V. This pressure increases as temperature increases, as shown by the increase in critical pressure of Nb at 500°C (0.03 MPa), primarily due to the decreasing solubility in the metals with increasing temperature.

An important note is that embrittlement at high hydrogen concentrations is not caused by β-hydride formation, which is a function of temperature, but brittleness within the α-hydride phase. The critical temperatures for β-hydride formation in BCC metals are shown in Table 6. Both α-hydride formation and the decrease in DBTC pressure as temperature decreases makes thermal cycling with BCC membranes difficult; membranes must be purged of H before any cycling to prevent fracture of the membrane.

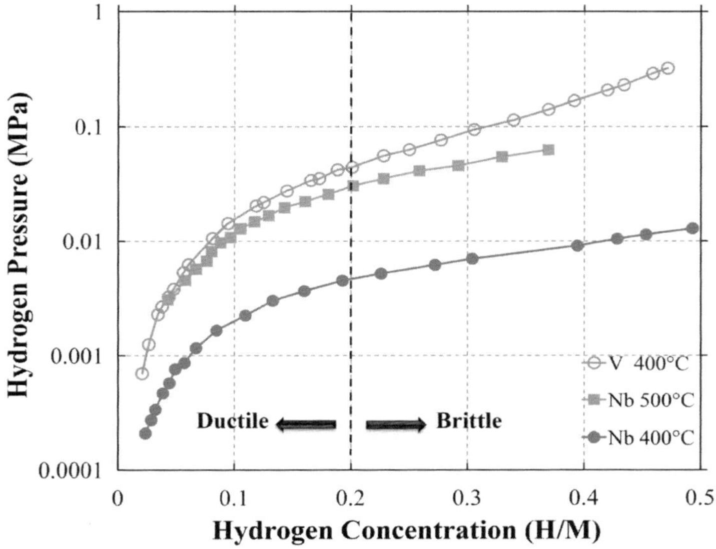

Fig. 5. PCT curve for Nb (blue circles) and V (orange circle) at 400°C[79] and Nb at 500°C[80] (green square).

Table 6. Critical temperatures (T_c) for β-hydride formation in select metals.

Metal	T_c (°C)
Pd[54]	293
V[81]	170
Nb[82]	171
Ta[83]	10

Applications such as ammonia decomposition and steam methane reforming require high hydrogen pressure to achieve high fluxes and significant shift of these reactions toward products (Sec. 3.1) which can cause BCC membranes to fail. Thus, alloys are primarily investigated to reduce hydrogen solubility and therefore prevent embrittlement caused by high hydrogen concentration in the lattice. Ideally, we desire alloys that are ductile, retain relatively high hydrogen permeability, and are resistant to embrittlement. These properties have been the central focus of research on alloys in this field for the past ca. 20 years.

To predict potential alloying metals for the BCC metals, Hume-Rothery rules can be used.[84] Four rules apply to the alloying element: (1) the atomic radii must differ by no more than 15%, (2) crystal structures must be similar, (3) the valency must be the same or higher, and (4) the electronegativities must be similar. Binary vanadium alloys have been the focus of research due to the smaller atomic radius of V (0.1316 Å) compared to Nb (0.1429 Å) and Ta (0.1430 Å) which allows V to alloy well with period 4 transition metals that have a similar atomic radius. Such metals include Cr,[80,85] Ni,[86–88] Fe,[80] and Co[80,85]; however, other metals have also been tested including Al,[85] Mo,[80,89,90] Pd,[45,91,92] and W.[89,90]

Unfortunately, not only does increasing the alloying-element composition decrease solubility, but it also has a negative impact on the material's ductility. This means a balancing act must be played while investigating these alloys for different applications. For instance, Paglieri *et al.*[45] conducted a study that screened numerous BCC alloys and found several, including V–7Fe, V-(4, 8, 14, and 33)Ru, V–9Co, and Nb–13Pd, were either too brittle to handle or embrittled in hydrogen and were unable to be tested. V–Pd has been shown to provide sufficient reduction in solubility while maintaining high permeability and ductility,[45,91,92] but this fails to eliminate the costly Pd metal from the membrane system. Alternatively, ternary alloys have also been studied by adding Ti into various binary alloys to increase mechanical strength and incorporate oxygen-scavenging properties.[93,94]

When considering Nb and Ta, these metals do not alloy well with other late transition metals that have been investigated with V due to their larger atomic radii. Yet, a limited number of studies do exist, including investigations of Nb with Mo and W,[90,95] and Ta with a smaller alloy concentration of Ni,[96] Fe,[97] Cr,[98] Al,[99] and Zr.[100] In addition, a different class of alloys has been developed with Ni and Nb wherein ternary or higher-order alloys are formed with Nb being only a minor fraction of the alloy, such as Ni–Nb–Zr[101] and Ni–Ti–Nb,[94] which can provide permeability similar to Pd at a fraction of the cost.

In addition to embrittlement severely affecting the range of applicable pressures and temperatures of operation, BCC metals also readily form oxides that have low hydrogen solubility and can severely limit hydrogen transport. Thus, coatings have been applied to feed and permeate sides of metal membranes in order to prevent surface oxidation and promote the dissociative-adsorption/reassociative-desorption of hydrogen. Pd has been the most studied and easiest choice because it exhibits high kinetics and strong chemical and environmental resistances.[71] Coatings are typically applied by electroless plating to produce 1–2 μm thick coatings[42] or by sputtering for thin films of 100–400 nm.[71] Surface oxides are removed before deposition by mechanical and chemical treatments in the case of electroless plating or by an Ar plasma etch when sputtering. Deposition of thin, 100 nm Pd coatings on BCC metals has shown to eliminate diffusion limitations in the Pd coatings and produce near theoretical results of hydrogen permeability for V, Nb, and Ta,[40] as shown in Fig. 3.

However, issues still persist with thin coatings including interdiffusion between the base BCC metal and the Pd coating that causes a non-Pd surface to be exposed and subsequently oxidized. This effect is most drastic between V and Pd, whereas poor surface interactions have been observed between Pd and either Nb or Ta. Pd is seen to flake off these BCC metals at 500°C,[40] presumably due to swelling of the Nb membrane from high hydrogen solubility. Additionally, the deposited Pd film has been found to change morphology as a function of temperature.[102,103] Films up to 250 nm formed pores at 420°C, and thicker films of 250–1000 nm formed porous structures at 600°C.[102]

Transition-metal carbides are known to perform similarly to platinum group metals and have been previously studied as replacements for catalyst applications.[104] On this topic, our group has recently applied a thin film of Mo_2C as a catalyst layer to V by RF sputtering[105,106] to provide both oxidation protection and catalytic ability to dissociate/associate hydrogen on the feed and permeate sides of the membrane. Given the low hydrogen solubility of Mo_2C,[105] the proposed transport mechanism is spillover wherein hydrogen surface diffuses down grain boundaries within the Mo_2C before entering the V lattice. Thicknesses of the carbide layer were optimized to 20 nm[106] which is dramatically thinner than typical Pd coatings. These membranes were able to exceed the permeability of pure Pd at high temperatures, with a measured permeability of 5.9×10^{-8} mol m m^{-2} s^{-1} Pa$^{-0.5}$ at 750°C.[106] The flux measurements at various trans-membrane pressures and temperatures for this system are shown in Fig. 6.

In this case, nanoscale coatings were sufficient because interdiffusion between the Mo_2C and V was not observed (Fig. 7), allowing V to be used in high temperature applications. Sieverts' constant analysis, shown in Fig. 6 as n-values, revealed that this system was dissociative adsorption/associated desorption limited at 600°C with

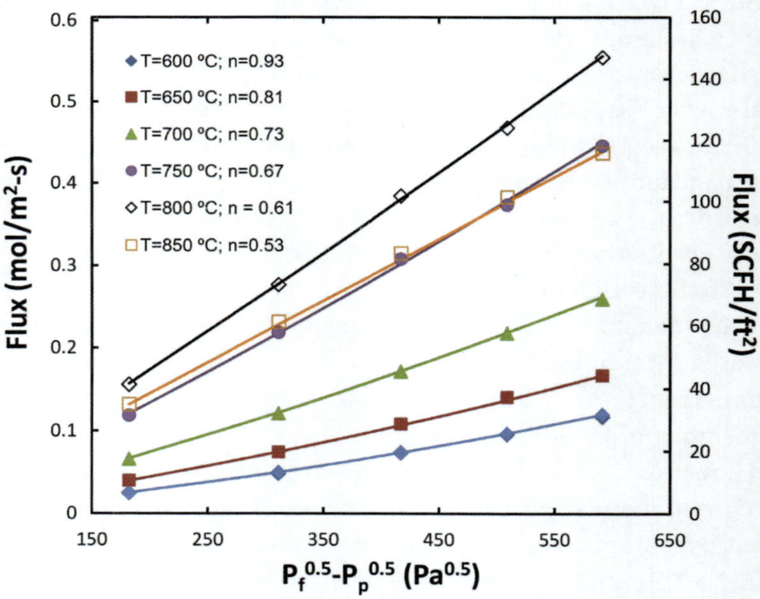

Fig. 6. Hydrogen flux versus driving force of 20 nm Mo_2C coated on 50 μm V membranes at selected operating temperatures. Extracted Sieverts' exponents are provided in the key next to the operating temperatures.

Source: Reprinted from Ref. 106 with permission from Elsevier.

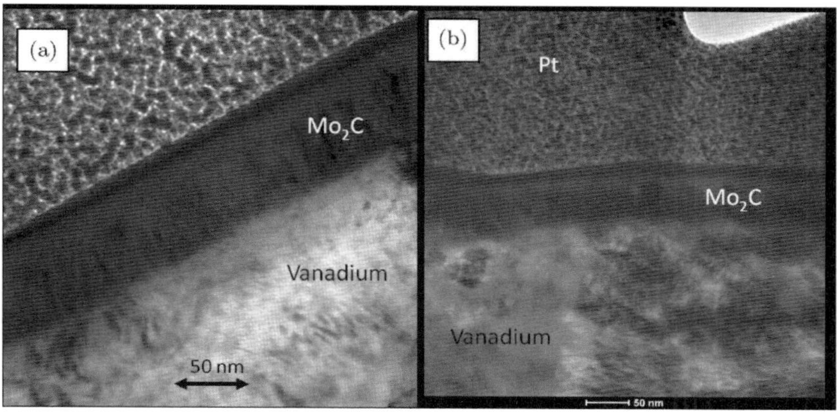

Fig. 7. Cross-sectional TEM images of the 65 nm Mo_2C–V interface (a) before testing and (b) after 168 hours of hydrogen-permeation testing. Note that Pt was deposited after permeation testing as part of the focused ion beam sample preparation. Scale bar on both images is 50 nm.

Source: Reprinted from Ref. 105.

a Sieverts' constant of 0.93. However, the membrane became diffusion limited as temperature increased, as indicated by the constant approaching 0.5. This is because lower temperatures are insufficient for hydrogen-dissociation kinetics due to the high activation energy of hydrogen dissociation on Mo_2C; however, this can protect the membrane from exceeding the DBTC discussed earlier. As an alternative to Mo_2C, Zhu *et al.*[107] conducted a first-principles (DFT) study to screen for potential new catalysts and proposed TiC. Yet, to our knowledge this has not been further explored experimentally.

Outside of using thin films as alternatives to Pd as a surface catalyst, barriers to prevent interdiffusion between Pd and BCC metals are another potential exciting field of which only limited literature has been published. Barriers could lead to longer lifetimes and allow high temperature applications that have previously been limited. Currently, oxides have been investigated as a Pd barrier[108–110] as well as Nb_2C.[111,112] However, upon realizing the ability of Mo_2C as a hydrogen permeable catalyst layer,[105,106] it is reasonable to assume that Mo_2C could also act as a potential Pd barrier. These barriers could be designed asymmetrically to provide transport limitations on the feed side of the membrane to alleviate issues with DBTC at high pressures. The use of new catalysts and designs as protective and transport barriers could provide exciting new breakthroughs to solve problems with the otherwise promising BCC metal membranes.

2.2. *Proton-Conducting Ceramic Membranes (Electrochemical Transport)*

The previous section described dense metal membranes that require a hydrogen pressure gradient to be applied across the membrane to drive the hydrogen transport. Dense ceramic membranes, however, can either be used in non-galvanic mode or coated with electrodes and operated in galvanic mode; the latter of which utilizes an electrical potential gradient across the membrane to drive the hydrogen transport rather than the chemical potential gradient of nongalvanic operation. As the Nernst equation shows (Eq. (4)),[113] pressure and electricity are equivalent forms of energy, with $RT/2F$ being equivalent to 0.042 V at 700°C. Thus, applying a 580 mV electrical potential gradient ($\Delta\varphi$) is equal to a hydrogen pressure gradient of one million times. Therefore, it is clearly easier to apply a potential gradient than to achieve an equivalent pressure gradient by compressing gases.

$$\Delta\phi = \frac{RT}{2F}\Delta\ln(P) = \frac{RT}{2F}\ln\left(\frac{P_{\text{Feed}}}{P_{\text{Permeate}}}\right). \tag{4}$$

In the case of ceramic proton conductors, hydrogen is transported purely as separate protons and electrons (Fig. 8). Thus, the electrons from the hydrogen gas need to be stripped on the upstream side of the membrane and supplied to the downstream side. Depending on the material used, the electrons can either pass through an outer circuit (galvanic mode), or, in the case of mixed protonic–electronic ceramics, can transport through the membrane by co-diffusion (passive membranes/non-galvanic mode similar to dense metallic membranes).[113] An electrode layer (like Pt, Pd, Cu, Ni, etc.) serves to dissociate the hydrogen gas on the upstream side and to reassociate the H species into hydrogen gas on the downstream side. Metal wires, usually of the same metal as the electrode to avoid interdiffusion, are used to complete the outer circuit and apply the potential gradient.

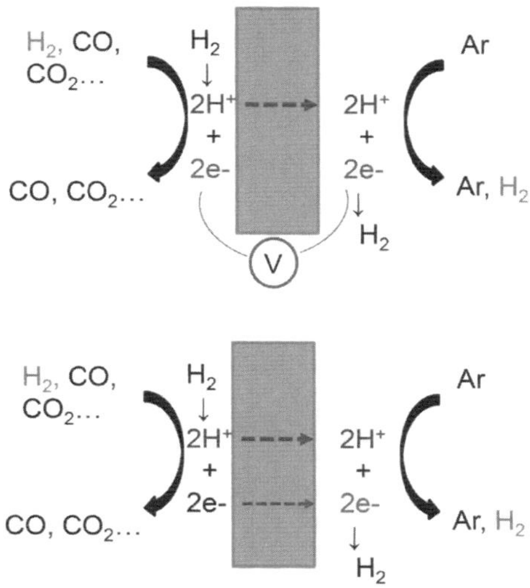

Fig. 8. Graphical illustration of galvanic mode (top) and nongalvanic mode (bottom) of operation.

For ceramic membranes with a second electronically percolating phase, a triple phase boundary (TPB) must exist between the electrode, proton-conducting ceramic, and electronically conducting phase. A large concentration of TPBs is necessary to ensure that the dissociation and reassociation reactions are not rate-limiting. The rate of hydrogen transport through the ceramic, then, is a function of the electrochemical potential gradient across the membrane (which is determined by the H_2 pressure gradient) and the rate of diffusion of the protons through the ceramic. In this case, the electromotive force (e.m.f.) across the membrane is limited to ≤ 1 V. However, when an external circuit is used, the e.m.f. can be increased by an external power source. Thus, the e.m.f. can be raised to increase the rate of hydrogen transport, with the only limitation being the degradation of the ceramic material.[113]

Proton conduction has been observed for ceramic materials with crystal structures of either the perovskite type or derivatives of the fluorite or pyrochlore structures.[114] Other types like acceptor-doped metal oxides, oxyacid salts, and mixed protonic electronic conductors are some of the newer types of proton-conducting ceramics that are being investigated.[114] For an in-depth discussion of the various types of proton-conducting oxide materials, the reader is directed to Chapter 2 of the book *Proton-Conducting Ceramics: From Fundamentals to Applied Research* by Mathieu Marrony.[114] Perovskite-based cerates and zirconates are the most studied proton-conducting systems with more than two decades of active research.[115] The progress of proton-conducting ceramics has been detailed in reviews by Iwahara,[116] Bonanos *et al.*,[117] Kreuer,[118] Fabbri *et al.*,[119] de Souza and Muccillo,[120] Malavasi *et al.*,[115] and Reijers and Haije.[121]

2.2.1. *Oxides Based on the Perovskite Structure*

Calcium titanate ($CaTiO_3$) was the first perovskite discovered in the Ural Mountains in 1839. Today, compounds with the chemical formula ABO_3 and a similar crystal structure are called perovskites. In the early 1980s pioneering research conducted by Iwahara *et al.*[10] revealed the proton-conducting nature of certain perovskites, creating a new field of proton conductors in which this property of simple perovskites is vital. The general perovskite structure, shown in regular cubic form in Fig. 9, contains red (corner) and grey (center)

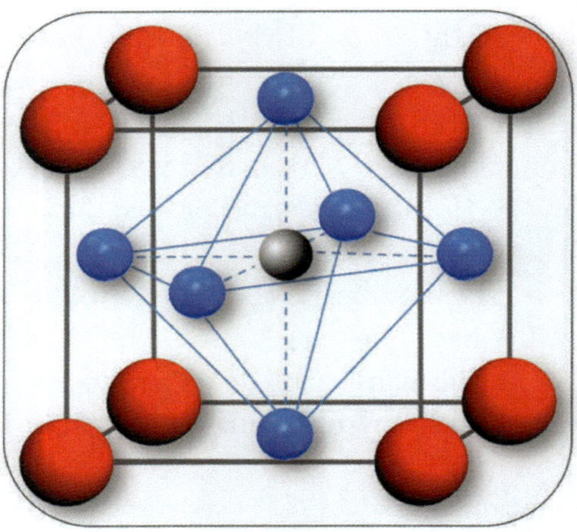

Fig. 9. Perovskite crystal structure (ABO_3). Large red (corner) spheres: A cation, small grey (center) sphere: B cation, and small blue (face) spheres: oxygen ions.

Source: Reproduced from Ref. 119 with permission from The Royal Society of Chemistry.

spheres also called site A for the large cation and site B for the smaller cation, respectively. Alkaline metals, alkaline earth metals, or rare earth metals, such as M^{+} (Na, K, etc.), M^{2+} (Ca, Sr, Ba, etc.), or M^{3+} (La, Fe, etc.), can occupy the A site while transition metals or rare earth elements like M^{5+} (Nb, W, etc.), M^{4+} (Ce, Ti, etc.), or M^{3+} (Mn, Fe, etc.) can occupy the B site. The following criteria need to be satisfied in order to form a stoichiometric perovskite oxide with the ion couple A^{m+} and B^{n+}:

(1) Sum of m and n of the oxidation numbers of the cations should be 6;
(2) Radius of the cation must be compatible with the geometry of the site it occupies.

In the structure shown in Fig. 9, the coordination number of A^{m+} is 12, whereas that of B^{n+} and O^{2-} is 6. The radius of the two cations must satisfy

$$r_a + r_o = t\sqrt{r_b + r_o}, \quad (5)$$

where r_a is the radius of A^{m+}, r_b is radius of B^{n+}, r_o is the radius of O^{2-}, and t is Goldschmidt tolerance factor.[114] The perovskite structure is inherently flexible as evidenced by the variety of distortions of the crystal lattice (cations displaced from the centers of their coordination polyhedral, tilting of the octahedra, and Jahn–Teller distortions) from the regular cubic structure.[114] Hence, perovskites can adopt different crystal systems like cubic, hexagonal, rhombohedral, orthorhombic, totetragonal, or monoclinic.[114]

2.2.2. *Protonic Defects, Doping, and Transport Mechanism*

Protonics is a common term which refers to the physics and chemistry of protons.[122] Although several compounds such as oxides, phosphates, borates, and silicates may conduct protons at temperatures above 300°C, protons are not considered structural components.[114] Instead, they appear as hydroxide defects, also referred to as protonic defects, with a positive charge denoted by $OH_O^{\bullet}$ in Kröger–Vink notation. The subscript O defines the site in the perovskite lattice (in this case the oxygen site) which is occupied by the specie OH, and the superscript is the difference between the real charge of the specie OH and the charge of the O specie ($\times$ is null, $\bullet$ is positive and $'$ is negative). Thus, a vacancy in an oxygen site in Kröger–Vink is given by $v_O^{\bullet\bullet}$. Protons from hydrogen and water vapor can be incorporated in the crystal structure when oxides are reduced via Reaction 1 and Reaction 2, respectively.[119] Reaction 1 occurs when the gas is dry hydrogen, but examples in the literature proving this reaction to be important are scarce.[123] Instead, Reaction 2, also known as the Stotz–Wagner equation,[7] is considered the most important reaction for the formation of protonic defects; water vapor dissociates into a proton which can form a covalent bond with the oxygen in the lattice and a hydroxide ion which fills an oxygen vacancy.[119]

$$\text{Reaction 1: } H_{2(g)} + 2O_O^{\times} \leftrightarrow 2OH_O^{\bullet} + 2e',$$

$$\text{Reaction 2: } H_2O_{(g)} + v_O^{\bullet\bullet} + O_O^{\times} \leftrightarrow 2OH_O^{\bullet} + 2e'.$$

As seen from Reaction 2, oxygen vacancies are necessary for the formation of protonic defects. Consequently, undoped oxides of the perovskite type, like $BaCeO_3$, show low proton conductivity. Extrinsic oxygen vacancies can be created in the $A^{2+}B^{4+}O_3$ perovskite by substituting the B^{4+} cation with a trivalent cation M^{3+} (Y, Gd, Sc, In, etc.).[114,115,119] Dopants which favor proton mobility in the oxide decrease in the order: Y > Gd > In > Sc.[118] The doping of the B cation with M^{3+} is given by Reaction 3 in Kröger–Vink notation.

$$\text{Reaction 3: } 2B_B^{\times} + O_O^{\times} + M_2O_3 \leftrightarrow 2M_B' + V_O^{\bullet\bullet} + 2BO_2.$$

Thus, the general formula can be modified as $AB_{1-x}M_xO_{3-\delta}$, where δ is the oxygen deficiency per unit cell. The concentration of the oxygen vacancies can be related to the dopant concentration by Eq. (6).[124,125] If the A^{2+} cation is doped with a trivalent cation, then oxygen vacancies are consumed instead of being formed.

$$[V_O^{\bullet\bullet}] = \frac{1}{2}[M_B^{3+}]. \tag{6}$$

The transport of protons through the oxide occurs in two steps: rotational motion of the proton around the oxygen followed by hopping to a neighboring oxide ion. In this process, the oxygen ion might undergo some displacement about its mean position in the crystal lattice, while only the proton is likely to diffuse longer distances. This mechanism of proton transport is termed as the Grotthuss mechanism[26] and is illustrated by quantum molecular dynamics simulations in Fig. 10. Experimental data as well as computer simulations have confirmed that the rotational motion of the proton is fast with a low activation energy barrier, with the proton transfer step being rate-limiting.[126–134]

2.2.3. *Cerates Versus Zirconates*

High-temperature proton conductors have been extensively studied during the past three decades. Yttrium-doped barium cerate ($BaCe_{1-x}Y_xO_{3-\delta}$, BCY), and yttrium-doped barium zirconate ($BaZr_{1-x}Y_xO_{3-\delta}$, BZY) continue to be considered amongst the most promising ones. The cerates are among those with the highest proton conductivity ($\sigma \geq 10^{-2}\,\mathrm{S\,cm^{-1}}$ at 650°C[135–143]). However, at

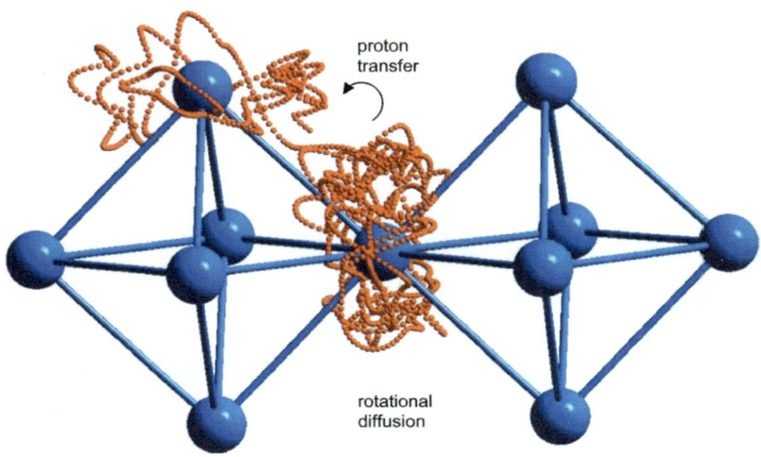

Fig. 10. The path of a proton (orange) in $BaCeO_3$ during the two steps of the Grotthuss mechanism. Reprinted from Ref. 126, Copyright (1996), with permission from Elsevier.

higher temperatures, they begin to conduct oxygen ions as well which reduces their proton-transference number (ratio of proton conductivity to total electrical conductivity).[135,143] Also, highly basic oxides like barium cerates and strontium cerates react with acidic gases like CO_2 and water vapor to form carbonates and hydroxides,[140,142,144–149] resulting in a loss in mechanical integrity of dense pellets.[150] $BaCeO_3$ reacts with pure CO_2 below 1150°C and with water vapor ($p_{H_2O} = 1$ atm) below 400°C.[140,144] $BaCeO_3$ doped with 10 mol% Y showed similar behavior in a pure CO_2 environment.[148]

On the other hand, zirconates offer greater chemical stability in the presence of acidic gases like CO_2 and water vapor. Nevertheless, they exhibit much lower proton conductivity as compared to the cerates. At the same temperature, sintered barium zirconate pellets generally have a conductivity which is one order of magnitude smaller than that of sintered barium cerate pellets.[118,138,142,151] The low conductivity can be explained by the poor grain boundary conduction. Space charge layers are present around the positively charged grain-boundary core, generating a depletion of the protonic defects.[152–157] Because of its refractory nature, barium zirconate requires very high sintering temperatures (above 1700°C), and its microstructure results in a small average grain size and consequently large grain-boundary volume content. Thus, poor grain-boundary conduction coupled with large grain-boundary volume is detrimental to its use as a proton conductor. One strategy investigated to increase the total conductivity of BZY was using sintering aids to promote densification; i.e., a small amount of an oxide such as ZnO,[158,159] NiO,[160] CuO,[161] etc. is added to the ceramic.

Increasing the concentration of the dopant Y in BCY (5–30 mol%) and BZY (5–15 mol%) results in an increase in the total conductivity of the material due to a larger oxygen-vacancy concentration.[114] But at temperatures > 400°C, as the Y concentration increases beyond 20 mol%, the total conductivity is found to decrease for both systems. Hindered mobility coupled with a decrease in the charge-carrier concentration outweighs the benefits of increased defect concentration.[162,163]

2.2.4. $BaZr_xCe_yY_{1-x-y}O_{3-\delta}$ (BZCY) and Solid-state Reactive Sintering (SSRS)

The lesson learned from the early research is that BCY and BZY have properties that are complementary to each other. Thus, one approach to develop new electrolyte compositions is using the properties of BZY and BCY in the same material. It is possible that a novel material consisting of some composition of $BaZr_xCe_yY_{1-x-y}O_{3-\delta}$ (BZCY) would benefit from the high chemical stability of BZY and the high conductivity of BCY, thereby offering a quality compromise between the two materials. BZCY has been recognized as the ideal proton-conducting ceramic for the past few years,[118,164] with several BZCY compositions having been investigated by different authors.[138,142,165–171] Yet, although a complete solid solution of barium cerate and barium zirconate can be fabricated for $0 \leq x \leq 1$, increasing the concentration of Zr beyond 50 mol% made it very difficult to sinter the BZCY; a dense polycrystalline ceramic could not be formed using traditional sintering techniques.[113]

One issue is that the use of very high sintering temperatures, required to form a dense ceramic, leads to barium evaporation. Stoichiometric differences induced by the barium deficiency resulted in a lower protonic conductivity for both bulk and grain-boundary conduction. Sacrificial powder with excess Ba precursor was shown to be an effective way to limit the barium evaporation.[172] Alternatively, these problems were alleviated with the development of solid-state reactive sintering (SSRS).[165,173,174]

The first step in the traditional sintering method consists of a solid-state reaction between carbonates and oxides used to form fine, single-phase BZCY powder (Reaction 4, in the case of a 10% Y doping on the B-site of the perovskite). This is then followed by milling the powder and then compacting to form green bodies ready for sintering at high temperatures. SSRS, on the other hand, simply combines all discrete steps of traditional synthesis and sintering into one processing step. The carbonate and oxide precursor powders are mixed with binders and a sintering additive (e.g., NiO, CuO, or ZnO, usually 1 wt.%) before being compacted into a green body and fired, causing the solid-state reaction to occur while the final product is simultaneously formed. Traditional sintering aids promote transport along the grain boundaries, whereas the metal oxides in SSRS play an important role in transport not only along the grain boundaries but also within the grains themselves. Different geometries including rods, tubes, and pellets can be fabricated by SSRS.[165,175,176]

$$\text{Reaction 4: } BaCO_{3(s)} + yCeO_{2(s)} + (0.9 - y)ZrO_{2(s)} + 0.05Y_2O_{3(s)} \xrightarrow{\geq 1300^\circ C} BaZr_{0.9-y}Ce_yY_{0.1}O_{3-\delta(s)} + CO_{2(g)}.$$

In most of the recent works on BZCY, 10 mol% of Y was used as a dopant.[113,165,176–178] Ricote *et al.*[165] studied the effect of varying the cerium content ($0 \leq y \leq 0.4$) on the conductivity of the BZCY between 300°C and 800°C when fabricated by SSRS at 1500°C using a mass fraction of 1% NiO as a sintering additive. The total conductivity was found to increase with cerium content due to the higher basicity of the Ce–O bond.[179,180] They also investigated using the non-toxic precursor barium sulfate instead of barium carbonate to fabricate BZCY and reported similar properties for the BZCY fabricated by using either of the Ba precursors.[165]

2.2.5. *Hydrogen Galvanic Flux and Its Measurement*

The H_2 flux dependence on the membrane thickness has been investigated for $SrCe_{0.95}Tm_{0.05}O_{3-\delta}$[181] and $SrCe_{0.95}Yb_{0.05}O_{3-\delta}$.[182] Cheng *et al.*[181] measured fluxes from 150 μm to 800 μm $SrCe_{0.95}Tm_{0.05}O_{3-\delta}$ membranes and found that the thinnest membrane had the highest flux. Hamakawa *et al.*[182] showed that for a supported $SrCe_{0.95}Yb_{0.05}O_{3-\delta}$ membrane, the H_2 flux was inversely proportional to thickness down to 2 μm. Song *et al.*[183] arrived at the same conclusion based on their investigation of Ni-supported $BaCe_{0.8}Y_{0.2}O_{3-\delta}$ membranes down to a thickness of 80 μm. This behavior is similar to that seen for thin, supported dense metal membranes and is indicative of bulk diffusion being the rate-limiting step of hydrogen transport.[183,184]

A summary of the H_2 fluxes for different proton-conducting ceramics was compiled in a table by Rosensteel *et al.*[185] that compared different ceramics to a palladium membrane doped with trace amounts of ruthenium. Most proton-conducting ceramic membranes are two to three orders of magnitude thicker than palladium-based membranes, but work is ongoing to reduce the thickness and improve the hydrogen flux through these membranes. Thin membranes (ca. 25 μm) lack the mechanical integrity of their thicker counterparts, which requires them to be supported. Coors and Manerbino[175] developed a cermet (ceramic–metal composite) consisting of 65 wt.% of reduced NiO and $BaZr_{0.7}Ce_{0.2}Y_{0.1}O_{3-\delta}$ on which a 30 μm $BaZr_{0.7}Ce_{0.2}Y_{0.1}O_{3-\delta}$ was deposited (Fig. 11, Ref. 175). They found that the Ni phase and the ceramic phase were interpenetrating and percolated throughout the cermet.

The theoretical value of hydrogen flux at a given current density (i, A cm^{-2}), also known as the theoretical Faradaic flux, can be

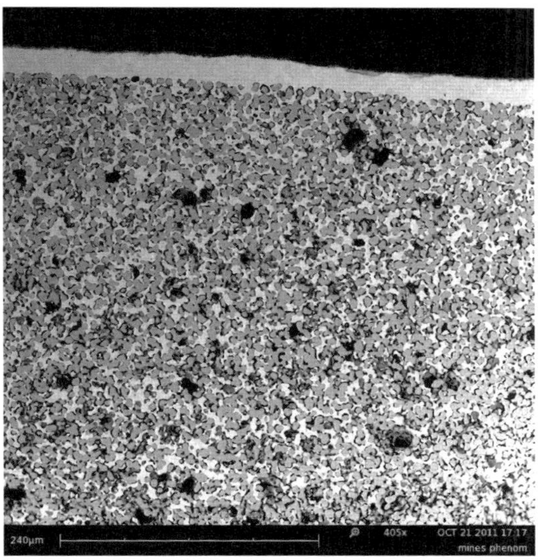

Fig. 11. Polished cross-sectional SEM image of reduced NiO and $BaZr_{0.7}Ce_{0.2}Y_{0.1}O_{3-\delta}$ cermet with a 30 μm $BaZr_{0.7}Ce_{0.2}Y_{0.1}O_{3-\delta}$ membrane. Light grey regions are $BaZr_{0.7}Ce_{0.2}Y_{0.1}O_{3-\delta}$, the dark grey regions are Ni, and the black regions are pores.

Source: Image courtesy of Dr. Anthony Manerbino — CoorsTek, Inc. Golden, CO, USA.

calculated from Eq. (7)[178]; where n is the number of electrons transferred per hydrogen molecule ($n = 2$), and F is Faraday's constant ($F = 96485\,\mathrm{C\ mol^{-1}}$). The Faradaic efficiency is the ratio of the measured flux to the theoretical Faradaic flux. $BaZr_{0.7}Ce_{0.2}Y_{0.1}O_{3-\delta}$ (BZCY72) fabricated by SSRS showed Faradaic efficiency as high as 98.6% of the theoretical value at 775°C and $0.8\,\mathrm{A\,cm^{-2}}$ during the galvanic pumping of hydrogen (i.e., hydrogen transport driven by a potential gradient) with no detectable impurities.[178]

$$j_{H_2} = \frac{i}{nF}. \tag{7}$$

The hydrogen flux for a $30\,\mu\mathrm{m}$ BZCY72 membrane at $0.8\,\mathrm{A\,cm^{-2}}$ and 98.6% faradaic efficiency is $0.04\,\mathrm{mol\,m^{-2}\,s^{-1}}$, which is an order of magnitude lower than that for thin (ca. $5\,\mu\mathrm{m}$) Pd-composite membranes. These membranes have been run for short periods at current densities up to $1\,\mathrm{A\,cm^{-2}}$, but their long-term performance still needs to be investigated. The limiting factors to increasing the current density lie in the range of power controllers used and the mechanical stability of the ceramic at high current densities. It is believed that the current density can be increased to $2\,\mathrm{A\,cm^{-2}}$ without compromising the ceramic. At higher current densities, mechanical stress causes fractures in the BZCY that shorts the circuit. Thus, there is scope for further improvements.

BZCY72 is a mixed ionic–electronic conductor (MIEC) and can have up to four different types of charge carriers depending on oxygen partial pressure and temperature.[113] Fortunately, electronic defects, which could result from oxidation of the ceramic, are not significant because most hydrogen-separation membranes are operated at low oxygen partial pressure. When BZCY72 is in a reducing environment, protons ($\mathrm{OH}_O^{\bullet}$), oxygen-ion vacancies ($v_O^{\bullet\bullet}$) and reduced cerium ions (Ce_B') are the charged species present. However, under milder reducing environments, such as a small amount of steam in hydrogen, reduced cerium cations exist only in extremely low concentrations. Hence, for practical purposes BZCY72 may be regarded as a co-ionic conductor (CIC).[186] Thus, protons and oxygen-ion vacancies are considered the only charge-carrying species present. A novel method called stoichiometric titration was developed[113] to measure the hydrogen flux through the BZCY72 but can be used for other ceramic proton conductors as well. BZCY72 has been studied for applications like methane steam reforming, ammonia synthesis, and gas-to-liquid fuels processing via methane dehydrogenation,[178,187–189] as will be discussed further in Sec. 3.

3. Hydrogen-Separation Applications

In general, hydrogen-transport membranes can be used anywhere it is desired to selectively transfer hydrogen from one gas stream to another. The reformation of hydrocarbons, for instance, is the primary industrial source of hydrogen, and significant work has been performed on the purification of hydrogen from such gas streams. Traditionally, the reforming occurs within a reactor before the gas stream flows into a separation process. This type of separation, however, results in direct competition with the existing pressure swing adsorption (PSA) technologies that have been developed over the last century. These PSA systems are highly refined and achieve both high energy efficiencies and hydrogen purities, making direct substitution unlikely to be economical. Moreover, despite their complexity, PSA processes produce hydrogen at near-feed pressure, reducing the need for recompression of the product hydrogen that reduces operating cost. For this reason, the primary future applications of hydrogen-transport membranes should focus on areas in which PSA cannot compete. In this section, three primary areas will be introduced that have a significant opportunity for efficiency and recovery enhancements via the implementation of hydrogen-transport membranes.

The first two sections will introduce the topic of catalytic membrane reactors (CMRs) in which the reaction and hydrogen-transport stages can be combined into one compact unit. CMR concepts using hydrogen-permeable membranes can generally be broken into two main subgroups: hydrogenation and dehydrogenation reactors. Specifically, hydrogenation reactors use membranes to supply hydrogen to a reacting environment whereas dehydrogenation reactors simultaneously remove the hydrogen from the reaction zone as it is formed. First, dehydrogenation membranes will be introduced, and the topics of steam methane reforming (SMR) and ammonia decomposition will be explored. Next, CMR designs involving hydrogenation reactions will be overviewed, with particular interest on the effects of permeating hydrogen on a reacting system. Lastly, the concept of using hydrogen-transport membranes for the separation of hydrogen isotopes will be explored. Membranes and CMRs play a critical role in the controlled fusion fuel cycle as well as being an available option for tritium separation and purification from typical fission nuclear power reactors.

3.1. *Dehydrogenation Membrane Reactors*

The vast majority of work on dense metal membrane reactors has been conducted on dehydrogenation reactions wherein a fuel, such as methane or ammonia, is cracked over a catalyst while the hydrogen produced is simultaneously removed from the system. Although a limited number of studies have used dense ceramic membranes on dehydrogenation reactions, the technique of applying a voltage has garnered far more attention for hydrogenation CMR concepts such as ammonia synthesis. Nevertheless, the CMR design for dehydrogenation, whether metal- or ceramic-based, offers three major benefits over traditional reactor configurations: (1) conversions can exceed equilibrium limitations, (2) similar conversions can be attained at milder operating conditions, and (3) capital cost can be reduced by combining the reaction and separation stages into a single unit.[190–192]

A generic CMR-design concept for dense metal membranes can be seen in Fig. 12. In this case, the concept of a shell-and-tube design for the decomposition of ammonia has been illustrated. Here, ammonia is fed to the catalyst bed where it is separated into hydrogen and nitrogen near the beginning of the permeable Pd-membrane region. As the ammonia decomposes over the catalyst bed, the hydrogen is simultaneously removed by permeation through the membrane and collected as a pure hydrogen product. The nitrogen, along with any unreacted ammonia or unseparated hydrogen, is meanwhile collected in the retentate stream.

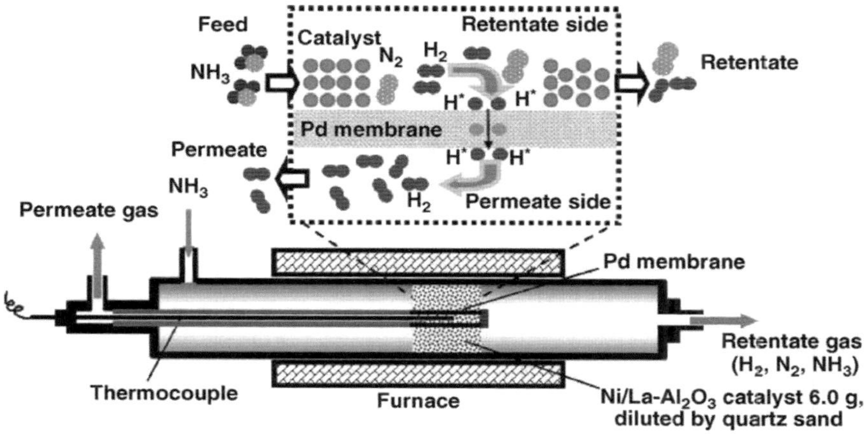

Fig. 12. Catalytic membrane reactor configuration for ammonia decomposition. *Source*: Reprinted from Ref. 193 with permission from Elsevier.

Although the feed here is ammonia, potentially any hydrogen-containing chemical can be fed into such a system. The chemical(s) can then be dehydrogenated while the hydrogen product is separated through any hydrogen-separation membrane, including any of the dense metal or dense ceramic membranes discussed above. At the end of the reaction/permeation zone, the by-product(s) exit the reactor on the shell side, and the hydrogen exits on the tube side. In this manner, the hydrogen from any particular feed source can be simultaneously generated and separated at high purity. The interested reader is referred to a recent review[194] on CMR designs which contains significantly more detail on several possible geometries and configurations.

Particular emphasis in this section will focus on methane and ammonia as hydrogen-generation fuels because of methane's natural abundance and ammonia's potential role in a carbon-free fuel economy. However, several recent reviews have detailed the use of such a reactor with Pd-based membranes using several alternative fuels including the reformation of methanol and ethanol,[195,196] as well as within water-gas shift (WGS) reactors.[197,198] In addition, information on the early dehydrogenation work using CMR processes can be found in the 2001 review by Dittmeyer *et al.*[199] This review includes several examples of dehydrogenation including cyclohexane to benzene, methylcyclohexane to toluene, alkanes to olefins, and ethylbenzene to styrene.

3.1.1. *Proton-conducting Ceramic Applications*

With regard to the proton-conducting ceramic cell reactors (PCCRs), only a limited number of studies have been performed; nevertheless, they include several notable reactions. For instance, PCCRs have been investigated as potential alternatives to traditional gas phase reactions between methane and oxygen,[200–203] wherein the use of proton-conducting ceramics allows dimerization of methane via dehydrogenation rather than partial oxidation. The reactions assumed to occur at the anode of a PCCR could be given by Reaction 5 or Reaction 6, while that at the cathode could be given by Reaction 7.

$$\text{Reaction 5: } 2CH_{4(g)} \leftrightarrow C_2H_{6(g)} + 2H^+ + 2e^-,$$

$$\text{Reaction 6: } 2CH_{4(g)} \leftrightarrow C_2H_{4(g)} + 4H^+ + 4e^-,$$

$$\text{Reaction 7: } 2H^+ + 2e^- \leftrightarrow H_{2(g)}$$

Mori and coworkers[204,205] were the first to report on the dimerization of methane in a PCCR in 1987, using $SrCe_{0.95}Yb_{0.05}O_{3-\alpha}$–Pt as the electrolyte–electrode combination, respectively. Iwahara *et al.*[206] passed a mixture of steam and methane over the anode of their PCCR that resulted in partial conversion of methane to C_2 hydrocarbons. The hydrogen reacted with oxygen at the cathode to be converted into water. Since then multiple research groups[207–216] have studied the conversion of methane to C_2 hydrocarbons using PCCRs, but in all cases the reported yields were less than 2%. The low yields could be partly explained by the methane pyrolysis reaction given in Reaction 8.

$$\text{Reaction 8: } CH_{4(g)} \leftrightarrow C_{(s)} + 2H_{2(g)},$$

$$\text{Reaction 9: } CO_{(g)} + H_2O_{(g)} \leftrightarrow CO_{2(g)} + H_{2(g)}.$$

Coors[217,218] studied steam methane reforming by operating a PCCR in fuel cell mode. Methane was fed on the anode side while steam was fed to the cathode side; permeation of steam from the cathode to the anode eliminated the need for co-fed steam and methane on the anode side. Another reaction of interest, the water-gas shift (WGS) reaction given by Reaction 9, is a key step in separating hydrogen from syngas ($CO+H_2$) streams obtained from either steam methane reforming or partial oxidation of hydrocarbons. The advantage of using PCCRs is two fold: selective removal of hydrogen from the reactant stream and shifting of equilibrium toward the products resulting in higher conversions.[218,219]

Dehydrogenation of alkanes to form alkenes with simultaneous separation of hydrogen has also been investigated. Iwahara *et al.*[206] studied the performance of a PCCR that converted ethane to ethylene. They found that addition of water vapor to the reactant stream resulted in better performance. Similar results were seen in the studies by Karagiannakis and coworkers[220,221] on a PCCR to convert propane to propene with hydrogen separation. Using a propane–steam mixture resulted in hydrogen recoveries of up to 90%, which was considered an improved performance over using pure propane as the reactant. This improved performance was believed to be a direct result of the increased faradaic efficiency due to the presence of steam.[221]

In the last couple of years, CoorsTek Inc., in collaboration with Protia AS, has evaluated the novel gas-to-liquid concept by utilizing the methane dehydro-aromatization (MDA) reaction (Reaction 10) in a catalytic membrane reactor (CMR) prototype. BZCY72 was used as the proton-conducting ceramic membrane to convert methane to benzene with simultaneous separation of hydrogen at 700°C and atmospheric pressure. The MDA-CMR was operated galvanically and achieved a faradaic efficiency of 90% while pumping $2.6 \times 10^{-6}\,\mathrm{mol\,m^{-2}\,s^{-1}}$ of hydrogen.[114] Work is ongoing to study the performance on a larger scale and to reduce the capital costs associated with manufacturing of the MDA-CMR module.

$$\text{Reaction 10: } 6CH_{4(g)} \leftrightarrow C_6H_{6(g)} + 6H_{2(g)}.$$

3.1.2. *Steam Methane Reforming*

Worldwide hydrogen production is estimated at around 50 million metric tons per year, with the majority being used on-site for ammonia synthesis and hydrotreatment.[222–224] It is estimated that about 50% of this hydrogen is produced via steam methane reformation (SMR), whereas water electrolysis accounts for only a minor fraction of production, and the remaining fraction is a combination of partial oxidation and coal gasification.[225,226] Although the U.S. Department of Energy (DOE) states that reformation of hydrocarbons should not be a long-term goal because it does not eliminate CO_2 emissions, the U.S. Energy Information Administration (EIA) predicts that the use of carbon-based fuels will continue to increase over the next 30 years. Specifically, the EIA predicts[227] that in 2040, natural gas will continue to fill about 23% of the world's energy demand and will simultaneously increase 70% in total quantity as compared to levels in 2010. This suggests that improvements in efficiency for methane reformation still have the potential for long-term impact on reducing the total quantity of fuel required — even if it is not the most desirable hydrogen source.

The conventional process for SMR consists of an upstream reactor run at 800°C or higher to convert methane to hydrogen and carbon monoxide. The endothermic reaction results in an increase in the number of moles of gas, meaning that the forward reaction favors high temperature and low pressure.

Reaction 11:

$$CH_{4(g)} + H_2O_{(g)} \leftrightarrow CO_{(g)} + 3H_{2(g)} \quad (\Delta H^0 = 206\,\text{kJ/mol}).$$

This product stream is then sent to additional reactors downstream to perform the exothermic water–gas shift (WGS) reaction at a lower temperature of 300°C–400°C.

Reaction 12:

$$CO_{(g)} + H_2O_{(g)} \leftrightarrow CO_{2(g)} + H_{2(g)} \quad (\Delta H^0 = -41\,\text{kJ/mol})$$

However, the output stream still consists of a gas mixture and must therefore be separated. A common method employed is pressure swing adsorption (PSA), but this system adds significant cost to the overall construction of a plant. In terms of using membranes as a replacement to PSA, recent publications by Iulianelli *et al.*[197] and Cornaglia *et al.*[198] provide detailed descriptions of the recent advancements and remaining challenges in the use of Pd-based membranes. Additionally, Basile *et al.*[228] include a description of the various catalysts being used in the recent SMR literature results.

As mentioned previously, one method for simplifying and improving the process is to run the reaction in a membrane reactor configuration. If conducted as a CMR, Reactions 11 and 12 take place simultaneously:

Reaction 13:

$$CH_{4(g)} + 2H_2O_{(g)} \leftrightarrow CO_{2(g)} + 4H_{2(g)} \quad (\Delta H^0 = 165\,\text{kJ/mol})$$

Because the hydrogen is continually removed from the system as it reacts, equilibrium is shifted toward the production of hydrogen, and overall conversions can approach 100%. If either a dense metal or proton-conducting ceramic membrane is used, then there is an added benefit that the permeate stream can potentially be 100% pure hydrogen because these materials are theoretically perfectly selective to hydrogen transport.

To understand the extent of the equilibrium shift, a summary of SMR results is shown in Fig. 13. Here, results for both catalytic membrane reactor (CMR) and traditional reactor (TR) configurations are given as either open or closed symbols, respectively. In addition, the equilibrium shift for a TR configuration was calculated for a reactor pressure of 0.1 MPa. Because the reaction increases the number of moles of gas, Le Châtelier's principle states that equilibrium conversion will decrease if the pressure is increased. This means that the line at 0.1 MPa can be considered an upper limit, regardless of reactor pressure for the plotted results, because any increase in pressure would reduce the expected conversion. As such, the benefits of CMR designs are clearly evident.

Whereas all of the TR results appear below the equilibrium shift line, the use of a CMR raises the conversion well above equilibrium

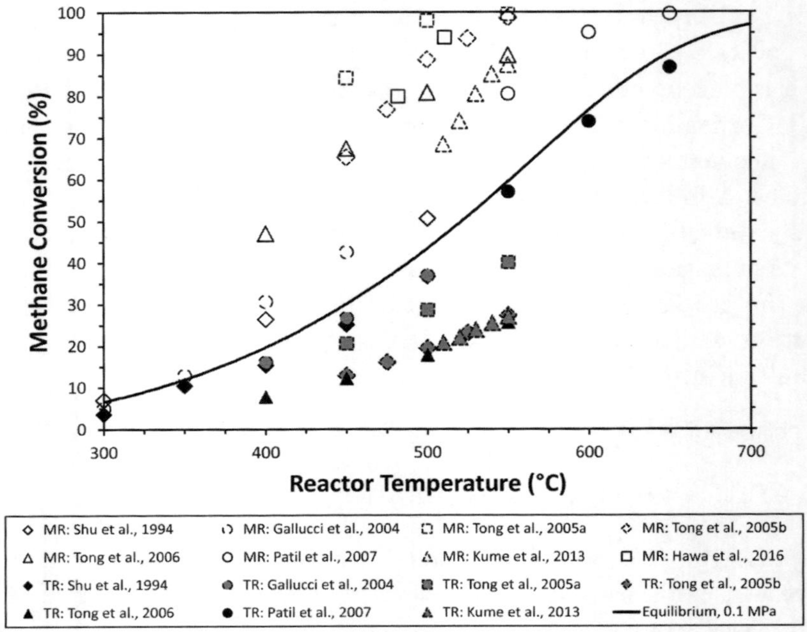

Fig. 13. Methane conversion vs. reactor temperature for both traditional (TR) catalytic and membrane (CMR) reactors.
Source: Data from Refs. 229–233.

for temperatures above 350°C — sometimes approaching 100%. In fact, conversions exceeding 90% can be seen at 500°C, whereas equilibrium would need to be above 650°C to match. This offers abundant evidence to conclude that CMR designs will allow for similar conversions at significantly lower operating temperatures than TR designs.

In addition to allowing the simultaneous recovery of a pure hydrogen stream, CMR designs also tend to follow the opposite trend as equilibrium in reference to pressure. Figure 14 shows the result of changing reactor pressure for several literature results, where it is evident that higher reactor pressures seem to increase, rather than decrease, methane conversion. In this case, results are included for temperatures less than 550°C, and the equilibrium limit again suggests an upper limit because decreasing reactor temperatures below

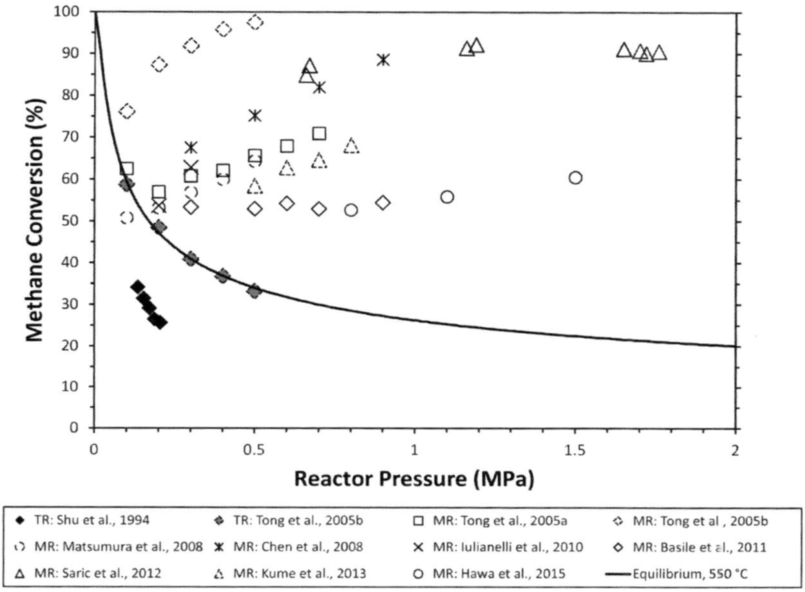

Fig. 14. Methane conversion vs. reactor pressure for both traditional (TR) and catalytic membrane (CMR) reactors.
Source: Data from Refs. 229, 233–241.

the calculated line will only depress the limit. However, in contrast to the decreasing equilibrium limit with pressure, literature results for CMR designs tend to see an increase in conversion with increasing pressure. This can be explained by the membrane's driving force. When the pressure of the reactor is increased, the driving force for hydrogen permeation increases, and thus the rate of hydrogen removal becomes faster. As hydrogen is removed, the reaction is pushed toward complete conversion, so increasing pressure actually offsets the decrease in local equilibrium conversion because the increase in hydrogen removal is faster than the decrease in the equilibrium conversion limitation.

Theoretically, once a full conversion has been achieved and all hydrogen has been removed, the effluent stream from the reactor consists of nothing but CO_2 and H_2O. This means that CO_2 sequestration could be simplified because the water can easily be condensed leaving nothing but a pure CO_2 stream. In essence, a well-optimized system could potentially lead from a CH_4 feed stream to both high purity H_2 and high purity CO_2.

An example of a well-optimized system for a Pd-alloy-based SMR configuration would be the system designed by Tokyo Gas.[242] In collaboration with NGK and Mitsubishi Heavy Industries, they developed a $40\,Nm^3\,h^{-1}$ hydrogen production plant with a purity >99.99% after 5000 h of testing. A methane conversion of 85% and overall conversion efficiency (HHV) of 81% were achieved, whereas the efficiency of a traditional SMR-PSA system is less than 70%.[242] While this system utilized a thick Pd-foil design, an alternative design, termed membrane-on-catalyst (MOC), has been under development which includes a Ni catalyst embedded in a YSZ support with a 6–10 μm Pd-alloy film deposited around the outside surface.[243] This design has been shown to produce >99.998% purity for >1000 h under SMR conditions. In addition, the effluent gas from the reactor contained 90% CO_2 when operated at 28% load, and a CO_2 separation and storage system attached to the reactor was able to reduce emissions by 50% with only a 3% loss in energy efficiency.[244]

Outside of the long-term durability test of Tokyo Gas, only a handful of other results are available. Recently, Iaquaniello *et al.*[245] tested a pilot-scale system designed for a production of about $20\,Nm^3\,h^{-1}$ for over 1300 h, but they did not report the specific purity of the permeate — merely mentioning that their system demonstrated reliability. Sarić *et al.*[240] reported on the long-term stability of a pure Pd CMR, but observed that purity declined during operation with a final value near 80% after 1100 h of testing. However, our recently published results[246] have suggested that Pd–Au may also be a viable candidate alloy for long-term stability in SMR environments. These results, shown in Fig. 15, found purity to remain above 99% for over 500 h of testing when using a 28% by mass Pd–Au membrane. Although a purity of 99% is still below the UHP quality desired, this result did suggest that improvements in initial fabrication might

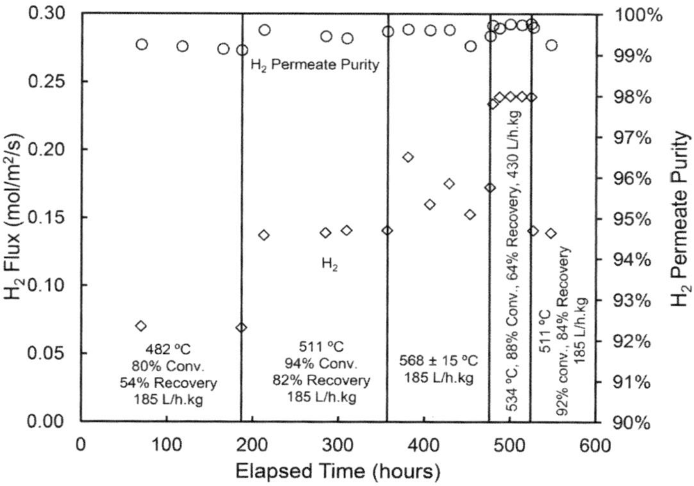

Fig. 15. SMR results for a PdAu CMR.
Source: Reprinted from Ref. 246 with permission from Elsevier.

provide a means of creating very high and stable purities with a Pd-based CMR.

Alternative membranes to Pd and its alloys have also recently been studied under SMR but represent only a small portion of the literature. Dolan *et al.*[247] attempted to use V-based membranes but encountered issues with the expansion of V when exposed to hydrogen. This expansion was blocked by the packed catalyst and caused the membranes to crack, suggesting future designs must allow for expansion of the vanadium if they are to survive the reactor startup. Meanwhile, Kyriakou *et al.*[189] were able to successfully operate a BZCY proton-conducting ceramic membrane in an SMR environment and saw an enhanced methane conversion and hydrogen recovery of up to 50%. The two important benefits of using BZCY membranes are that it limits the use of noble metals — primarily Pd — and these membranes enhance hydrogen recovery through overpotentials rather than pressure gradients, so no pressurization of the reactor is required. However, the hydrogen productivity of the membrane appears to have been less than 0.004 mol m^{-2} s^{-1} for their best case, which remains about an order of magnitude lower than some results for recent Pd-based studies,[248,249] and nearly two orders of magnitude lower than some of the higher Pd-based studies.[240,241] It is possible, however, that even a low productivity would be cost-beneficial given the lack of expensive metals and low compression costs. Thus, further study is necessary, but it will be of great interest to see how proton-conducting membranes develop within this field.

3.1.3. *Ammonia Decomposition*

As a primary ingredient in the fertilizers which help sustain the world's current population,[250] ammonia is arguably one of the most important chemicals produced in modern society. However, there is a second potential application for ammonia which could prove substantial — that of an energy storage material. Ammonia has been suggested as a potential storage medium for hydrogen that is easier to transport, less costly to produce and has a higher energy density than pure hydrogen.[251–254] With a mass fraction of 0.176 hydrogen and volumetric density about 40% of gasoline, ammonia presents itself as a potentially competitive fuel — especially when compared to using compressed hydrogen. Even when compressed to 70 MPa, compressed hydrogen is less than 17% of the energy density of gasoline. Other fuels, such as methanol and ethanol, offer better volumetric energy densities than ammonia (Table 7) but ultimately remain carbon-based. And as far as mass implementation of the ammonia network, the infrastructure and knowledge required for transporting ammonia has been considered advanced enough to meet the demands of an ammonia-based energy economy for decades.[255]

Additionally, in recent advances in ammonia production,[257,258] water, rather than hydrogen, is used as the feedstock for electrolysis which can also be performed at much lower pressures than the traditional Haber–Bosch process. If methane were to be eliminated from the feed requirements, ammonia synthesis and decomposition would represent a complete carbon-free route for fuel cells with storage benefits over direct hydrogen production. However, ammonia-based fuel cells still have a number of limitations and are far from being commercialized.[254] Thus, once synthesized and transported,

Table 7. Higher heating values and energy densities for several potential hydrogen-storage fuels. Calculations based on heat of formation values from NIST webbook[256] with storage at 298 K.

Hydrogen source	Storage pressure (MPa)	HHV (kJ/mol)	H_2 content (g/g)	Gravimetric energy density (MJ/kg)	Volumetric energy density (MJ/L)
Natural gas (methane)	2.5	889	0.250	55.5	0.938
Hydrogen	70.	286	1.00	143	5.61
Ammonia	1.0	383	0.176	22.5	13.5
Methanol	0.1	726	0.125	22.7	18.0
Ethanol	0.1	1300	0.130	28.2	22.2
Gasoline (octane)	0.1	5512	0.158	48.3	33.7

ammonia must be decomposed and separated for the hydrogen to be used efficiently in fuel cells.[253,259]

Compared to other hydrogen-production methods, such as methane reformation, ammonia decomposition is only represented by a handful of literature studies. Chiuta *et al.*[253] published a thorough review on ammonia decomposition from the previous decade in 2013. Yet, they only cite 35 articles for the period 2002 to 2012 that involve ammonia decomposition experiments. These publications were divided into four major devices for ammonia decomposition: microreactors, monolithic reactors, membrane reactors and electrolyzers.

Although the microreactors and monolithic reactors were the most heavily studied, both of these devices result in an output stream of nitrogen and hydrogen even upon full conversion. This means that either the hydrogen needs to be separated or the fuel cell using the hydrogen will suffer an efficiency loss due to the diluted hydrogen stream. As with SMR, however, one method for simplifying and improving the process is to run the reaction in a membrane reactor configuration. In addition to the benefits of a CMR mentioned in the opening of Sec. 3.1, a dense hydrogen membrane reactor (assuming theoretical infinite selectivity) can also eliminate trace amounts of unreacted ammonia which cause degradation in fuel cells even at concentrations as low as 10 ppm.[260]

In this method, a hydrogen-transport membrane is surrounded by catalyst just as in Fig. 12. As the ammonia reaches the catalyst, it is decomposed into hydrogen and nitrogen via the endothermic reaction

Reaction 14:

$$2NH_{3(g)} \leftrightarrow N_{2(g)} + 3H_{2(g)} \quad (\Delta H^0 = 92.4\,\text{kJ/mol})$$

while the hydrogen is simultaneously separated by the membrane. This allows for the production of a pure hydrogen stream for use in a fuel cell, while the nitrogen and trace unreacted ammonia are simply exhausted into the atmosphere.

One of the primary issues with the current literature database is that the ammonia experiments, both experimental and modeled, often include either a sweep gas or vacuum in the permeate.[19,261–266] This is because a sweep stream or vacuum effectively lowers the hydrogen partial pressure on the permeate side of the membrane, maximizing the partial pressure difference and thus increasing the hydrogen-flux. This hydrogen flux increase in turn also increases ammonia conversion. However, this also means that either the hydrogen needs to be separated from the sweep gas in another stage or the reactor system needs to have an additional vacuum system attached which would add cost and complexity to the system. The only simple option is to use steam as a sweep gas, which is easily condensed out of the hydrogen permeate. However, even a steam sweep would add complexity to the system and may not offer significant benefits worth the cost. Therefore, any future industrial application would be unlikely to use such an aid.

Although many studies involved the use of sweep- or vacuum-assisted permeation, several studies exist which collected the permeate stream at atmospheric pressure.[193,267–269] For instance, Zhang *et al.*[193] used a 3 μm thick pure Pd-composite membrane with a Ni catalyst and obtained 93% conversion with 92% recovery at 773 K and 0.4 MPa. However, Zhang *et al.* suggested the membranes were not fully utilized in the combined membrane reactor; instead, they recommended using a higher temperature cracker followed by a separator. In this configuration, they performed a 500 h stability test and achieved 99.95% conversion, 83% recovery, and 99.97% purity with an H_2 productivity of $0.37\,\mathrm{mol\,m^{-2}\,s^{-1}}$ for the duration. The reason for this increase in performance was that the ammonia could be fully converted in the high-temperature cracker, allowing the membrane to see the maximum 75% hydrogen at its inlet rather than needing to operate in parallel with the catalyst. Yet, this may still be undesirable because it again requires a separate high-temperature reactor before the membrane. The concept of the CMR design is to allow for a lower overall temperature, which may increase the life span of the components and allow the use of less costly materials of construction.

A potentially better designed geometry was used by Rizzuto, Palange, and Del Prete[269] who tested a membrane module from REB Research and Consulting and achieved a 99.8% conversion with 87.2% hydrogen yield at 723 K and 0.6 MPa. The authors also stated that the membrane achieved higher than 99.9999% purity under any condition, although specific measurements were not provided. In contrast to the results of Zhang *et al.*[193] and Israni *et al.*,[267] they reported little difference in hydrogen yield above 0.5 MPa. Unfortunately, the description of their proprietary module did not include a membrane area, making the hydrogen productivity of the system impossible to calculate.

Another issue with ammonia decomposition lies in the low vapor pressure at ambient temperature. With a vapor pressure of only 0.8 MPa at 298 K, ammonia reactors are unable to increase pressures to similar levels as reactors for SMR without heating the ammonia reservoir. This essentially prevents a reactor from increasing in pressure to increase hydrogen recovery and potentially the ammonia conversion. Zhang *et al.*,[193] for example, saw a significant enhancement in operation between 0.2 MPa and 0.4 MPa, but they did not go above this pressure, presumably because higher pressures would lower the pressure differential from their source tank and cause issue with mass flow controller operations.

In fact, none of the experimental studies have reported results for feed pressures above 1.0 MPa due to the limitation of the vapor pressure of pure ammonia around 298 K. Israni, Nair, and Harold[267] reported little pressure dependence on conversion for their Pd hollow fiber membrane reactor between 0.1 MPa and 0.5 MPa at 773 K but did see an increase in hydrogen flux through the membrane of about 60% at several flow rates. Di Carlo, Dell'Era, and Del Prete[268] conducted a 3D simulation on a membrane reactor at 1.0 MPa and from 773 K to 873 K and found an increase in conversion over a traditional reactor configuration of 18% at 823 K and 1.0 MPa, but they did not consider simulations for other pressures.

One final note on the CMR references in this section is that several groups studied microporous silica membranes.[19,21,262] Given the limited number of results using Pd-based membranes, microporous silica membrane results were included to aid in the understanding of the benefits of a CMR in terms of a one-step reaction and conversion reactor. However, unlike the solution-diffusion mechanism of dense membranes, microporous silica membranes function based on molecular sieving. Thus, it is only when the pore size is very finely controlled that hydrogen selectivities over nitrogen and ammonia can be high. Of those that reported purity, microporous silica membranes had hydrogen purities of only 71%[19] and 84%,[263] whereas Pd-based reactors obtained purities of 99.2%[267] and 99.6%.[193] Additionally, the 500 h stability test by Zhang *et al.*[193] suggested no performance degradation for Pd-membrane performance, and our group recently reported[270] on a lack of hydrogen flux inhibition by either ammonia- or nitrogen-containing gas streams. Because Pd-based membranes offer much higher purities and have no degradation issues, the current literature suggests them to be the better option to produce very high purity hydrogen from an ammonia decomposition CMR.

3.2. *Hydrogenation Membrane Reactors*

In addition to removal of hydrogen from a system, dense hydrogen-permeable membranes can also be considered as an alternative means of feeding hydrogen to a reacting environment. In fact, several groups have found that permeating hydrogen is significantly more reactive than when co-fed over a Pd membrane — even when this was an undesired result. Okazaki and co-workers,[271–273] for example, observed that their α-alumina supports were reduced to metallic aluminum by permeating hydrogen while testing thin-film Pd membranes as low as 650°C. Considering alumina is often used as a catalyst support, it was not expected that alumina would reduce in hydrogen at this temperature, especially given the lack of an electropotential. Nevertheless, under permeation conditions, Okazaki and coworkers observed that Al alloyed with their Pd-based membranes, which could only occur if the Al_2O_3 was reduced.

One of the potential reasons for this increase in reactivity is thought to be due to the free energy associated with the atomic hydrogen. Dissociation of hydrogen on the feed side provides about 100 kcal mol^{-1} of free energy that is simply lost to heat upon recombination on the permeate side. However, this means that the permeating hydrogen has significant free energy to react with the first surface species they contact on the permeate side of the membrane. The hydrogenation membrane reactor concept, then, is an attempt to force this atomic hydrogen to react with a precursor other than a second hydrogen atom.

Yet, it is not necessary to rely on the free energy of the hydrogen if an electrical bias is used to supply any necessary energy requirement. In the case of proton-conducting ceramics, the applied bias can be modified to drive a reaction at the interface. This section will explore the research on hydrogenation reactions using membranes pertaining to both the nongalvanic dense metal membrane as well as the galvanic proton-conducting ceramic setups.

3.2.1. *Dense Metal Membrane Applications*

As far back as 1966, Rennard and Kokes[274] studied ethylene and propylene hydrogenation over Pd and observed reaction kinetics to be of zero order for gas-phase hydrogen but of first order on the hydride concentration of PdH. They stipulated that, under their conditions, ethylene and propylene dominated the surface coverage of the Pd catalyst, making it difficult for the gas-phase hydrogen to adsorb onto the surface. Because of this, neither the olefin nor the hydrogen gas-phase pressures had an effect on the rate of reaction. However, it was observed that increasing the hydride content of the initial Pd had a first-order effect on the kinetics of hydrogenation. Although the authors at the time attributed this to possible electronic changes in the Pd sample, an alternative explanation is that the desorbing hydrogen from the Pd bulk simply did not need to compete with the olefin for a surface site and may have reacted from the bottom up. Essentially, this method of providing hydrogen atoms from the interior of the Pd has the capability of being more reactive than that of gas-phase hydrogen because it bypasses the competitive adsorption on the surface.

To begin, a schematic of a generic CMR for hydrogenation can be explored (Fig. 16). First, hydrogen is fed to the membrane causing it to adsorb, dissociate, and diffuse through the membrane. Once it diffuses to the far surface, however, the novelty of the CMR design is that hydrogen may be coerced into reacting with any prevalent surface species. Here, the surface acts as a catalyst for the hydrogenation of a precursor that was simultaneously fed on the catalytic side of the membrane. Once the hydrogen atoms diffuse through the bulk, they essentially run into any adsorbed species on the surface and, in their highly reactive atomic form, immediately hydrogenate any species on the surface.

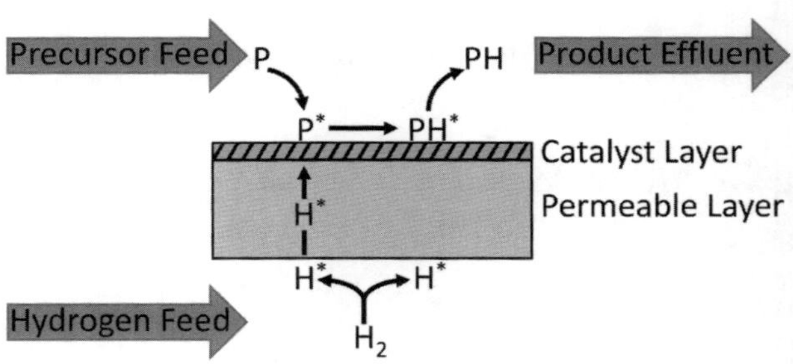

Fig. 16. Schematic of hydrogenation mechanism on a dense metal membrane. (P) is the precursor, (H) is a hydrogen atom, and (PH) is the hydrogenated product. (*) denotes an adsorbed or absorbed species.

Although several hydrogenation reactions proceed quickly using a pure Pd surface, the permeate side of the membrane could potentially be covered with a thin-film of nearly any metal to provide a good catalyst for a specified reaction. Much of the early work occurred before supported thin-film membranes were readily being fabricated in laboratories, so pure Pd (or a Pd alloy) was often the catalyst of choice. However, current sputtering technologies would allow for very thin (<100 nm) coatings of nearly any metal to be deposited on a Pd surface.

Within the Pd-CMR field, Gryaznov and co-workers are often attributed as being the first major contributors, having published multiple studies on hydrogenation reactors.[53] Within their 1981 publication,[275] for instance, they discussed the ability of Pd-based membranes to selectively hydrogenate from triple to double bonds. And in 1983, they performed hydrogenation of benzene, cyclohexadiene, cyclopentadiene, cyclo-octa-1,5-diene, and naphthalene using several Pd-alloy membranes.[276] From their results, it was concluded that permeating hydrogen enhanced both the rate and selectivity for partial hydrogenation of these reactants over any similar co-fed reactor.

Further work on a Pd-CMR was performed by Nagamoto and Inoue[277] in 1985 where they hydrogenated propylene, 1-butene, and 1,3-butadiene. In this study, the reactivities of both gas-phase and permeate hydrogen were compared for the hydrogenation reactions using a Pd membrane with Pd-black electrodeposited as a surface catalyst. The authors observed that propylene and 1-butene had only a slight effect on the hydrogen adsorption from the gas phase, meaning that the surface was well covered in hydrogen. This resulted in only a small change in the reaction kinetics from using the permeating hydrogen. However, 1,3-butadiene affected the hydrogen adsorption strongly, and the use of permeating hydrogen was able to increase the hydrogenation rate by more than an order of magnitude. Essentially, permeating hydrogen was only able to increase the reaction rate for a hydrogenation reaction where the precursor dominated the surface coverage of the Pd catalyst.

This can be understood by considering four principle reactor conditions (Fig. 17). Essentially, the concept of a precursor and hydrogen

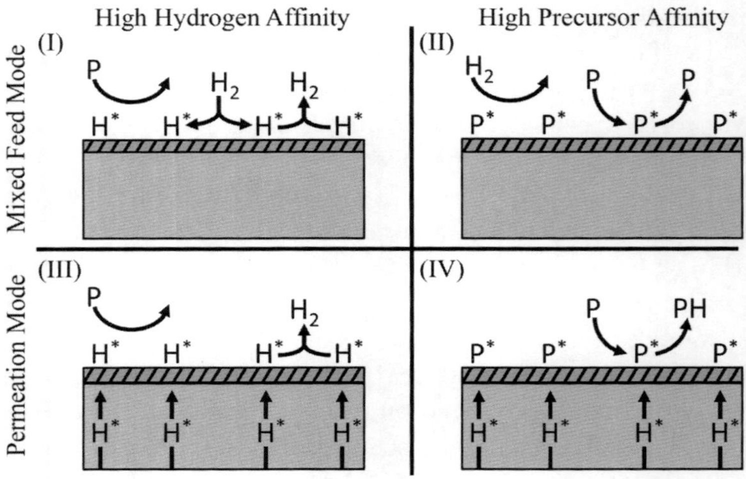

Fig. 17. Schematic of four extreme adsorption/reaction conditions — I and II represent mixed feed conditions while III and IV represent permeation modes.

species can be broken into four extreme cases: (I) co-fed P (precursor reactant) and H_2 with surface dominated in H species, (II) co-fed P and H_2 with surface covered in P, (III) H_2 permeated with surface covered in H species, and (IV) H_2 permeated with surface covered in P. In each case, it is assumed that the PH product will not remain on the surface, and it is only the P and H_2 competing for surface sites. From this, it becomes evident that the most likely hydrogenation reaction will only occur from case IV because this is the only case where hydrogen atoms and the precursor can meet on the surface. In each of the first three cases, the P and H species do not successfully occupy the surface together and would thus suffer a low rate of hydrogenation.

Although the permeating hydrogen appears to increase the reaction rates, it is interesting that several early studies saw no thermodynamic change in the created products. Arai, Wada, and Nishiyama[278] used a Pd-CMR for hydrodesulfurization (HDS) of thiophene and found permeating hydrogen to provide a five times better conversion than that of pre-mixed hydrogen, but they observed the product distributions to be identical for both permeation and pre-mixed hydrogen. In the work of Itoh and Xu,[279] a palladium-based membrane was used to supply hydrogen for the hydrogenation of phenol to cyclohexanone. A primary issue with this reaction when performed in a traditional reactor is that cyclohexanone will further hydrogenate to form cyclohexanol.

$$\text{Reaction 15: } C_6H_5OH_{(g)} + 2H_{2(g)} \leftrightarrow C_6H_{10}O_{(g)}$$

$$\text{Reaction 16: } C_6H_{10}O_{(g)} + H_{2(g)} \leftrightarrow C_6H_{11}OH_{(g)}$$

The result of varying hydrogen flows, shown in Fig. 18, indicated that a significant increase in overall conversion was obtained when permeating hydrogen was used, but selectivity to cyclohexanone decreased for the permeating hydrogen, as it was equilibrium limited for both mixed feed and permeation modes.

Shortly after their work on phenol to cyclohexanone conversion, Itoh, Xu and Sathe[280] published results on the capability of permeated hydrogen to enhance the hydrogenation of acetylene to ethene (Fig. 19). For this reaction scheme, the authors observed

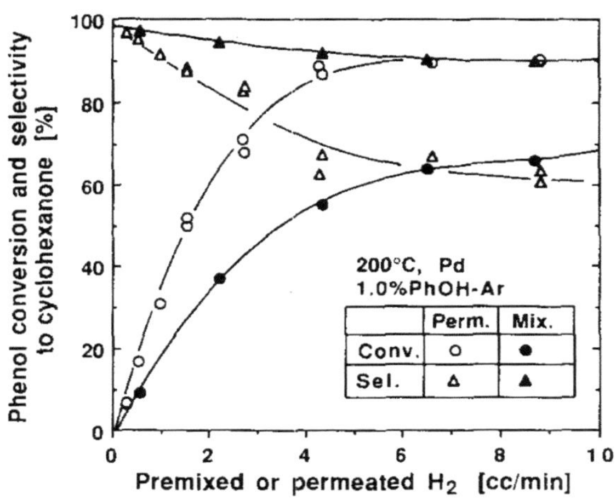

Fig. 18. Phenol conversion and selectivity to cyclohexanone for various hydrogen mixtures supplied via permeation or pre mixing. Selectivity to cyclohexanone was equilibrium limited for both permeated and pre-mixed conditions.

Source: Reprinted from Itoh and Xu[279] with permission from Elsevier.

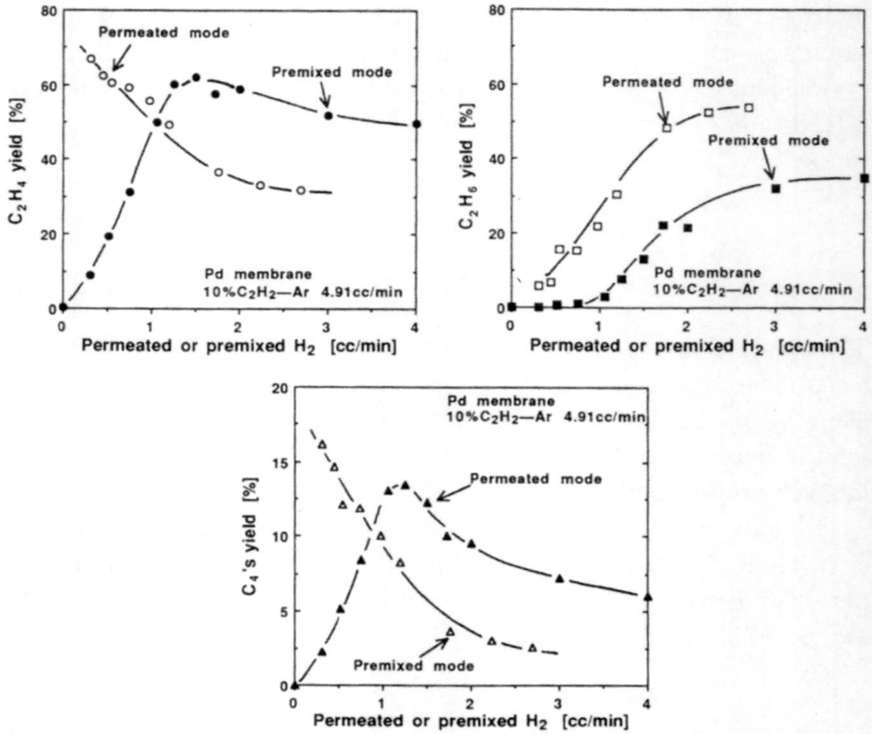

Fig. 19. Conversion of acetylene to various products versus hydrogen flow rates. Hydrogen was supplied either by permeation through a Pd membrane or premixed and co-fed over the membrane surface.

Source: Reprinted with permission from Ref. 280, Copyright (1993), American Chemical Society.

that all hydrogenation reaction rates, including the creation of C_4 compounds, were low under pre-mixed conditions for lower flows of hydrogen. Under permeation conditions, however, the reactivity of the hydrogen was observed to be greatly enhanced. At low hydrogen flows, conversion to both ethene and C_4 compounds was observed to be greatly enhanced, whereas at higher hydrogen flows, the conversion to ethane began to dominate over the pre-mixed feed condition. This can easily be explained by an increase in hydrogen reactivity at both low and high hydrogen flows. At low hydrogen flows, the partial hydrogenation to either ethene or a C_4 compound is preferential to the formation of ethane due to a limited amount of hydrogen available, but at higher flows, the highly reactive hydrogen begins to hydrogenate preferentially to ethane. Although product distributions did not vary significantly between ethene and C_4 compounds, permeation mode allowed for a much more efficient use of the hydrogen fed to the system.

In all, dozens of studies were conducted on hydrogenation reactors during the 1980s and 1990s. Yet, despite their promising results of higher hydrogen efficiencies, most of this work died out by the year 2000. Several reviews[281–284] of the early CMR work involving hydrogenation may offer insight into this decline in research interest. For one, the initial work had to be conducted on thick Pd foils, often 1 mm. This meant that the material cost of the Pd versus the number of active surface sites was enormous as compared to any supported heterogeneous catalyst. Moreover, given that Pd is one of the most expensive metals, this aspect was heavily criticized due to the seemingly prohibitive costs associated with the use of CMRs. Another issue was the poisoning of Pd by certain gases — primarily CO and H_2S. Since this research was conducted, much has been learned concerning how to alloy Pd to reduce poisoning. Yet, at the time, it was a significant cause for concern in the implementation of Pd under any oil- or natural gas-based feed. A third potential reason for the decline was that tough environmental regulations on the fuels industry suddenly required heavy hydrocracking and hydrotreating of fuels. This caused refineries to quickly change from being hydrogen producers to hydrogen consumers during the 1990s.

Despite these issues, however, the highly reactive atomic hydrogen available has been of continued interest. Certain applications, such as ammonia synthesis or hydrogen peroxide synthesis, have neither the carbon-containing feeds that may coke nor the surface-poisoning gases CO or H_2S. Inoue *et al.*,[285] for example, utilized a Pd-CMR in 2010 in the production of hydrogen peroxide. In this case, the permeate (reactor) side was placed in a water solution containing sulfuric acid, phosphoric acid, and sodium bromide to stabilize the H_2O_2 while O_2 was bubbled over the surface of the membrane. With regard to the use of this atomic hydrogen for ammonia synthesis, Machida and coworkers employed a Pd–Ag membrane with a slurry coated Ru/Al_2O_3 catalyst and observed a 50% decrease in activation energy for ammonia synthesis[286] and an ammonia synthesis rate increase of 28%[287] over co-fed conditions. Other recent studies using a Pd-CMR include the hydrogenation of benzene to cyclohexanone,[288] a coupled ammonia decomposition and CO_2 methanation reactor,[289] the direct hydroxylation of methyl benzoate to methyl salicylate,[290] and several studies on the hydroxylation of benzene to phenol.[291–294]

However, some issues do arise when attempting to use Pd-CMR designs. Concerning the hydroxylation of benzene to phenol, a review by Shu *et al.*[295] suggested the membrane concept to be ineffective due to both the low catalytic activity of the Pd surface and the high rate of membrane degradation due to oxidation. This lack of activity for the amount of Pd required has a huge economic impact because even with higher reactivities on a surface area basis, the cost of Pd is such that the reactivity per mass basis makes such systems uneconomical. Therefore, the use of alternative materials may be the best way to overcome these cost issues. In 2016, Morimoto *et al.*[296] successfully coated a V-10Fe membrane with Pd–Ag co catalyst sections to produce a small quantity of ammonia, suggesting novel surface catalyst designs can enhance reactivities and reduce the catalyst cost. Perhaps future research will reveal new and exciting membrane–catalyst combinations, but for now, it seems that the most likely membrane-synthesis techniques will need to involve the use of a galvanic cell, as will now be detailed.

3.2.2. *Dense Ceramic Membrane Applications*

As discussed in Sec. 3.1.3, ammonia is widely used in the production of fertilizers. More recently, interest has developed in ammonia as an effective and safe compound to store large quantities of hydrogen. Yet, commercial ammonia synthesis relies on the Haber–Bosch process which has remained relatively unchanged for more than a century.[297] Although minor improvements in catalysts have been achieved, the core process still requires a temperature of at least 400°C due to catalyst kinetic limitations and pressures above

10 MPa for reasonable thermodynamic conversions. The reaction for ammonia synthesis is exothermic and can be written as

Reaction 17:

$$N_{2(g)} + 3H_{2(g)} \leftrightarrow 2NH_{3(g)} \quad (\Delta H^0 = -92.4\,\text{kJ/mol})$$

Although the mechanism for ammonia synthesis over a Fe or Ru catalyst has been elucidated using both experimental and theoretical studies,[297–301] the challenge to form ammonia at reduced temperatures and pressures remains. This is where electrochemical cells based on dense ceramic membranes have begun to shine.

Figure 20 is a graphical representation of a solid electrolyte cell, which consists of a solid-state proton conductor serving the dual purpose of being an electrolyte as well as a dense proton-conducting membrane. A metal electrode on either side of the solid-state proton conductor is connected by metal wires to an external power source, allowing gas-phase hydrogen to be converted into protons at the anode. These protons are transported through the proton conductor to the cathode where they react with adsorbed nitrogen to form ammonia. In 1998, Marnellos and Stoukides[302] demonstrated the feasibility of the process shown in Fig. 20, and since then many research groups have studied different aspects of this system in an effort to improve conversion. More than 30 solid electrolytes and 15 catalysts were investigated. The results were summarized in the 2011 review by Amar *et al.*[303]

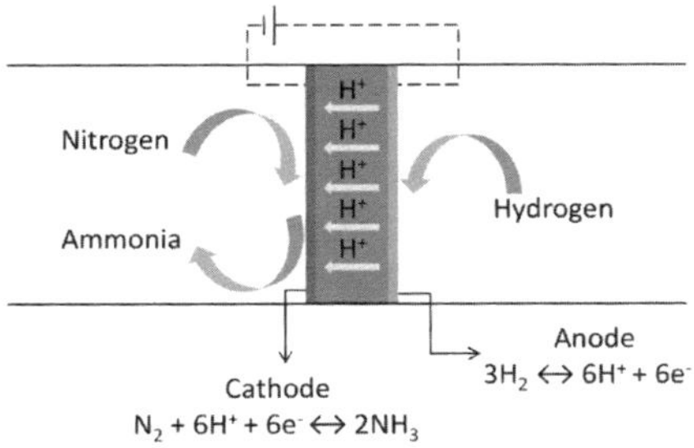

Fig. 20. Graphical representation of the principle of operation of an electrochemical cell reactor.

Source: Adapted from Ref. 114.

Researchers have considered N_2-dissociative adsorption onto the catalyst surface as the rate-limiting step in ammonia synthesis since the work of Emmett and Brunauer[304] in 1934, wherein they observed that the rate of N_2 dissociation over a Ni surface was of the same order of magnitude as NH_3 synthesis. And although Fe- or Ru-based compounds are still considered the best conventional ammonia synthesis catalysts,[305] solid-state ammonia synthesis (SSAS) experiments have used Pd, Pt, Ru, Fe, Pd–Ag, as well as some conductive oxides and composite materials over the last 15 years. Table 8 summarizes the catalytic activity (ammonia synthesis rate, r_{NH3}) and faradaic efficiency (defined as the percentage of protons transported that are converted into ammonia) of several SSAS studies.

It is interesting to note that Pd–Ag has appeared as the cathode in 50% of all SSAS studies, even though it is not the best conventional catalyst. Yet, at temperatures between 450°C and 600°C, Pd–Ag electrodes have demonstrated high catalytic activity and above average faradaic efficiencies (40% to 80%). This then lends itself to the notion that catalyst improvements might even improve these current synthesis rates.

According to a recent DFT study by Song and Hu,[322] one of the primary reasons that Ru is a good conventional catalyst is because the binding energies of N(ads) and H(ads) surface species cross. In fact, as shown in Fig. 21, they calculated the binding energies of N(ads) and H(ads) species to be identical near the 1/4 monolayer coverage of nitrogen, which is quite similar to the 1/3 monolayer coverage observed experimentally by Dietrich *et al.*[323] In contrast, they calculated that the surface of Pd strongly prefers H(ads) species whereas Zr surfaces strongly prefer N(ads) species.

To explore the competing nitrogen dissociation and hydrogenation reactions further, Song and Hu plotted simplified mathematical models for the rates of reaction (Fig. 21(b)).[322] The parabolic curves represent NH_3 synthesis rates, whereas the linear lines represent the ease of hydrogenation on the surface. The intersection of these lines is predicted to be the overall turnover frequency (TOF), and it is

Table 8. Summary of catalytic activities of different electrode–electrolyte combinations for SSAS.

Electrode	Electrolyte	T (°C)	r_{NH3} (mol m^{-2} s^{-1})	Faradaic efficiency (%)	Refs.
Pd–Ag	$Ce_{0.8}Y_{0.2}O_{2-\delta} - [Ca_3(PO_4)_2 - K_3PO_4]$	650	9.50×10^{-5}	40	306
Pd	$SrCe_{0.95}Yb_{0.05}O_{3-\alpha}$	570	4.50×10^{-5}	78	302, 307
Pd–Ag	$La_{0.9}Sr_{0.1}Ga_{0.8}Mg_{0.2}O_{3-\delta}$	550	3.37×10^{-5}	70	308
Pd–Ag	$BaCe_{0.85}Dy_{0.15}O_{3-\alpha}$	530	3.50×10^{-5}	80	309
$Ba_{0.5}Sr_{0.5}Co_{0.8}\ Fe_{0.2}O_{3-\delta}$	$BaCe_{0.85}Y_{0.15}O_{3-\alpha}$	530	4.10×10^{-5}	60	310
Pd–Ag	$La_{0.9}Ba_{0.1}Ga_{0.8}Mg_{0.2}O_{3-\delta}$	520	1.89×10^{-5}	60	311
Pd–Ag	$BaCe_{0.85}Y_{0.15}O_{3-\alpha}$	500	2.10×10^{-5}	60	312, 313
Pd–Ag	$Ce_{0.8}Sm_{0.2}O_{2-\delta}$	500	2.67×10^{-5}	80	314
Pd–Ag	$BaCe_{0.9}Ca_{0.1}O_{3-\alpha}$	480	2.69×10^{-5}	50	315
$Ba_{0.5}Sr_{0.5}Co_{0.8}\ Fe_{0.2}O_{3-\delta}$	$BaCe_{0.85}Y_{0.15}O_{3-\alpha}$	450	5.39×10^{-5}	3.5	316
$CoFe_2O_4$–Ag	Carbonate–$LiAlO_2$	400	2.32×10^{-6}	2.5	317
Ru/C	Nafion®	90	2.12×10^{-7}	1.0	318
$Sm_{1.5}Sr_{0.5}NiO_4$	Sulfonated polysulfone	80	1.05×10^{-4}	0.8	319
$SmFe_{0.7}Cu_{0.1}\ Ni_{0.2}O_3$	Sulfonated polysulfone	80	1.13×10^{-4}	90.4	320
SmBaCuNiO	Nafion®	80	8.70×10^{-5}	0.7	321
Pt	$BaZr_{0.7}Ce_{0.2}Y_{0.1}O_{3-\delta}$	620	4.10×10^{-5}	3.0	188

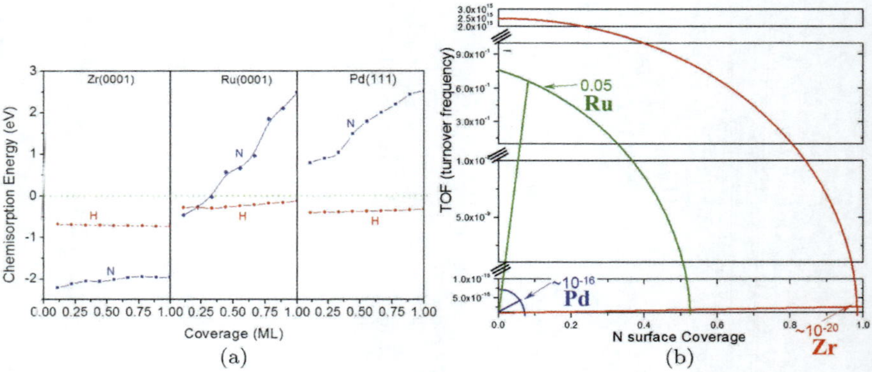

Fig. 21. (a) Chemisorption energy versus surface coverage for both N and H adatoms on Zr, Ru, and Pd surfaces. (b) Theoretical turnover frequency versus N-adatom surface coverage for Zr (red), Ru (green), and Pd (blue). The parabola represents the theoretical rate of NH_3 formation while the linear line is the rate of hydrogenation.

Source: Reprinted from Ref. 322 with permission from AIP publishing.

seen that the rates predict Ru > Pd > Zr. However, note that Zr has the potential to be a much better catalyst than Ru if the rate of hydrogenation (linear line) could be greatly enhanced, thus lowering N adatom surface coverage and intersecting the parabolic curve at a higher TOF. Ru is simply a better catalyst due to a good match between the rate of hydrogenation and the rate of dissociation of nitrogen. In sharp contrast, a Pd surface is predicted to have no potential to have a higher TOF than Ru or Zr because the rate of nitrogen dissociation is always low.

This improvement based on differences in catalytic surfaces was also suggested by the DFT work of Skúlason *et al.*[311] The authors argue that based on their results Sc, Y, Ti, and Zr warrant further research as potential catalysts in electrochemical cells primarily because these metals lie in the N adatom dominated regime, rather than other metals that will likely be dominantly covered in H adatoms. And because electrochemical cell reactors will provide the H species from within, the predictions are that easing nitrogen-dissociation kinetics should provide significant benefits to the overall rate.

In addition to increasing the catalytic rate of ammonia synthesis, PCCRs have even been shown to exceed gas-phase thermodynamic conversion at higher temperatures. The concept of using PCCRs to improve the conversion dictated by thermodynamics of reversible hydrogenation reactions was suggested in the mid-1990s.[324–326] An early modeling study by Panagos, Voudouris, and Stoukides[325] showed that using electrochemically supplied H^+ could actually produce more NH_3 than gas-phase thermodynamics predicts because the partial pressure of hydrogen is not relevant when protons are supplied separately. The authors showed that below about 330°C, the gas-phase reactions are insignificant, and it is the activity of the surface-adsorbed atomic hydrogen, rather than the partial pressure of H_2 gas, that decides the amount of NH_3 formed. Because the activity of protons is significantly higher than H_2 gas, the system shifts toward NH_3 production, and this effect was later confirmed by the same group experimentally.[307] In their experimental study, Marnellos, Zisekas, and Stoukides[307] used a proton-conducting ceramic electrochemical cell (Fig. 22), and obtained conversions about two

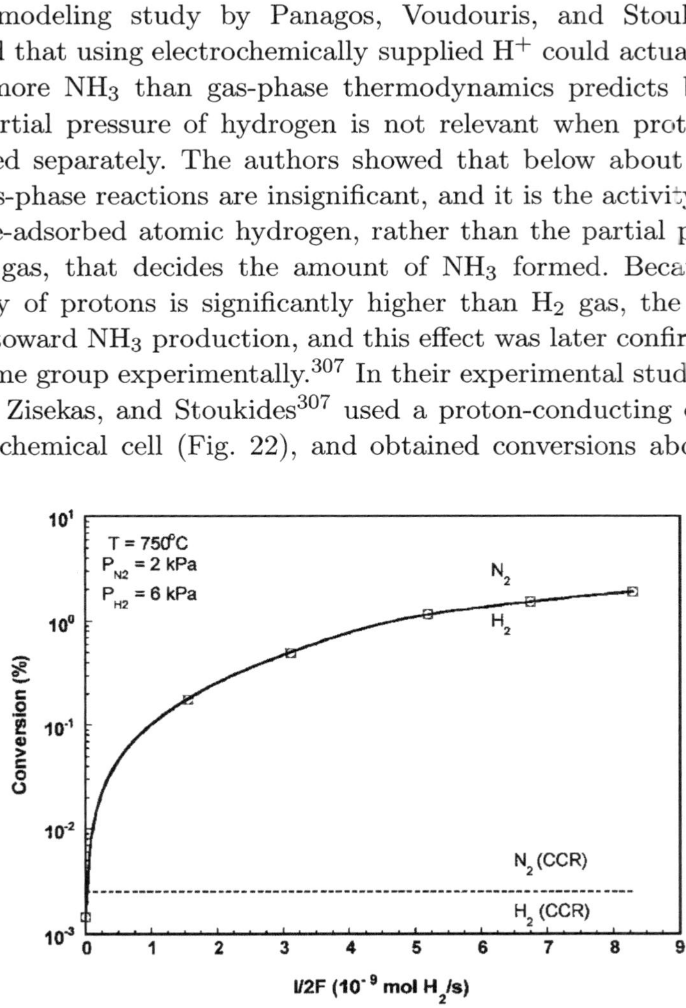

Fig. 22. Conversion of feed to ammonia versus supplied H_2-driving force. The conversion of the electrochemical cell exceeds that of a CCR.

Source: Reprinted from Ref. 307 with permission from Elsevier.

orders of magnitude higher for the electrochemical cell than for a conventional catalytic reactor (CCR).

It is important to note, however, that Fig. 22 also suggests that the proton conduction was rate-limiting in this experiment. As the current was increased, and thus the rate of proton conduction through the material, the rate of ammonia production increased. Yet, in a separate study where N_2 pressure was varied, no effects on the rate of ammonia formation was observed. This suggests that the system was either proton limited due to flux through the material or that the hydrogenation reaction was rate-limiting. It may then follow that using a better proton-conducting material may further enhance the ammonia synthesis rate.

The various electrochemical cell types and results were recently reviewed by Giddey, Badwal, and Kulkarni,[327] including liquid electrolytes, molten salts, composite membranes, and solid-state electrolytes. Thus far, the highest reported synthesis rate[320] used a Nafion polymer electrolyte membrane to obtain $1.13 \times 10^{-4}\,\mathrm{mol\,m^{-2}\,s^{-1}}$ at ambient pressure and 80°C. It is likely that this rate was achieved because Nafion also has the highest proton conductivity amongst the various materials.

However, Giddey *et al.*[327] noted that the field is still underdeveloped, with production rates being about two orders of magnitude low for industrial viability and long-term stability yet to be demonstrated. Indeed, the best result thus far was likely from a short-term experiment due to the fact that conductivity in Nafion is known to fall dramatically with NH_3 exposure.[328] Additionally, the use of aqueous electrolytes like Nafion requires low-temperature operation, which limits the kinetic reaction rates, which would suggest that a higher proton-conductivity material at higher temperature might do better. Solid-state proton conductors are more stable than aqueous electrolytes and can be operated at higher temperatures, but their proton conductivities remain much lower than aqueous electrolytes. Nevertheless, the mechanism of reaction in these cells does allow for formation of NH_3 at ambient pressure, which suggests that utilizing this strategy could eventually offer significant benefits over the conventional Haber–Bosch process.

3.3. *Hydrogen Isotope Separations*

Tritium (T) is a hydrogen isotope with a nucleus containing two neutrons and a proton. Although naturally occurring by interactions between atmospheric N and O with cosmic neutrons and protons, its half-life of 12.26 years limits significant build-up in nature.[329] Artificial sources of T, on the other hand, include thermonuclear detonations and nuclear reactors wherein neutrons interact with elements in the reactor core which can cause significant build-up of T in the localized environment. Due to issues with water table contamination, the U.S. Nuclear Regulatory Commission (NRC) regularly inspects nuclear plants to ensure large amounts of T are not released into the environment. Yet, not only is prevention of T leakage necessary for environmental protection, but collecting and concentrating T is also critical for nuclear weapon programs as well as fuel for fusion reactors. The fusion reaction with the highest cross-section and energy release (Reaction 18) occurs when deuterium (D) and T undergo nuclear fusion in a high-temperature plasma tokamak reactor to produce He, a neutron, and 17.6 MeV of energy.[330]

$$\text{Reaction 18: D} + \text{T} \leftrightarrow \text{He} + n.$$

As opposed to T, D has a natural abundance of 0.015% which allows it to be separated from water by conventional methods such as chemical exchange or water distillation.[331] T, however, is more difficult to produce because it must come from breeding reactions or by reprocessing spent fuel from fission reactors. Methods of separating T include water electrolysis, water distillation, thermal diffusion, selective membranes, porous membranes, adsorption/chromatography, and laser isotope separation.[329] A dense metal membrane, falling under the selective membrane category, is one of the materials used to separate H isotopes from spent fuel.[332] A commercial example of a Pd membrane for this application is the hydrogen separators produced by power + energy, INC which are marketed for this exact application.

However, the main application for dense metallic membranes is in the fusion fuel cycle. Here, T must be managed in the breeder blanket as well as concentrated and recycled in the tokamak exhaust processing (TEP) cycle in order to conserve fuel and decontaminate exhaust streams.[333] Generally, Pd–Ag membranes have been the most heavily studied and implemented for this application[334]; however, other dense membranes such as BCC metals and proton-conducting ceramics could certainly be implemented as permeators.

The current state-of-the-art configuration in the TEP cycle to recover and recycle T is the CAPER (catalyst permeation) process (Fig. 23), which was developed for use in the ITER reactor.[333] The CAPER process consists of permeators (Pd–Ag membranes), a catalyst bed, and a PERMCAT reactor. The feed into the CAPER process is a mixture of hydrogenated species consisting of different isotopes Q (H, D, or T) along with carbon and oxygen impurities that arise from contamination within the fusion system. In order to extract the isotopes into a pure gas form, isotope exchange reactions must occur in the catalyst bed and the PERMCAT reactor.[335] The PERMCAT reactor consists of a Pd–Ag membrane packed with isotope exchange catalyst — nominally Nikki 111, which is a commercially available Ni catalyst on a kieselguhr support. Pure H_2 enters the reactor countercurrent to the tritiated species, permeates through the membrane, and undergoes isotope exchange interactions on the catalyst surface that allows tritiated species to then permeate back through the membrane.

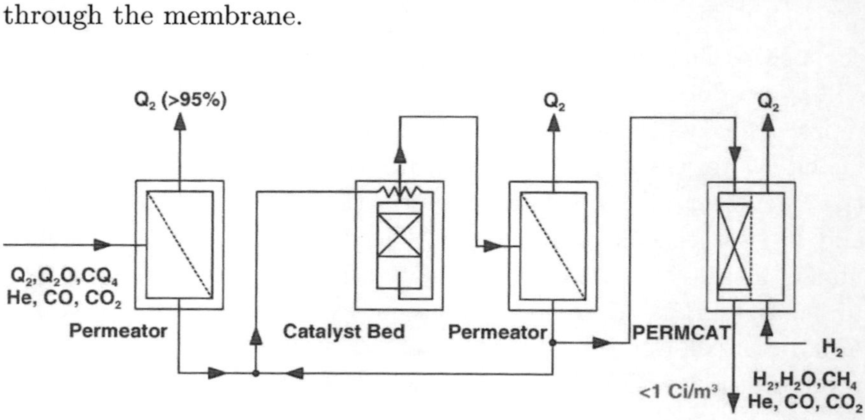

Fig. 23. Process flow diagram of CAPER process where Q is H, D, or T. *Source*: Reprinted from Ref. 333 with permission from Elsevier.

The second process — the catalyst bed followed by the permeator — can be operated in either batch or continuous mode. In batch mode, the loop is filled with gas and is cycled until the tritium level is lowered to a specified value, but this is more difficult to control and has lower throughput than continuous operation. Alternatively, gas can be supplied continuously through a recycle loop with a portion of the downstream gas withdrawn and fed into the PERMCAT. In general, this process is able to decontaminate off-gas to levels below $1\,Ci\,m^{-3}$, which relates to a decontamination factor (DF) of 10^8 (i.e. 99.999999% of T is recovered), and meets the requirements for ITER.[333]

Current research at the Tritium Laboratory Karlsruhe (TLK) consists of PERMCAT process optimization. For example, Demange *et al.*[335] provide an overview of the research and development done at TLK which includes improved mechanical designs for the PERMCAT as well as testing under various operating conditions. Bükki-Deme *et al.*[336] investigated high He-carrier flowrates but concluded that He-carrier gas decreased the DF by reducing the residence time and the water vapor partial pressure. They also suggested that new catalysts could potentially improve performance.

The potential future of fusion-generated electricity can be considered to lie in the ITER and DEMO projects. Whereas the goal of ITER (latin for "the way") is to prove controlled fusion for energy generation is possible, DEMO (standing for DEMOnstration power station) plans to take the next step and demonstrate the large-scale viability by producing 500 MW of electricity from controlled fusion. ITER is planned to consume the relatively minor quantity of $100\,g\,yr^{-1}$ of T, with the necessary T supplied externally via extraction from CANDU reactors. In contrast, DEMO consumption will exceed $100\,kg\,yr^{-1}$. This will require T to be produced inside the fusion reactor in a breeder blanket using ^6Li and neutrons from Reaction 18; however, T management in the breeder blanket remains challenging with the current process relying on packed beds.[337]

Demange and coworkers[338,339] have studied combing zeolite membranes and a PERMCAT reactor for application in the breeder blanket fuel cycle. They found this process to be challenging due to the necessary high He-sweep flow rates diluting the tritiated species, which reduces the driving force for membrane separation. In general, dense inorganic membranes are critical to this field, and improvements in such membranes are beneficial to both advancing fusion energy as well as safe handling of radioactive T in fission reactors. Improvements can come from developing novel materials for composite membranes as well as mechanical design for membrane reactors. This is an exciting field where despite over a century of research and development, constant breakthroughs and advances are ever improving the state-of-the-art.

4. Summary and Outlook

Overall, hydrogen-permeable membranes are still within the research stage and may have another decade of development before they are considered viable for industrial purposes. Yet, it can also be said that the progress over the last two decades has been extraordinary. Continuous improvements in the hydrogen flux, purity, and membrane durability have increased significantly since the early 1990s. Pd films are now consistently producing greater than 99% purity with high fluxes and good durability, while being fabricated with 10 μm thickness. This means that the overall Pd inventory is now low enough to begin to be competitive and not a large hindrance on the implementation of the technology. Unfortunately, annealing and chemical tolerance remains an issue, with lifetime durability only tested over several months. Thus, the technology still needs to improve, and tests must be performed on the year-long scale before an industrial scale plant can be considered.

Alternative metals to Pd, the so-called BCC metals (V, Nb, and Ta), have also seen an increase in viability due to both surface catalyst and metal-alloy configurations. Historically, the biggest issue facing the BCC metals has been their embrittlement when operated at high partial pressures of hydrogen. Although the hydrogen flux is large, and a thicker foil can be used to ensure 100% purity, they often crack and degrade due to swelling and embrittlement. This has partially been solved by alloying them with other metals to supress the hydrogen solubility and prevent embrittlement, but the surface-catalyst issue remains. Because these metals require a secondary catalyst to dissociate hydrogen, the biggest issue facing the material today is in the development of a stable and effective surface catalyst. Thin films of Pd have been fabricated and shown to be good at transporting hydrogen for short durations, but the Pd quickly alloys with the BCC metals, and they lose their ability to transport hydrogen within a few days. Alternative catalysts, such as molybdenum carbide, have shown promise in pure-gas configurations, but there is a limited amount of research shown for gas-mixture performance. Thus, BCC metals will remain a promising, but underdeveloped field until an effective surface catalyst is found.

The recent advances in proton-conducting ceramics may actually show the best promise for future applications. Although these materials have low hydrogen flux when compared to the dense metal membranes, they remain perfectly selective and will potentially avoid several issues. The materials are typically cheap, so reducing the inventory is not as critical as it is for the thin-film Pd membranes, and because the support is ceramic, alloying of the surface catalyst layer will not pose a large issue as it does with BCC metals. Thus, fabrication of the membrane itself is likely to result in good long-term durability. However, the low hydrogen flux of most proton-conducting ceramics limits their use and should be the primary concern for future research.

Much of this chapter has emphasized the various potential applications within the dense hydrogen-permeable membrane categories, primarily focusing on membrane reactor configurations. Although these membranes can potentially separate hydrogen from any gas mixture, the most interesting field is the one in which separation and reaction are combined to produce a compact multi functional unit. The two biggest fields in the research community for hydrogen generation have been steam methane reforming (SMR) and ammonia decomposition. SMR has proven to be interesting because of the large thermodynamic shift in conversion possible by operation in a membrane reactor configuration. This has obvious cost-saving benefits if a cheap membrane can be used because it not only eliminates the need for a separate reactor and separator but also shifts equilibrium and allows for operation at a lower temperature — something that can potentially reduce the cost of the necessary reactor materials. Ammonia decomposition, on the other hand, does not necessarily see an increase in overall conversion but does combine the reaction and separation stages. And given the recent interest in alternative carbon-free fuels, ammonia has been suggested as a superior chemical for hydrogen storage and transport due to its low vapor pressure and high hydrogen content. Yet, as with each of the membrane types, both ammonia decomposition and SMR reactors lack long-term studies necessary for industrial implementation. In particular, the exposure to CO and potentially H_2S in an SMR environment can cause severe degradation on certain membrane materials. These issues continue to be addressed by finding alloys less susceptible to poisoning, and further research will undoubtedly solve these problems over the coming years.

Finally, hydrogenation membrane reactors were discussed for both Pd and proton-conducting ceramic membranes. These were broken into two groups because the Pd-based reactors run in a nongalvanic mode whereas proton-conducting ceramic membranes typically have an applied potential. Although Pd-based membranes were shown to increase the reaction rate on a surface area basis, this rate increase was not enough to off set the cost per mass. The requirement to use a foil of Pd in the reactor is such that the increase in kinetic reaction rate is not an overall economic benefit. Yet, proton-conducting ceramics are much cheaper and may thus achieve a cost benefit sooner. Proton-conducting ceramics have primarily been used in electrochemical cells to produce ammonia and have actually performed extremely well compared to traditional reactor designs. However, the rates remain too low to effectively compete with the well-established Haber–Bosch process. These cells appear to often be limited by a low flux of hydrogen, so it is quite possible that newer proton-conducting materials will be developed that will allow for an increase in the hydrogen flux, leading to an increase in the ammonia synthesis rate. Perhaps ammonia will be entirely generated by these galvanically operated membrane reactors in another 20 years.

Acknowledgments

The authors gratefully acknowledge support for this project from the U.S. National Science Foundation under Grant number 1512172. Thomas F. Fuerst acknowledges support under a U.S. Department of Energy Nuclear Energy University Programs Graduate Fellowship. Any opinions, findings, and conclusions or recommendations expressed in this material are those of the author(s) and do not necessarily reflect the views of the National Science Foundation or the Department of Energy Office of Nuclear Energy.

References

1. Deville, H. S.-C., Troost, L. Sur la perméabilité du fer à haute température. *Comptes Rendus.* 1863, 57, 965.
2. Deville, H. S.-C. Sur le passage des gaz à travers des corps solides homogènes. *Comptes Rendus.* 1864, 59, 102.
3. Graham, T. On the absorption and dialytic separation of gases by colloid septa. *Philos. Trans. R. Soc. Lond.* 1866, 156(0), 399–439.
4. Grashoff; Pilkington; Corti. The Purification of Hydrogen: A Review of the Technology Emphasising the Current Status of Palladium Membrane Diffusion. *Platin. Met. Rev.* 1983, 27(4), 157.
5. Marrony, M. *Proton-Conducting Ceramics: From Fundamentals to Applied Research.* Pan Stanford: Singapore, 2015.
6. Forrat, F., Dauge, G., Trevoux, P., Danner, G., Christen, M. Electrolyte solide à base de $AlLaO_3$. Application aux piles à combustible. *Comptes Rendus L'Académie Sci.* 1964, 259, 2813–2816.
7. Stotz, S., Wagner, C. Die Löslichkeit von Wasserdampf und Wasserstoff in festen Oxiden. *Berichte Bunsenges. Für Phys. Chem.* 1966, 70(8), 781–788.
8. Takahashi, T., Iwahara, H. Ionic conduction in perovskite-type oxide solid solution and its application to the solid electrolyte fuel cell. *Energy Convers.* 1971, 11(3), 105–111.
9. Takahashi, T., Iwahara, H. Solid state ionics: Proton conduction in perovskite type oxide solid solutions. *Rev. Chim. Miner.* 1980, 17, 243.
10. Iwahara, H., Esaka, T., Uchida, H., Maeda, N. Proton conduction in sintered oxides and its application to steam electrolysis for hydrogen production. *Solid State Ion.* 1981, 3–4, 359–363.
11. Uemiya, S., Kajiwara, M., Kojima, T. Composite membranes of group VIII metal supported on porous alumina. *AIChE J.* 1997, 43(S11), 2715–2723.
12. Kajiwara, M., Uemiya, S., Kojima, T., Kikuchi, E. Rhodium- and iridium-dispersed porous alumina membranes and their hydrogen permeation properties. *Catal. Today* 2000, 56(1–3), 83–87.

13. Kajiwara, M., Uemiya, S., Kojima, T., Kikuchi, E. Hydrogen permeation properties through composite membranes of platinum supported on porous alumina. *Catal. Today* 2000, 56(1–3), 65–73.
14. Tsuru, T., Igi, R., Kanezashi, M., Yoshioka, T., Fujisaki, S., Iwamoto, Y. Permeation properties of hydrogen and water vapor through porous silica membranes at high temperatures. *AIChE J.* 2011, 57(3), 618–629.
15. Lee, D., Zhang, L., Oyama, S. T., Niu, S., Saraf, R. F. Synthesis, characterization, and gas permeation properties of a hydrogen permeable silica membrane supported on porous alumina. *J. Membr. Sci.* 2004, 231(1–2), 117–126.
16. Gu, Y., Ted Oyama, S. Ultrathin, hydrogen-selective silica membranes deposited on alumina-graded structures prepared from size-controlled boehmite sols. *J. Membr. Sci.* 2007, 306(1–2), 216–227.
17. Kanezashi, M., Yamamoto, A., Yoshioka, T., Tsuru, T. Characteristics of ammonia permeation through porous silica membranes. *AIChE J.* 2009.
18. Kanezashi, M., Sasaki, T., Tawarayama, H., Yoshioka, T., Tsuru, T. Hydrogen permeation properties and hydrothermal stability of sol-gel-derived amorphous silica membranes fabricated at high temperatures. *J. Am. Ceram. Soc.* 2013, *96*(9), 2950–2957.
19. Li, G., Kanezashi, M., Yoshioka, T., Tsuru, T. Ammonia decomposition in catalytic membrane reactors: Simulation and experimental studies. *AIChE J.* 2013, 59(1), 168–179.
20. Gu, Y., Oyama, S. T. High molecular permeance in a poreless ceramic membrane. *Adv. Mater.* 2007, 19(12), 1636–1640.
21. Meng, L., Tsuru, T. Hydrogen production from energy carriers by silica-based catalytic membrane reactors. *Catal. Today* 2016, 268, 3–11.
22. Hirschfelder, C. O., Curtiss, C. F., Bird, R. B. *The Molecular Theory of Gases and Liquids* (Updated Edition). Wiley: New York, 1964.
23. Tables of Physical & Chemical Constants (16th edition). 2.2.4 Mean velocity, free path and size of molecules. Kaye & Laby Online, 1995.
24. van Leeuwen, M. E. Derivation of Stockmayer potential parameters for polar fluids. *Fluid Phase Equilibria.* 1994, 99, 1–18.
25. Wijenayake, S. N., Panapitiya, N. P., Nguyen, C. N., Huang, Y., Balkus, K. J., Musselman, I. H., Ferraris, J. P. Composite membranes with a highly selective polymer skin for hydrogen separation. *Sep. Purif. Technol.* 2014, 135, 190–198.
26. Agmon, N. The Grotthuss mechanism. *Chem. Phys. Lett.* 1995, 244(5–6), 456–462.
27. Li, P., Wang, Z., Qiao, Z., Liu, Y., Cao, X., Li, W., Wang, J., Wang, S. Recent developments in membranes for efficient hydrogen purification. *J. Membr. Sci.* 2015, 495, 130–168.
28. Al-Mufachi, N. A., Rees, N. V., Steinberger-Wilkens, R. Hydrogen selective membranes: A review of palladium-based dense metal membranes. *Renew. Sustain. Energy Rev.* 2015, 47, 540–551.
29. Uemiya, S. Brief review of steam reforming using a metal membrane reactor. *Top. Catal.* 2004, 29(1/2), 79–84.

30. Gillespie, L. J., Galstaun, L. S. The palladium-hydrogen equilibrium and new palladium hydrides. *J. Am. Chem. Soc.* 1936, 58(12), 2565–2573.
31. Sieverts, A. *Z. Met.* 1929, 21(44).
32. Brüning; Sieverts. *Z. Für Phys. Chem.* 1933, 163A(409).
33. Sieverts, A., Zapf, G. *Z. Für Phys. Chem.* 1935, A174(359).
34. Lewis, F. A. *The Palladium-Hydrogen System.* Academic Press: London, 1967.
35. Caravella, A., Scura, F., Barbieri, G., Drioli, E. Sieverts law empirical exponent for Pd-based membranes: Critical analysis in pure H_2 permeation. *J. Phys. Chem. B.* 2010, 114(18), 6033–6047.
36. Caravella, A., Hara, S., Drioli, E., Barbieri, G. Sieverts law pressure exponent for hydrogen permeation through Pd-based membranes: Coupled influence of non-ideal diffusion and multicomponent external mass transfer. *Int. J. Hydrog. Energy* 2013, 38(36), 16229–16244.
37. Deveau, N. D., Ma, Y. H., Datta, R. Beyond Sieverts' law: A comprehensive microkinetic model of hydrogen permeation in dense metal membranes. *J. Membr. Sci.* 2013, 437, 298–311.
38. Vadrucci, M., Borgognoni, F., Moriani, A., Santucci, A., Tosti, S. Hydrogen permeation through Pd–Ag membranes: Surface effects and Sieverts' law. *Int. J. Hydrog. Energy* 2013, 38(10), 4144–4152.
39. Steward, S. A. *Review of Hydrogen Isotope Permeability Through Materials.* Lawrence Livermore National Lab., CA (USA), 1983.
40. Cooney, D. A., Way, J. D., Wolden, C. A. A comparison of the performance and stability of Pd/BCC metal composite membranes for hydrogen purification. *Int. J. Hydrog. Energy* 2014, 39(33), 19009–19017.
41. Holleck, G. L. Diffusion and solubility of hydrogen in palladium and palladium–silver alloys. *J. Phys. Chem.* 1970, 74(3), 503–511.
42. Buxbaum, R. E., Marker, T. L. Hydrogen transport through non-porous membranes of palladium-coated niobium, tantalum and vanadium. *J. Membr. Sci.* 1993, 85(1), 29–38.
43. Buxbaum, R. E., Kinney, A. B. Hydrogen transport through tubular membranes of palladium-coated tantalum and niobium. *Ind. Eng. Chem. Res.* 1996, 35(2), 530–537.
44. Nambu, T., Shimizu, K., Matsumoto, Y., Rong, R., Watanabe, N., Yukawa, H., Morinaga, M., Yasuda, I. Enhanced hydrogen embrittlement of Pd-coated niobium metal membrane detected by in situ small punch test under hydrogen permeation. *J. Alloys Compd.* 2007, 446–447, 588–592.
45. Paglieri, S. N., Wermer, J. R., Buxbaum, R. E., Ciocco, M. V., Howard, B. H., Morreale, B. D. Development of membranes for hydrogen separation: Pd coated V–10Pd. *Energy Mater.* 2008, 3(3), 169–176.
46. Cotterill, P. The hydrogen embrittlement of metals. *Prog. Mater. Sci.* 1961, 9(4), 205–301.
47. Owen, C. V., Scott, T. E. Relation between hydrogen embrittlement and the formation of hydride in the group V transition metals. *Metall. Trans.* 1972, 3(7), 1715–1726.

48. Dolan, M. D. Non-Pd BCC alloy membranes for industrial hydrogen separation. *J. Membr. Sci.* 2010, 362(1–2), 12–28.
49. Sakamoto, Y., Hirata, S., Nishikawa, H. Diffusivity and solubility of hydrogen in Pd–Ag and Pd–Au alloys. *J. Common Met.* 1982, 88(2), 387–395.
50. Hunter, J. B. Silver-palladium film for separation and purification of hydrogen, 1956.
51. Hunter, J. B. A New hydrogen purification process: Commercial development of palladium alloy diffusion cells. *Platin. Met. Rev.* 1960, 4(4), 130.
52. McKinley, D. L. Metal alloy for hydrogen separation and purification. 1967.
53. Gryaznov, V. Metal containing membranes for the production of ultrapure hydrogen and the recovery of hydrogen isotopes. *Sep. Purif. Rev.* 2000, 29(2), 171–187.
54. *Ga-Gd — Hf-Zr*, Madelung, O., Ed., Landolt-Börnstein — Group IV Physical Chemistry, Vol. f. Springer-Verlag: Berlin/Heidelberg, 1996.
55. *Noble Metal Systems. Selected Systems from Ag-Al-Zn to Rh-Ru-Sc*; Effenberg, G., Ilyenko, S., Eds., Landolt-Börnstein — Group IV Physical Chemistry, Vol. 11B. Springer. Berlin Heidelberg, 2006.
56. Lewis, A. E., Kershner, D. C., Paglieri, S. N., Slepicka, M. J., Way, J. D. Pd–Pt/YSZ composite membranes for hydrogen separation from synthetic water–gas shift streams. *J. Membr. Sci.* 2013, 437, 257–264.
57. Coulter, K. E., Way, J. D., Gade, S. K., Chaudhari, S., Alptekin, G. O., DeVoss, S. J., Paglieri, S. N., Pledger, B. Sulfur tolerant PdAu and PdAuPt alloy hydrogen separation membranes. *J. Membr. Sci.* 2012, 405–406, 11–19.
58. Okamoto, H., Massalski, T. B. The Au-Pd (Gold-Palladium) system. *Bull. Alloy Phase Diagr.* 1985, 6(3), 229–235.
59. Shi, L., Goldbach, A., Xu, H. High-flux H_2 separation membranes from (Pd/Au)n nanolayers. *Int. J. Hydrog. Energy* 2011, 36(3), 2281–2284.
60. Tarditi, A., Gerboni, C., Cornaglia, L. PdAu membranes supported on top of vacuum-assisted ZrO_2-modified porous stainless steel substrates. *J. Membr. Sci.* 2013, 428, 1–10.
61. Patki, N. S., Lundin, S.-T., Way, J. D. Rapid annealing of sequentially plated Pd–Au composite membranes using high pressure hydrogen. *J. Membr. Sci.* 2016, 513, 197–205.
62. Gade, S. K., Coulter, K. E., Way, J. D. Effects of fabrication technique upon material properties and permeation characteristics of palladium-gold alloy membranes for hydrogen separations. *Gold Bull.* 2010, 43(4), 287–297.
63. Chen, C.-H., Ma, Y. H. The effect of H_2S on the performance of Pd and Pd/Au composite membrane. *J. Membr. Sci.* 2010, 362(1–2), 535–544.
64. Gade, S. K., DeVoss, S. J., Coulter, K. E., Paglieri, S. N., Alptekin, G. O., Way, J. D. Palladium–gold membranes in mixed gas streams with hydrogen sulfide: Effect of alloy content and fabrication technique. *J. Membr. Sci.* 2011, 378(1–2), 35–41.
65. Braun, F., Miller, J. B., Gellman, A. J., Tarditi, A. M., Fleutot, B., Kondratyuk, P., Cornaglia, L. M. PdAgAu alloy with high resistance to corrosion by H_2S. *Int. J. Hydrog. Energy* 2012, 37(23), 18547–18555.

66. Braun, F., Tarditi, A. M., Miller, J. B., Cornaglia, L. M. Pd-based binary and ternary alloy membranes: morphological and perm-selective characterization in the presence of H_2S. *J. Membr. Sci.*
67. Peters, T. A., Kaleta, T., Stange, M., Bredesen, R. Development of ternary Pd–Ag–TM alloy membranes with improved sulphur tolerance. *J. Membr. Sci.* 2013, 429, 448–458.
68. Løvvik, O. M., Peters, T. A., Bredesen, R. First-principles calculations on sulfur interacting with ternary Pd–Ag-transition metal alloy membrane alloys. *J. Membr. Sci.* 2014, 453, 525–531.
69. Hagen, H., Sieverts, A. Germanium, Indium, Niob, Titan und Wasserstoff. *Z. Für Anorg. Allg. Chem.* 1929, 185(1), 225–238.
70. Veleckis, E., Edwards, R. K. Thermodynamic properties in the systems vanadium-hydrogen, niobium-hydrogen, and tantalum-hydrogen. *J. Phys. Chem.* 1969, 73(3), 683–692.
71. Makrides, A. C., Wright, M. A., Jewett, D. N. Separation of hydrogen by permeation, 1967.
72. Lee, K., Yuan, M., Wilcox, J. Understanding deviations in hydrogen solubility predictions in transition metals through first-principles calculations. *J. Phys. Chem. C* 2015, 119(34), 19642–19653.
73. Matsumoto, Y., Yukawa, H., Nambu, T. In situ quantitative evaluation of hydrogen embrittlement in group 5 metals used for hydrogen separation and purification. In *Advances in Hydrogen Production, Storage and Distribution*, Elsevier: Amsterdam, 2014, pp. 317–340.
74. Sholl, D. S. Using density functional theory to study hydrogen diffusion in metals: A brief overview. *J. Alloys Compd.* 2007, 446–447, 462–468.
75. Völkl, J., Alefeld, G. Diffusion of hydrogen in metals. In *Hydrogen in Metals I*, Alefeld, P. D. G., Völkl, D. J. (Eds.), Topics in Applied Physics, Springer. Berlin Heidelberg, 1978, pp. 321–348.
76. Fukai, Y. *The Metal-Hydrogen System.* Springer Series in Materials Science, Vol. 21. Springer-Verlag: Berlin/Heidelberg, 2005.
77. Gahr, S., Birnbaum, H. K. Hydrogen embrittlement of niobium—III. High temperature behavior. *Acta Metall.* 1978, 26(11), 1781–1788.
78. Gahr, S., Grossbeck, M. L., Birnbaum, H. K. Hydrogen embrittlement of Nb I — Macroscopic behavior at low temperatures. *Acta Metall.* 1977, 25(2), 125–134.
79. Yukawa, H., Nambu, T., Matsumoto, Y. 13 — Design of group 5 metal-based alloy membranes with high hydrogen permeability and strong resistance to hydrogen embrittlement. In *Advances in Hydrogen Production, Storage and Distribution*, Woodhead Publishing, Cambridge, 2014, pp. 341–367.
80. Suzuki, A., Yukawa, H., Ijiri, S., Nambu, T., Matsumoto, Y., Murata, Y. Alloying effects on hydrogen solubility and hydrogen permeability for V-based alloy membranes. *Mater. Trans.* 2015, 56(10), 1688–1692.
81. Griffiths, R., Pryde, J. A., Righini-Brand, A. Phase diagram and thermodynamic data for the hydrogen/vanadium system. *J. Chem. Soc. Faraday Trans. 1 Phys. Chem. Condens. Phases* 1972, 68(0), 2344–2349.

82. Zabel, H., Peisl, J. The incoherent phase transitions of hydrogen and deuterium in niobium. *J. Phys. F Met. Phys.* 1979, 9(7), 1461.
83. San-Martin, A., Manchester, F. D. The H-Ta (hydrogen-tantalum) system. *J. Phase Equilibria.* 1991, 12(3), 332–343.
84. Hume-Rothery, W. Atomic diameters, atomic volumes and solid solubility relations in alloys. *Acta Metall.* 1966, 14(1), 17–20.
85. Dolan, M. D., Kellam, M. E., McLennan, K. G., Liang, D., Song, G. Hydrogen transport properties of several vanadium-based binary alloys. *Int. J. Hydrog. Energy* 2013, 38(23), 9794–9799.
86. Ozaki, T., Zhang, Y., Komaki, M., Nishimura, C. Preparation of palladium-coated V and V–15Ni membranes for hydrogen purification by electroless plating technique. *Int. J. Hydrog. Energy* 2003, 28(3), 297–302.
87. Nishimura, C., Komaki, M., Hwang, S., Amano, M. V–Ni alloy membranes for hydrogen purification. *J. Alloys Compd.* 2002, 330, 902–906.
88. Dolan, M. D., McLennan, K. G., Way, J. D. Diffusion of atomic hydrogen through V–Ni alloy membranes under nondilute conditions. *J. Phys. Chem. C* 2012, 116(1), 1512–1518.
89. Yukawa, H., Tsukada, C., Nambu, T., Matsumoto, Y. Hydrogen solubility and permeability of V–W–Mo alloy membrane for hydrogen separation and purification. *J. Alloys Compd.* 2013, 580, *Supplement 1*, S386–S390.
90. Yukawa, H., Nambu, T., Matsumoto, Y. 13 — Design of group 5 metal-based alloy membranes with high hydrogen permeability and strong resistance to hydrogen embrittlement. In *Advances in Hydrogen Production, Storage and Distribution*, Woodhead Publishing: Cambridge, 2014, 341–367.
91. Alimov, V. N., Busnyuk, A. O., Notkin, M. E., Peredistov, E. Y., Livshits, A. I. Substitutional V–Pd alloys for the membranes permeable to hydrogen: Hydrogen solubility at 150–400°C. *Int. J. Hydrog. Energy* 2014, 39(34), 19682–19690.
92. Alimov, V. N., Busnyuk, A. O., Notkin, M., Peredistov, E. Y., Livshits, A. I. Hydrogen transport through V–Pd alloy membranes: Hydrogen solution, permeation and diffusion. *J. Membr. Sci.* 2015, 481, 54–62.
93. Dolan, M. D., Song, G., McLennan, K. G., Kellam, M. E., Liang, D. The effect of Ti on the microstructure, hydrogen absorption and diffusivity of V–Ni alloy membranes. *J. Membr. Sci.* 2012, 415–416, 320–327.
94. Ohtsu, N., Ishikawa, K., Kobori, Y. Hydrogen permeability degradation of Pd-coated Nb–TiNi alloy caused by its interfacial diffusion. *Appl. Surf. Sci.* 2016, 360, 566–571.
95. Tsuchimoto, K., Yukawa, H., Nambu, T., Matsumoto, Y., Murata, Y. Design of Nb–W–Mo alloy membrane for hydrogen separation and purification. *J. Alloys Compd.* 2013, 580, *Supplement 1*, S391–S396.
96. Taxak, M., Kumar, S., Kalekar, B. B., Krishnamurthy, N. Effect of nickel addition on the solubility of hydrogen in tantalum. *Int. J. Hydrog. Energy* 2013, 38(18), 7561–7568.
97. Taxak, M., Kumar, S., Sheelvantra, S., Krishnamurthy, N. Effect of iron on the solubility of hydrogen in tantalum. *J. Mater. Sci.* 2014, 49(24), 8471–8477.

98. Taxak, M., Kumar, S., Krishnamurthy, N. Thermodynamic parameters for the Ta–Cr–H solid solution from equilibrium (P–C–T) data. *J. Chem. Thermodyn.* 2013, 67, 48–54.
99. Taxak, M., Krishnamurthy, N. Effect of aluminum on hydrogen absorption kinetics of tantalum. *J. Alloys Compd.* 2015, 623, 121–126.
100. Kumar, S., Singh, A., Jain, U., Dey, G. K. Development of tantalum–zirconium alloy for hydrogen purification. *Fusion Eng. Des.*
101. Yamaura, S., Shimpo, Y., Okouchi, H., Nishida, M., Kajita, O., Kimura, H., Inoue, A. Hydrogen permeation characteristics of melt-spun Ni-Nb-Zr amorphous alloy membranes. *Mater. Trans.* 2001, 42(9), 1885–1890.
102. Busnyuk, A. O., Notkin, M. E., Grigoriadi, I. P., Alimov, V. N., Lifshitz, A. I. Thermal degradation of a palladium coating on hydrogen-tight niobium membranes. *Tech. Phys.* 2010, 55(1), 117–124.
103. Sasaki, K., Hattori, M., Tsuchimoto, K., Yukawa, H., Arai, S., Tokunaga, T., Murata, Y., Yamamoto, T. Microstructural analysis of thermal degradation of palladium-coated niobium membrane. *J. Alloys Compd* 2013, 573, 192–197.
104. Lee, J. S., Oyama, S. T., Boudart, M. Molybdenum carbide catalysts: I. Synthesis of unsupported powders. *J. Catal.* 1987, 106(1), 125–133.
105. Gade, S. K., Chmelka, S. J., Parks, S., Way, J. D., Wolden, C. A. Dense carbide/metal composite membranes for hydrogen separations without platinum group metals. *Adv. Mater.* 2011, 23(31), 3585–3589.
106. Wolden, C. A., Adeyemo, K., Vervuurt, R. H. J., Ostwal, M., Douglas Way, J. Mechanistic studies of hydrogen transport through Mo_2C/V composite membranes. *J. Membr. Sci.* 2013, 427, 150–154.
107. Zhu, R., Hao, S., Sholl, D. S. Selection of surface coatings for high H_2 Permeability group 5 metal membranes using first-principles calculations. *J. Phys. Chem. C* 2015, 119(14), 7848–7855.
108. Yepes, D., Cornaglia, L. M., Irusta, S., Lombardo, E. A. Different oxides used as diffusion barriers in composite hydrogen permeable membranes. *J. Membr. Sci.* 2006, 274(1–2), 92–101.
109. Nam, S.-E., Lee, K.-H. Hydrogen separation by Pd alloy composite membranes: Introduction of diffusion barrier. *J. Membr. Sci.* 2001, 192(1), 177–185.
110. David J. Edlund; Jack McCarthy. The relationship between intermetallic diffusion and flux decline in composite-metal membranes: implications for achieving long membrane lifetime. *J. Membr. Sci.* 1995, 107(1–2), 147–153.
111. Hatano, Y., Ishiyama, K., Homma, H., Watanabe, K. Improvement of high temperature stability of Pd coating on Nb by intermediate layer comprising NbC and Nb_2C. *J. Alloys Compd.* 2007, 446–447, 539–542.
112. Hatano, Y., Ishiyama, K., Homma, H., Watanabe, K. Improvement in high temperature stability of Pd coating on Nb by Nb_2C intermediate layer. *Int. J. Hydrog. Energy* 2007, 32(5), 615–619.
113. Grover Coors, W. A stoichiometric titration method for measuring galvanic hydrogen flux in ceramic hydrogen separation membranes. *J. Membr. Sci.* 2014, 458, 245–253.

114. Mathieu Marrony. Proton-Conducting Ceramics: From Fundamentals to Applied Research.
115. Malavasi, L., Fisher, C. A. J., Islam, M. S. Oxide-ion and proton conducting electrolyte materials for clean energy applications: structural and mechanistic features. *Chem. Soc. Rev.* 2010, 39(11), 4370.
116. Iwahara, H. Proton conducting ceramics and their applications. *Solid State Ion.* 1996, 86–88, Part 1, 9–15.
117. Bonanos, N., Knight, K. S., Ellis, B. Perovskite solid electrolytes: Structure, transport properties and fuel cell applications. *Solid State Ion.* 1995, 79, 161–170.
118. Kreuer, K. D. Proton-Conducting Oxides. *Annu. Rev. Mater. Res.* 2003, 33(1), 333–359.
119. Fabbri, E., Pergolesi, D., Traversa, E. Materials challenges toward proton-conducting oxide fuel cells: A critical review. *Chem. Soc. Rev.* 2010, 39(11), 4355–4369.
120. de Souza, E. C. C., Muccillo, R. Properties and applications of perovskite proton conductors. *Mater. Res.* 2010, 13(3), 385–394.
121. Reijers, R., Haije, W. Literature review on high temperature proton conducting materials. *Energy Res. Cent. Neth.* 2008.
122. Philippe Colomban. Proton Conductors: Solids, Membranes and Gels — Materials and Devices.
123. Marius Wideroe, Nadezhda Kochetova, Truls Norby. Transport numbers from hydrogen concentration cells over different oxides under oxidising and reducing conditions. *Dalton Trans.* 2004, 3147–3151.
124. Wu, J., Davies, R. A., Islam, M. S., Haile, S. M. Atomistic study of doped $BaCeO_3$: Dopant site-selectivity and cation nonstoichiometry. *Chem. Mater.* 2005, 17(4), 846–851.
125. Kumar, R. V. Electrical conducting properties of rare earth doped perovskites. *J. Alloys Compd.* 2006, 408–412, 463–467.
126. Münch, W., Seifert, G., Kreuer, K. D., Maier, J. A quantum molecular dynamics study of proton conduction phenomena in $BaCeO_3$. *Solid State Ion.* 1996, 86–88, Part 1, 647–652.
127. Shimojo, F., Hoshino, K., Okazaki, H. First-principles molecular-dynamics simulation of proton diffusion in Sc-doped $SrTiO_3$. *J. Phys. Soc. Jpn.* 1997, 66(1), 8–10.
128. Shimojo, F., Hoshino, K. Microscopic mechanism of proton conduction in perovskite oxides from ab initio molecular dynamics simulations. *Solid State Ion.* 2001, 145(1–4), 421–427.
129. Pionke, M., Mono, T., Schweika, W., Springer, T., Schober, H. Investigation of the hydrogen mobility in a mixed perovskite: $Ba[Ca_{(1+x)/3}Nb_{(2-x)/3}]O_{3-x/2}$ by quasielastic neutron scattering. *Solid State Ion.* 1997, 97(1–4), 497–504.
130. Matzke, T., Stimming, U., Karmonik, C., Soetratmo, M., Hempelmann, R., Güthoff, F. Quasielastic thermal neutron scattering experiment on the proton conductor $SrCe_{0.95}Yb_{0.05}H_{0.02}O_{2.985}$. *Solid State Ion.* 1996, 86–88, Part 1, 621–628.

131. Hempelmann, R., Soetratmo, M., Hartmann, O., Wäppling, R. Muon diffusion and trapping in proton conducting oxides. *Solid State Ion.* 1998, 107(3–4), 269–280.
132. Mather, G. C., Islam, M. S. Defect and dopant properties of the $SrCeO_3$-based proton conductor. *Chem. Mater.* 2005, 17(7), 1736–1744.
133. Islam, M. S., Slater, P. R., Tolchard, J. R., Dinges, T. Doping and defect association in $AZrO_3$ (A = Ca, Ba) and $LaMO_3$ (M = Sc, Ga) perovskite-type ionic conductors. *Dalton Trans.* 2004, (19), 3061–3066.
134. Cherry, M., Islam, M. S., Gale, J. D., Catlow, C. R. A. Computational studies of proton migration in perovskite oxides. *Solid State Ion.* 1995, 77, 207–209.
135. Iwahara, H., Yajima, T., Hibino, T., Ushida, H. Performance of solid oxide fuel cell using proton and oxide ion mixed conductors based on $BaCe_{1-x}Sm_xO_{3-\alpha}$. *J. Electrochem. Soc.* 1993, 140(6), 1687–1691.
136. Stevenson, D. A., Jiang, N., Buchanan, R. M., Henn, F. E. G. Characterization of Gd, Yb and Nd doped barium cerates as proton conductors. *Solid State Ion.* 1993, 62(3), 279–285.
137. Li, A., Liang, W., Hughes, R. Characterisation and permeation of palladium/stainless steel composite membranes. *J. Membr. Sci.* 1998, 149(2), 259–268.
138. Katahira, K., Kohchi, Y., Shimura, T., Iwahara, H. Protonic conduction in Zr-substituted $BaCeO_3$. *Solid State Ion.* 2000, 138(1–2), 91–98.
139. Hibino, T., Hashimoto, A., Suzuki, M., Sano, M. Proton conduction at the surface of Y-doped $BaCeO_3$. *J. Phys. Chem. B.* 2001, 105(46), 11399–11401.
140. Matsumoto, H., Kawasaki, Y., Ito, N., Enoki, M., Ishihara, T. Relation between electrical conductivity and chemical stability of $BaCeO_3$-based proton conductors with different trivalent dopants. *Electrochem. Solid-State Lett.* 2007, 10(4), B77–B80.
141. Melnik, J., Luo, J., Chuang, K. T., Sanger, A. R. Stability and electric conductivity of barium cerate perovskites co-doped with praseodymium. *Open Fuels Energy Sci. J.* 2008, 1(1).
142. Fabbri, E., D'Epifanio, A., Di Bartolomeo, E., Licoccia, S., Traversa, E. Tailoring the chemical stability of $Ba(Ce_{0.8-x}Zr_x)Y_{0.2}O_{3-\delta}$ protonic conductors for Intermediate Temperature Solid Oxide Fuel Cells (IT-SOFCs). *Solid State Ion.* 2008, 179(15–16), 558–564.
143. Oishi, M., Akoshima, S., Yashiro, K., Sato, K., Mizusaki, J., Kawada, T. Defect structure analysis of B-site doped perovskite-type proton conducting oxide $BaCeO_3$: Part 2: The electrical conductivity and diffusion coefficient of $BaCe_{0.9}Y_{0.1}O_{3-\delta}$. *Solid State Ion.* 2008, 179(39), 2240–2247.
144. Scholten, M. J., Schoonman, J., van Miltenburg, J. C., Oonk, H. A. J. Synthesis of strontium and barium cerate and their reaction with carbon dioxide. *Solid State Ion.* 1993, 61(1), 83–91.
145. Chen, F., Sørensen, O. T., Meng, G., Peng, D. Chemical stability study of $BaCe_{0.9}Nd_{0.1}O_{3-\alpha}$ high-temperature proton-conducting ceramic. *J. Mater. Chem.* 1997, 7(3), 481–485.

146. Bhide, S. V., Virkar, A. V. Stability of $BaCeO_3$-based proton conductors in water-containing atmospheres. *J. Electrochem. Soc.* 1999, 146(6), 2038–2044.
147. Haile, S. M., Staneff, G., Ryu, K. H. Non-stoichiometry, grain boundary transport and chemical stability of proton conducting perovskites. *J. Mater. Sci.* 2001, 36(5), 1149–1160.
148. Zakowsky, N., Williamson, S., Irvine, J. T. S. Elaboration of CO_2 tolerance limits of $BaCe_{0.9}Y_{0.1}O_{3-\delta}$ electrolytes for fuel cells and other applications. *Solid State Ion.* 2005, 176(39–40), 3019–3026.
149. Taniguchi, N., Nishimura, C., Kato, J. Endurance against moisture for protonic conductors of perovskite-type ceramics and preparation of practical conductors. *Solid State Ion.* 2001, 145(1–4), 349–355.
150. Shima, D., Haile, S. M. The influence of cation non-stoichiometry on the properties of undoped and gadolinia-doped barium cerate. *Solid State Ion.* 1997, 97(1–4), 443–455.
151. Snijkers, F. M. M., Buekenhoudt, A., Cooymans, J., Luyten, J. J. Proton conductivity and phase composition in $BaZr_{0.9}Y_{0.1}O_{3-\delta}$. *Scr. Mater.* 2004, 50(5), 655–659.
152. Kim, S., Fleig, J., Maier, J. Space charge conduction: Simple analytical solutions for ionic and mixed conductors and application to nanocrystalline ceria. *Phys. Chem. Chem. Phys.* 2003, 5(11), 2268–2273.
153. Iguchi, F., Sata, N., Yugami, H. Proton transport properties at the grain boundary of barium zirconate based proton conductors for intermediate temperature operating SOFC. *J. Mater. Chem.* 2010, 20(30), 6265–6270.
154. Kjølseth, C., Fjeld, H., Prytz, Ø., Dahl, P. I., Estournès, C., Haugsrud, R., Norby, T. Space–charge theory applied to the grain boundary impedance of proton conducting $BaZr_{0.9}Y_{0.1}O_{3-\delta}$. *Solid State Ion.* 2010, 181(5–7), 268–275.
155. Chen, C.-T., Danel, C. E., Kim, S. On the origin of the blocking effect of grain-boundaries on proton transport in yttrium-doped barium zirconates. *J. Mater. Chem.* 2011, 21(14), 5435–5442.
156. Shirpour, M., Merkle, R., Maier, J. Evidence for space charge effects in Y-doped $BaZrO_3$ from reduction experiments. *Solid State Ion.* 2012, 216, 1–5.
157. Ricote, S., Bonanos, N., Manerbino, A., Sullivan, N. P., Coors, W. G. Effects of the fabrication process on the grain-boundary resistance in $BaZr_{0.9}Y_{0.1}O_{3-\delta}$. *J. Mater. Chem. A.* 2014, 2(38), 16107–16115.
158. Babilo, P., Haile, S. M. Enhanced sintering of yttrium-doped barium zirconate by addition of ZnO. *J. Am. Ceram. Soc.* 2005, 88(9), 2362–2368.
159. Tao, S., Irvine, J. T. S. Conductivity studies of dense yttrium-doped $BaZrO_3$ sintered at 1325°C. *J. Solid State Chem.* 2007, 180(12), 3493–3503.
160. Bi, L., Fabbri, E., Sun, Z., Traversa, E. Sinteractive anodic powders improve densification and electrochemical properties of $BaZr_{0.8}Y_{0.2}O_{3-\delta}$ electrolyte films for anode-supported solid oxide fuel cells. *Energy Environ. Sci.* 2011, 4(4), 1352.

161. Gao, D., Guo, R. Structural and electrochemical properties of yttrium-doped barium zirconate by addition of CuO. *J. Alloys Compd.* 2010, 493(1–2), 288–293.
162. Taillades, G., Jacquin, M., Khani, Z., Jones, D., Marrony, M., Roziere, J. Development of proton conducting thin films from nanoparticulate precursors. *ECS Trans.* 2007, 7(1), 2291–2298.
163. Fabbri, E., Pergolesi, D., Licoccia, S., Traversa, E. Does the increase in Y-dopant concentration improve the proton conductivity of $BaZr_{1-x}Y_xO_{3-\delta}$ fuel cell electrolytes? *Solid State Ion.* 2010, 181(21–22), 1043–1051.
164. Kreuer, K.-D., Paddison, S. J., Spohr, E., Schuster, M. Transport in proton conductors for fuel-cell applications: Simulations, elementary reactions, and phenomenology. *Chem. Rev.* 2004, 104(10), 4637–4678.
165. Ricote, S., Bonanos, N., Manerbino, A., Coors, W. G. Conductivity study of dense $BaCe_xZr_{(0.9-x)}Y_{0.1}O_{(3-\delta)}$ prepared by solid state reactive sintering at 1500°C. *Int. J. Hydrog. Energy* 2012, 37(9), 7954–7961.
166. Zuo, C., Zha, S., Liu, M., Hatano, M., Uchiyama, M. $Ba(Zr_{0.1}Ce_{0.7}Y_{0.2})O_{3-\delta}$ as an Electrolyte for low-temperature solid-oxide fuel cells. *Adv. Mater.* 2006, 18(24), 3318–3320.
167. Sawant, P., Varma, S., Wani, B. N., Bharadwaj, S. R. Synthesis, stability and conductivity of $BaCe_{0.8-x}Zr_xY_{0.2}O_{3-\delta}$ as electrolyte for proton conducting SOFC. *Int. J. Hydrog. Energy* 2012, 37(4), 3848–3856.
168. Guo, Y., Lin, Y., Ran, R., Shao, Z. Zirconium doping effect on the performance of proton-conducting $BaZr_yCe_{0.8-y}Y_{0.2}O_{3-\delta}$ ($0.0 \leq y \leq 0.8$) for fuel cell applications. *J. Power Sources* 2009, 193(2), 400–407.
169. Ricote, S., Bonanos, N., Caboche, G. Water vapour solubility and conductivity study of the proton conductor $BaCe_{(0.9-x)}Zr_xY_{0.1}O_{(3-\delta)}$. *Solid State Ion.* 2009, 180(14–16), 990–997.
170. Barison, S., Battagliarin, M., Cavallin, T., Doubova, L., Fabrizio, M., Mortalò, C., Boldrini, S., Malavasi, L., Gerbasi, R. High conductivity and chemical stability of $BaCe_{1-x-y}Zr_xY_yO_{3-\delta}$ proton conductors prepared by a sol–gel method. *J. Mater. Chem.* 2008, 18(42), 5120–5128.
171. Nasani, N., Dias, P. A. N., Saraiva, J. A., Fagg, D. P. Synthesis and conductivity of $Ba(Ce,Zr,Y)O_{3-\delta}$ electrolytes for PCFCs by new nitrate-free combustion method. *Int. J. Hydrog. Energy* 2013, 38(20), 8461–8470.
172. Babilo, P., Uda, T., Haile, S. M. Processing of yttrium-doped barium zirconate for high proton conductivity. *J. Mater. Res.* 2007, 22(5), 1322–1330.
173. Coors, W. G. Co-Ionic Conduction in Protonic Ceramics of the Solid Solution, $BaCe_{(x)}Zr_{(y-x)}Y_{(1-y)}O_{(3-d)}$ Part I: Fabrication and Microstructure. *Adv. Ceram. Synth. Charact. Process. Specif. Appl. InTech Croat.* 2011.
174. Tong, J., Clark, D., Hoban, M., O'Hayre, R. Cost-effective solid-state reactive sintering method for high conductivity proton conducting yttrium-doped barium zirconium ceramics. *Solid State Ion.* 2010, 181(11–12), 496–503.
175. Coors, W. G., Manerbino, A. Characterization of composite cermet with 68 wt.% NiO and $BaCe_{0.2}Zr_{0.6}Y_{0.2}O_{3-\delta}$. *J. Membr. Sci.* 2011, 376(1–2), 50–55.

176. Babiniec, S. M., Ricote, S., Sullivan, N. P. Characterization of ionic transport through $BaCe_{0.2}Zr_{0.7}Y_{0.1}O_{3-\delta}$ membranes in galvanic and electrolytic operation. *Int. J. Hydrog. Energy* 2015, 40(30), 9278–9286.
177. Dippon, M., Babiniec, S. M., Ding, H., Ricote, S., Sullivan, N. P. Exploring electronic conduction through $BaCe_xZr_{0.9-x}Y_{0.1}O_{3-d}$ proton-conducting ceramics. *Solid State Ion.* 2016, 286, 117–121.
178. Robinson, S., Manerbino, A., Grover Coors, W. Galvanic hydrogen pumping in the protonic ceramic perovskite. *J. Membr. Sci.* 2013, 446, 99–105.
179. Song, S.-J., Wachsman, E. D., Dorris, S. E., Balachandran, U. Defect chemistry modeling of high-temperature proton-conducting cerates. *Solid State Ion.* 2002, 149(1–2), 1–10.
180. Norby, T., Widerøe, M., Glöckner, R., Larring, Y. Hydrogen in oxides. *Dalton Trans.* 2004, (19), 3012–3018.
181. Cheng, S., Gupta, V. K., Lin, J. Y. S. Synthesis and hydrogen permeation properties of asymmetric proton-conducting ceramic membranes. *Solid State Ion.* 2005, 176(35–36), 2653–2662.
182. Hamakawa, S., Li, L., Li, A., Iglesia, E. Synthesis and hydrogen permeation properties of membranes based on dense $SrCe_{0.95}Yb_{0.05}O_{3-\alpha}$ thin films. *Solid State Ion.* 2002, 148(1–2), 71–81.
183. Song, S.-J., Moon, J.-H., Lee, T. H., Dorris, S. E., Balachandran, U. Thickness dependence of hydrogen permeability for Ni–$BaCe_{0.8}Y_{0.2}O_{3-\delta}$. *Solid State Ion.* 2008, 179(33–34), 1854–1857.
184. Fontaine, M., Norby, T., Larring, Y., Grande, T., Bredesen, R. Oxygen and Hydrogen Separation Membranes Based on Dense Ceramic Conductors. In *Membrane Science and Technology*, Vol. 13, Elsevier: Amsterdam, 2008, pp. 401–458.
185. Rosensteel, W. A., Ricote, S., Sullivan, N. P. Hydrogen permeation through dense $BaCe_{0.8}Y_{0.2}O_{3-\delta}$ — $Ce_{0.8}Y_{0.2}O_{2-\delta}$ composite-ceramic hydrogen separation membranes. *Int. J. Hydrog. Energy* 2016, 41(4), 2598–2606.
186. Coors, W. G. Co-ionic conduction in protonic ceramics of the solid solution, $BaCe_{(x)}Zr_{(y-x)}Y_{(1-y)}O_{(3-d)}$ Part II: Co-ionic conduction. *Adv. Ceram. Synth. Charact. Process. Specif. Appl. InTech Croat.* 2011.
187. Vasileiou, E., Kyriakou, V., Garagounis, I., Vourros, A., Manerbino, A., Coors, W. G., Stoukides, M. Reaction rate enhancement during the electrocatalytic synthesis of ammonia in a $BaZr_{0.7}Ce_{0.2}Y_{0.1}O_{2.9}$ Solid Electrolyte Cell. *Top. Catal.* 2015, 58(18–20), 1193–1201.
188. Vasileiou, E., Kyriakou, V., Garagounis, I., Vourros, A., Manerbino, A., Coors, W. G., Stoukides, M. Electrochemical enhancement of ammonia synthesis in a $BaZr_{0.7}Ce_{0.2}Y_{0.1}O_{2.9}$ solid electrolyte cell. *Solid State Ion.* 2016, 288, 357–362.
189. Kyriakou, V., Garagounis, I., Vourros, A., Vasileiou, E., Manerbino, A., Coors, W. G., Stoukides, M. Methane steam reforming at low temperatures in a $BaZr_{0.7}Ce_{0.2}Y_{0.1}O_{2.9}$ proton conducting membrane reactor. *Appl. Catal. B Environ.* 2016, 186, 1–9.

190. De Falco, M., Iaquaniello, G., Palo, E., Cucchiella, B., Palma, V., Ciambelli, P. Palladium-based membranes for hydrogen separation: preparation, economic analysis and coupling with a water gas shift reactor. In *Handbook of Membrane Reactors*, Elsevier: Amsterdam, 2013, pp. 456–486.
191. Mendes, D., Mendes, A., Madeira, L. M., Iulianelli, A., Sousa, J. M., Basile, A. The water-gas shift reaction: from conventional catalytic systems to Pd-based membrane reactors-a review. *Asia-Pac. J. Chem. Eng.* 2010, 5(1), 111–137.
192. Saracco, G., Neomagus, H. W. J. P., Versteeg, G. F., Swaaij, W. P. M. va. High-temperature membrane reactors: potential and problems. *Chem. Eng. Sci.* 1999, 54(13–14), 1997–2017.
193. Zhang, J., Xu, H., Li, W. High-purity COx-free H_2 generation from NH_3 via the ultra permeable and highly selective Pd membranes. *J. Membr. Sci.* 2006, 277(1–2), 85–93.
194. Lu, N., Xie, D. Novel membrane reactor concepts for hydrogen production from hydrocarbons: A Review. *Int. J. Chem. React. Eng.* 2016, 14(1), 1–31.
195. Iulianelli, A., Ribeirinha, P., Mendes, A., Basile, A. Methanol steam reforming for hydrogen generation via conventional and membrane reactors: A review. *Renew. Sustain. Energy Rev.* 2014, 29, 355–368.
196. Basile, A., Iulianelli, A., Tong, J. Membrane reactors for the conversion of methanol and ethanol to hydrogen. In *Membrane Reactors for Energy Applications and Basic Chemical Production*, Elsevier: Amsterdam, 2015, pp. 187–208.
197. Iulianelli, A., Pirola, C., Comazzi, A., Galli, F., Manenti, F., Basile, A. Water gas shift membrane reactors. In *Membrane Reactors for Energy Applications and Basic Chemical Production*, Elsevier: Amsterdam, 2015, pp. 3–29.
198. Cornaglia, L., Múnera, J., Lombardo, E. Recent advances in catalysts, palladium alloys and high temperature WGS membrane reactors. *Int. J. Hydrog. Energy* 2015, 40(8), 3423–3437.
199. Dittmeyer, R., Höllein, V., Daub, K. Membrane reactors for hydrogenation and dehydrogenation processes based on supported palladium. *J. Mol. Catal. Chem.* 2001, 173(1–2), 135–184.
200. Iwahara, H. Technological challenges in the application of proton conducting ceramics. *Solid State Ion.* 1995, 77, 289–298.
201. Liu, S., Tan, X., Li, K., Hughes, R. Methane coupling using catalytic membrane reactors. *Catal. Rev.* 2001, 43(1–2), 147–198.
202. Kokkofitis, C., Ouzounidou, M., Skodra, A., Stoukides, M. High temperature proton conductors: Applications in catalytic processes. *Solid State Ion.* 2007, 178(7–10), 507–513.
203. Garagounis, I., Kyriakou, V., Anagnostou, C., Bourganis, V., Papachristou, I., Stoukides, M. Solid electrolytes: Applications in heterogeneous catalysis and chemical cogeneration. *Ind. Eng. Chem. Res.* 2011, 50(2), 431–472.
204. Mori, K., Lida, K. Method for decomposing methane, 1987.
205. Mori, K., Imai, T. Method for Reforming Methane, 1987.

206. Iwahara, H., Uchida, H., Tanaka, S. High temperature-type proton conductive solid oxide fuel cells using various fuels. *J. Appl. Electrochem.* 1986, 16(5), 663–668.
207. Iwahara, H., Uchida, H., Morimoto, K., Hosogi, S. High-temperature C_1-gas fuel cells using proton-conducting solid electrolytes. *J. Appl. Electrochem.* 1989, 19(3), 448–452.
208. Woldman, L. S., Sokolovskii, V. D. Electrocatalytic methane coupling in the absence of oxygen on a high-temperature proton-conducting electrolyte. *Catal. Lett.* 1991, 8(1), 61–66.
209. Chiang, P.-H., Eng, D., Stoukides, M. Electrocatalytic methane dimerization with a Yb-doped $SrCeO_3$ solid electrolyte. *J. Electrochem. Soc.* 1991, 138(6), L11–L12.
210. Chiang, P. H., Eng, D., Alqahtany, H., Stoukides, M. Nonoxidative methane coupling with the aid of solid electrolytes. *Solid State Ion.* 1992, 53, 135–141.
211. Hamakawa, S., Hibino, T., Iwahara, H. Electrochemical methane coupling using protonic conductors. *J. Electrochem. Soc.* 1993, 140(2), 459–462.
212. Hamakawa, S., Hibino, T., Iwahara, H. Electrochemical Hydrogen Permeation in a Proton-Hole Mixed Conductor and Its Application to a Membrane Reactor. *J. Electrochem. Soc.* 1994, 141(7), 1720–1725.
213. Langguth, J., Dittmeyer, R., Hofmann, H., Tomandl, G. Studies on oxidative coupling of methane using high-temperature proton-conducting membranes. *Appl. Catal. Gen.* 1997, 158(1–2), 287–305.
214. Terai, T., Li, X., Tomishige, K., Fujimoto, K. Development of catalyst-membrane system for oxidative coupling of methane by water. *Chem. Lett.* 1999, (4), 323–324.
215. Kyriakou, V., Athanasiou, C., Garagounis, I., Skodra, A., Stoukides, M. Production of C_2 hydrocarbons and H_2 from CH_4 in a proton conducting cell. *Solid State Ion.* 2012, 225, 219–222.
216. Kyriakou, V., Athanasiou, C., Garagounis, I., Skodra, A., Stoukides, M. Production of H_2 and C_2 hydrocarbons from methane in a proton conducting solid electrolyte cell using a Au–5Ce–$5Na_2WO_4/SiO_2$ anode. *Int. J. Hydrog. Energy* 2012, 37(21), 16636–16641.
217. Coors, W. G. Protonic ceramic fuel cells for high-efficiency operation with methane. *J. Power Sources* 2003, 118(1–2), 150–156.
218. Coors, W. G. Steam reforming and water-gas shift by steam permeation in a protonic ceramic fuel cell. *J. Electrochem. Soc.* 2004, 151(7), A994–A997.
219. Matsumoto, H., Okubo, M., Hamajima, S., Katahira, K., Iwahara, H. Extraction and production of hydrogen using high-temperature proton conductor. *Solid State Ion.* 2002, 152–153, 715–720.
220. Karagiannakis, G., Kokkofitis, C., Zisekas, S., Stoukides, M. Catalytic and electrocatalytic production of H_2 from propane decomposition over Pt and Pd in a proton-conducting membrane-reactor. *Catal. Today* 2005, 104(2–4), 219–224.

221. Karagiannakis, G., Zisekas, S., Kokkofitis, C., Stoukides, M. Effect of H_2O presence on the propane decomposition reaction over Pd in a proton conducting membrane reactor. *Appl. Catal. Gen.* 2006, 301(2), 265–271.
222. Lu, G. Q., Diniz da Costa, J. C., Duke, M., Giessler, S., Socolow, R., Williams, R. H., Kreutz, T. Inorganic membranes for hydrogen production and purification: A critical review and perspective. *J. Colloid Interface Sci.* 2007, 314(2), 589–603.
223. Rostrup-Nielsen, J. R., Rostrup-Nielsen, T. Large-scale hydrogen production. *Cattech.* 2002, 6(4), 150–159.
224. *Report of the hydrogen production expert panel: A subcommittee of the hydrogen and fuel cell technical advisory committee.* Department of Energy, 2013.
225. Yun, S., Ted Oyama, S. Correlations in palladium membranes for hydrogen separation: A review. *J. Membr. Sci.* 2011, 375(1–2), 28–45.
226. Schoots, K., Ferioli, F., Kramer, G., Vanderzwaan, B. Learning curves for hydrogen production technology: An assessment of observed cost reductions. *Int. J. Hydrog. Energy* 2008, 33(11), 2630–2645.
227. Sieminski, A. International Energy Outlook 2013.
228. Basile, A., Liguori, S., Iulianelli, A. Membrane reactors for methane steam reforming (MSR). In *Membrane Reactors for Energy Applications and Basic Chemical Production*, Elsevier: Amsterdam, 2015, pp. 31–59.
229. Shu, J., Grandjean, B. P. A., Kaliaguine, S. Methane steam reforming in asymmetric Pd- and Pd–Ag/porous SS membrane reactors. *Appl. Catal. Gen.* 1994, 119(2), 305–325.
230. Gallucci, F., Paturzo, L., Famà, A., Basile, A. Experimental Study of the Methane Steam Reforming Reaction in a Dense Pd/Ag Membrane Reactor. *Ind. Eng. Chem. Res.* 2004, 43(4), 928–933.
231. Tong, J., Su, L., Kashima, Y., Shirai, R., Suda, H., Matsumura, Y. Simultaneously depositing Pd–Ag thin membrane on asymmetric porous stainless steel tube and application to produce hydrogen from steam reforming of methane. *Ind. Eng. Chem. Res.* 2006, 45(2), 648–655.
232. Patil, C. S., van Sint Annaland, M., Kuipers, J. A. M. Fluidised bed membrane reactor for ultrapure hydrogen production via methane steam reforming: Experimental demonstration and model validation. *Chem. Eng. Sci.* 2007, 62(11), 2989–3007.
233. Kume, T., Ikeda, Y., Iseki, T., Yakabe, H., Tanaka, H., Hikosaka, H., Takagi, Y., Ito, M. Performance evaluation of membrane on catalyst module for hydrogen production from natural gas. *Int. J. Hydrog. Energy* 2013, 38(14), 6079–6084.
234. Tong, J., Matsumura, Y. Effect of catalytic activity on methane steam reforming in hydrogen-permeable membrane reactor. *Appl. Catal. Gen.* 2005, 286(2), 226–231.
235. Tong, J., Matsumura, Y., Suda, H., Haraya, K. Experimental study of steam reforming of methane in a thin ($6\,\mu$M) Pd-based membrane reactor. *Ind. Eng. Chem. Res.* 2005, 44(5), 1454–1465.

236. Matsumura, Y., Tong, J. Methane steam reforming in hydrogen-permeable membrane reactor for pure hydrogen production. *Top. Catal.* 2008, 51(1–4), 123–132.
237. Chen, Y., Wang, Y., Xu, H., Xiong, G. Efficient production of hydrogen from natural gas steam reforming in palladium membrane reactor. *Appl. Catal. B Environ.* 2008, 81(3–4), 283–294.
238. Iulianelli, A., Manzolini, G., De Falco, M., Campanari, S., Longo, T., Liguori, S., Basile, A. H_2 production by low pressure methane steam reforming in a Pd–Ag membrane reactor over a Ni-based catalyst: Experimental and modeling. *Int. J. Hydrog. Energy* 2010, 35(20), 11514–11524.
239. Basile, A., Campanari, S., Manzolini, G., Iulianelli, A., Longo, T., Liguori, S., De Falco, M., Piemonte, V. Methane steam reforming in a Pd–Ag membrane reformer: An experimental study on reaction pressure influence at middle temperature. *Int. J. Hydrog. Energy* 2011, 36(2), 1531–1539.
240. Sarić, M., van Delft, Y. C., Sumbharaju, R., Meyer, D. F., de Groot, A. Steam reforming of methane in a bench-scale membrane reactor at realistic working conditions. *Catal. Today* 2012, 193(1), 74–80.
241. Abu El Hawa, H. W., Paglieri, S. N., Morris, C. C., Harale, A., Douglas Way, J. Application of a Pd–Ru composite membrane to hydrogen production in a high temperature membrane reactor. *Sep. Purif. Technol.* 2015, 147, 388–397.
242. Shirasaki, Y., Tsuneki, T., Ota, Y., Yasuda, I., Tachibana, S., Nakajima, H., Kobayashi, K. Development of membrane reformer system for highly efficient hydrogen production from natural gas. *Int. J. Hydrog. Energy* 2009, 34(10), 4482–4487.
243. Kume, T., Ikeda, Y., Nakagawa, Y., Tominaga, R., Iseki, T., Tamura, M., Yakabe, H., Tanaka, H., Hikosaka, H., Takagi, Y., *et al.* Development of compact and efficient hydrogen production module with membrane on catalyst. *J. Nat. Gas Sci. Eng.* 2011, 3(5), 591–600.
244. Shirasaki, Y., Yasuda, I. Membrane reactor for hydrogen production from natural gas at the Tokyo Gas Company: A case study. In *Handbook of Membrane Reactors*, Elsevier: Amsterdam, 2013, pp. 487–507.
245. Iaquaniello, G., Palo, E., Salladini, A., Cucchiella, B. Using palladium membrane reformers for hydrogen production. In *Palladium Membrane Technology for Hydrogen Production, Carbon Capture and Other Applications*, Elsevier: Amsterdam, 2015, pp. 287–301.
246. Abu El Hawa, H. W., Lundin, S.-T. B., Patki, N. S., Douglas Way, J. Steam methane reforming in a PdAu membrane reactor: Long-term assessment. *Int. J. Hydrog. Energy*
247. Dolan, M. D., Hla, S. S., Morpeth, L. D. Design and operational considerations for a catalytic membrane reactor incorporating a vanadium-based membrane. *Sep. Purif. Technol.* 2015, 147, 398–405.
248. Vigneault, A., Grace, J. R. Hydrogen production in multi-channel membrane reactor via steam methane reforming and methane catalytic combustion. *Int. J. Hydrog. Energy* 2015, 40(1), 233–243.

249. Patrascu, M., Sheintuch, M. On-site pure hydrogen production by methane steam reforming in high flux membrane reactor: Experimental validation, model predictions and membrane inhibition. *Chem. Eng. J.* 2015, 262, 862–874.
250. Erisman, J. W., Sutton, M. A., Galloway, J., Klimont, Z., Winiwarter, W. How a century of ammonia synthesis changed the world. *Nat. Geosci.* 2008, 1(10), 636–639.
251. Klerke, A., Christensen, C. H., Nørskov, J. K., Vegge, T. Ammonia for hydrogen storage: challenges and opportunities. *J. Mater. Chem.* 2008, 18(20), 2304.
252. Lan, R., Irvine, J. T. S., Tao, S. Ammonia and related chemicals as potential indirect hydrogen storage materials. *Int. J. Hydrog. Energy* 2012, 37(2), 1482–1494.
253. Chiuta, S., Everson, R. C., Neomagus, H. W. J. P., van der Gryp, P., Bessarabov, D. G. Reactor technology options for distributed hydrogen generation via ammonia decomposition: A review. *Int. J. Hydrog. Energy* 2013, 38(35), 14968–14991.
254. Afif, A., Radenahmad, N., Cheok, Q., Shams, S., Kim, J. H., Azad, A. K. Ammonia-fed fuel cells: a comprehensive review. *Renew. Sustain. Energy Rev.* 2016, 60, 822–835.
255. Green Jr., L. An ammonia energy vector for the hydrogen economy. *Int. J. Hydrog. Energy* 1982, 7(4), 355–359.
256. Burgess, D. R. Thermochemical data. In *NIST Chemistry WebBook, NIST Standard Reference Database Number 69*, Linstrom, P. J., Mallard, W. G. (Eds.), National Institute of Standards and Technology. Gaithersburg MD.
257. Lan, R., Irvine, J. T. S., Tao, S. Synthesis of ammonia directly from air and water at ambient temperature and pressure. *Sci. Rep.* 2013, 3.
258. Amar, I. A., Petit, C. T. G., Mann, G., Lan, R., Skabara, P. J., Tao, S. Electrochemical synthesis of ammonia from N_2 and H_2O based on $(Li,Na,K)_2CO_3$–$Ce_{0.8}Gd_{0.18}Ca_{0.02}O_{2-\delta}$ composite electrolyte and $CoFe_2O_4$ cathode. *Int. J. Hydrog. Energy* 2014, 39(9), 4322–4330.
259. Lan, R., Tao, S. Ammonia as a suitable fuel for fuel cells. *Front. Energy Res.* 2014, 2.
260. Halseid, R., Vie, P. J. S., Tunold, R. Effect of ammonia on the performance of polymer electrolyte membrane fuel cells. *J. Power Sources* 2006, 154(2), 343–350.
261. Rahimpour, M. R., Asgari, A. Production of hydrogen from purge gases of ammonia plants in a catalytic hydrogen-permselective membrane reactor. *Int. J. Hydrog. Energy* 2009, 34(14), 5795–5802.
262. Li, G., Kanezashi, M., Tsuru, T. Highly enhanced ammonia decomposition in a bimodal catalytic membrane reactor for COx-free hydrogen production. *Catal. Commun.* 2011, 15(1), 60–63.
263. Li, G., Kanezashi, M., Lee, H. R., Maeda, M., Yoshioka, T., Tsuru, T. Preparation of a novel bimodal catalytic membrane reactor and its application to ammonia decomposition for COx-free hydrogen production. *Int. J. Hydrog. Energy* 2012, 37(17), 12105–12113.

264. Gómez-García, M. á., Dobrosz-Gómez, I., Fontalvo, J., Rynkowski, J. M. Membrane reactor design guidelines for ammonia decomposition. *Catal. Today* 2012, 191(1), 165–168.
265. García-García, F. R., Ma, Y. H., Rodríguez-Ramos, I., Guerrero-Ruiz, A. High purity hydrogen production by low temperature catalytic ammonia decomposition in a multifunctional membrane reactor. *Catal. Commun.* 2008, 9(3), 482–486.
266. Itoh, N., Oshima, A., Suga, E., Sato, T. Kinetic enhancement of ammonia decomposition as a chemical hydrogen carrier in palladium membrane reactor. *Catal. Today* 2014, 236, 70–76.
267. Israni, S. H., Nair, B. K. R., Harold, M. P. Hydrogen generation and purification in a composite Pd hollow fiber membrane reactor: Experiments and modeling. *Catal. Today* 2009, 139(4), 299–311.
268. Di Carlo, A., Dell'Era, A., Del Prete, Z. 3D simulation of hydrogen production by ammonia decomposition in a catalytic membrane reactor. *Int. J. Hydrog. Energy* 2011, 36(18), 11815–11824.
269. Rizzuto, E., Palange, P., Del Prete, Z. Characterization of an ammonia decomposition process by means of a multifunctional catalytic membrane reactor. *Int. J. Hydrog. Energy* 2014, 39(22), 11403–11410.
270. Lundin, S.-T. B., Yamaguchi, T., Wolden, C. A., Oyama, S. T., Way, J. D. The role (or lack thereof) of nitrogen or ammonia adsorption-induced hydrogen flux inhibition on palladium membrane performance. *J. Membr. Sci.* 2016, 514, 65–72.
271. Okazaki, J., Ikeda, T., Tanaka, D. A. P., Tanco, M. A. L., Wakui, Y., Sato, K., Mizukami, F., Suzuki, T. M. Strong interaction at the palladium/alumina interface of membrane during hydrogen permeation at elevated temperature. *Chem. Lett.* 2008, 37(9), 1004–1005.
272. Okazaki, J., Ikeda, T., Pacheco Tanaka, D. A., Llosa Tanco, M. A., Wakui, Y., Sato, K., Mizukami, F., Suzuki, T. M. Importance of the support material in thin palladium composite membranes for steady hydrogen permeation at elevated temperatures. *Phys. Chem. Chem. Phys.* 2009, 11(38), 8632.
273. Okazaki, J., Ikeda, T., Tanaka, D. A. P., Sato, K., Suzuki, T. M., Mizukami, F. An investigation of thermal stability of thin palladium–silver alloy membranes for high temperature hydrogen separation. *J. Membr. Sci.* 2011, 366(1–2), 212–219.
274. Rennard, R. J., Kokes, R. J. Hydrogenation of ethylene and propylene over palladium hydride. *J. Phys. Chem.* 1966, 70(8), 2543–2549.
275. Gryaznov, V. M., Slin'ko, M. G. Selectivity in catalysis by hydrogen-porous membranes. *Faraday Discuss. Chem. Soc.* 1981, 72, 73.
276. Gryaznov, V. M., Ermilova, M. M., Morozova, L. S., Orekhova, N. V., Polyakova, V. P., Roshan, N. R., Savitsky, E. M., Parfenova, N. I. Palladium alloys as hydrogen permeable catalysts in hydrogenation and dehydrogenation reactions. *J. Common Met.* 1983, 89(2), 529–535.
277. Nagamoto, H., Inoue, H. A reactor with catalytic membrane permeated by hydrogen. *Chem. Eng. Commun.* 1985, 34(1–6), 315–323.

278. Arai, M., Wada, Y., Nishiyama, Y. Thiophene hydrodesulfurization by catalytic palladium membrane systems. *J. Jpn. Pet. Inst.* 1993, 36(1), 44–49.
279. Itoh, N., Xu, W.-C. Selective hydrogenation of phenol to cyclohexanone using palladium-based membranes as catalysts. *Appl. Catal. Gen.* 1993, 107(1), 83–100.
280. Itoh, N., Xu, W. C., Sathe, A. M. Capability of permeate hydrogen through palladium-based membranes for acetylene hydrogenation. *Ind. Eng. Chem. Res.* 1993, 32(11), 2614–2619.
281. Armor, J. N. Catalysis with permselective inorganic membranes. *Appl. Catal.* 1989, 49(1), 1–25.
282. Shu, J., Grandjean, B. P. A., Neste, A. V., Kaliaguine, S. Catalytic palladium-based membrane reactors: A review. *Can. J. Chem. Eng.* 1991, 69(5), 1036–1060.
283. Saracco, G., Specchia, V. Catalytic inorganic-membrane reactors: present experience and future opportunities. *Catal. Rev.* 1994, 36(2), 305–384.
284. Armor, J. N. Membrane catalysis: Where is it now, what needs to be done? *Catal. Today* 1995, 25(3–4), 199–207.
285. Inoue, T., Tanaka, Y., Pacheco Tanaka, D. A., Suzuki, T. M., Sato, K., Nishioka, M., Hamakawa, S., Mizukami, F. Direct production of hydrogen peroxide from oxygen and hydrogen applying membrane-permeation mechanism. *Chem. Eng. Sci.* 2010, 65(1), 436–440.
286. Itoh, M., Machida, K., Adachi, G. Ammonia production characteristics of Ru/Al_2O_3 catalysts using hydrogen permeable membrane. *Chem. Lett.* 2000, (10), 1162–1163.
287. Itoh, M., Saito, M., Tajima, N., Machida, K. Ammonia synthesis using atomic hydrogen supplied from silver-palladium alloy membrane. *Mater. Sci. Forum* 2007, 561–565, 1597–1600.
288. Sato, K., Natsui, M., Hasegawa, Y. Preparation of double layer membrane combined with palladium metal and FAU zeolite for catalytic membrane reactor. *Mater. Trans.* 2015, 56(4), 473–478.
289. Miyamoto, M., Hayakawa, R., Makino, Y., Oumi, Y., Uemiya, S., Asanuma, M. CO_2 methanation combined with NH_3 decomposition by in situ H_2 separation using a Pd membrane reactor. *Int. J. Hydrog. Energy* 2014, 39(19), 10154–10160.
290. Sato, K., Niwaa, S., Hanaoka, T., Komura, K., Nambab, T., Mizukami, F. Direct hydroxylation of methyl benzoate to methyl salicylate by using new Pd membrane reactor. *Catal. Lett.* 2004, 96(1/2), 107–112.
291. Niwa, S., Eswaramoorthy, M., Nair, J., Raj, A., Itoh, N., Shoji, H., Namba, T., Mizukami, F. A one-step conversion of benzene to phenol with a palladium membrane. *Science* 2002, 295(5552), 105–107.
292. Orita, H., Itoh, N. Simulation of phenol formation from benzene with a Pd membrane reactor: ab initio periodic density functional study. *Appl. Catal. Gen.* 2004, 258(1), 17–23.
293. Dittmeyer, R., Bortolotto, L. Modification of the catalytic properties of a Pd membrane catalyst for direct hydroxylation of benzene to phenol in a

double-membrane reactor by sputtering of different catalyst systems. *Appl. Catal. Gen.* 2011, 391(1–2), 311–318.
294. Wang, X., Meng, B., Tan, X., Zhang, X., Zhuang, S., Liu, L. Direct hydroxylation of benzene to phenol using palladium–titanium silicalite zeolite bifunctional membrane reactors. *Ind. Eng. Chem. Res.* 2014, 53(14), 5636–5645.
295. Shu, S., Huang, Y., Hu, X., Fan, Y., Xu, N. On the membrane reactor concept for one-step hydroxylation of benzene to phenol with oxygen and hydrogen. *J. Phys. Chem. C* 2009, 113(45), 19618–19622.
296. Morimoto, S., Yukawa, H., Nambu, T., Murata, Y. Investigation of new ammonia synthesis process utilizing vanadium-based hydrogen permeable alloy membrane. *Mater. Trans.* 2016, 57(3), 423–427.
297. Ertl, G. Primary steps in catalytic synthesis of ammonia. *J. Vac. Sci. Technol. Vac. Surf. Films* 1983, 1(2), 1247.
298. Ertl, G., Huber, M., Lee, S. B., Paál, Z., Weiss, M. Interactions of nitrogen and hydrogen on iron surfaces. *Appl. Surf. Sci.* 1981, 8(4), 373–386.
299. Dahl, S., Sehested, J., Jacobsen, C., Tornqvist, E., Chorkendorff, I. Surface science based microkinetic analysis of ammonia synthesis over ruthenium catalysts. *J. Catal.* 2000, 192(2), 391–399.
300. Rod, T. H., Logadottir, A., Nørskov, J. K. Ammonia synthesis at low temperatures. *J. Chem. Phys.* 2000, 112(12), 5343.
301. Honkala, K., Hellman, A., Remediakis, I. N., Logadottir, A., Carlsson, A., Dahl, S., Christensen, C. H., Norskov, J. K. Ammonia synthesis from first-principles calculations. *Science.* 2005, 307(5709), 555–558.
302. Marnellos, G., Stoukides, M. Ammonia synthesis at atmospheric pressure. *Science* 1998, 282(5386), 98–100.
303. Amar, I. A., Lan, R., Petit, C. T. G., Tao, S. Solid-state electrochemical synthesis of ammonia: A review. *J. Solid State Electrochem.* 2011, 15(9), 1845–1860.
304. Emmett, P. H., Brunauer, S. The adsorption of nitrogen by iron synthetic ammonia catalysts. *J. Am. Chem. Soc.* 1934, 56(1), 35–41.
305. Satterfield, C. *Heterogeneous Catalysis in Practice.* McGraw-Hill Inc.: New York, 1980.
306. Wang, J.-D., Xie, Y.-H., Zhang, Z.-F., Liu, R.-Q., Li, Z.-J. Protonic conduction in Ca^{2+}-doped $La_2M_2O_7$ (M = Ce, Zr) with its application to ammonia synthesis electrochemically. *Mater. Res. Bull.* 2005, 40(8), 1294–1302.
307. Marnellos, G., Zisekas, S., Stoukides, M. Synthesis of ammonia at atmospheric pressure with the use of solid state proton conductors. *J. Catal.* 2000, 193(1), 80–87.
308. Zhang, F., Yang, Q., Pan, B., Xu, R., Wang, H., Ma, G. Proton conduction in $La_{0.9}Sr_{0.1}Ga_{0.8}Mg_{0.2}O_{3-\alpha}$ ceramic prepared via microemulsion method and its application in ammonia synthesis at atmospheric pressure. *Mater. Lett.* 2007, 61(19–20), 4144–4148.

309. Wang, W. B., Liu, J. W., Li, Y. D., Wang, H. T., Zhang, F., Ma, G. L. Microstructures and proton conduction behaviors of Dy-doped $BaCeO_3$ ceramics at intermediate temperature. *Solid State Ion.* 2010, 181(15–16), 667–671.
310. Wang, W. B., Cao, X. B., Gao, W. J., Zhang, F., Wang, H. T., Ma, G. L. Ammonia synthesis at atmospheric pressure using a reactor with thin solid electrolyte $BaCe_{0.85}Y_{0.15}O_{3-\alpha}$ membrane. *J. Membr. Sci.* 2010, 360(1–2), 397–403.
311. Skúlason, E., Bligaard, T., Gudmundsdóttir, S., Studt, F., Rossmeisl, J., Abild-Pedersen, F., Vegge, T., Jónsson, H., Nørskov, J. K. A theoretical evaluation of possible transition metal electro-catalysts for N_2 reduction. *Phys. Chem. Chem. Phys.* 2012, 14(3), 1235.
312. Guo, X., Sigle, W., Maier, J. Blocking grain boundaries in yttria-doped and undoped ceria ceramics of high purity. *J. Am. Ceram. Soc.* 2003, 86(1), 77–87.
313. Murakami, T., Nishikiori, T., Nohira, T., Ito, Y. Electrolytic synthesis of ammonia in molten salts under atmospheric pressure. *J. Am. Chem. Soc.* 2003, 125(2), 334–335.
314. Murakami, T., Nohira, T., Goto, T., Ogata, Y. H., Ito, Y. Electrolytic ammonia synthesis from water and nitrogen gas in molten salt under atmospheric pressure. *Electrochimica Acta.* 2005, 50(27), 5423–5426.
315. Liu, J., Li, Y., Wang, W., Wang, H., Zhang, F., Ma, G. Proton conduction at intermediate temperature and its application in ammonia synthesis at atmospheric pressure of $BaCe_{1-x}Ca_xO_{3-\alpha}$. *J. Mater. Sci.* 2010, 45(21), 5860–5864.
316. Amar, I. A., Petit, C. T. G., Zhang, L., Lan, R., Skabara, P. J., Tao, S. Electrochemical synthesis of ammonia based on doped-ceria-carbonate composite electrolyte and perovskite cathode. *Solid State Ion.* 2011, 201(1), 94–100.
317. Amar, I. A., Lan, R., Petit, C. T. G., Arrighi, V., Tao, S. Electrochemical synthesis of ammonia based on a carbonate-oxide composite electrolyte. *Solid State Ion.* 2011, 182(1), 133–138.
318. Kordali, V., Kyriacou, G., Lambrou, C. Electrochemical synthesis of ammonia at atmospheric pressure and low temperature in a solid polymer electrolyte cell. *Chem. Commun.* 2000, (17), 1673–1674.
319. Liu, R., Xu, G. Comparison of electrochemical synthesis of ammonia by using sulfonated polysulfone and nafion membrane with $Sm_{1.5}Sr_{0.5}NiO_4$. *Chin. J. Chem.* 2010, 28(2), 139–142.
320. Xu, G., Liu, R., Wang, J. Electrochemical synthesis of ammonia using a cell with a Nafion membrane and $SmFe_{0.7}Cu_{0.3-x}Ni_xO_3$ (x = 0–0.3) cathode at atmospheric pressure and lower temperature. *Sci. China Ser. B Chem.* 2009, 52(8), 1171–1175.
321. Zhang, Z., Zhong, Z., Liu, R. Cathode catalysis performance of $SmBaCuMO_{5+\delta}$ (M=Fe, Co, Ni) in ammonia synthesis. *J. Rare Earths* 2010, 28(4), 556–559.

322. Song, T., Hu, P. Insight into the adsorption competition and the relationship between dissociation and association reactions in ammonia synthesis. *J. Chem. Phys.* 2007, 127(23), 234706.
323. Dietrich, H., Jacobi, K., Ertl, G. Coverage, lateral order, and vibrations of atomic nitrogen on Ru(0001). *J. Chem. Phys.* 1996, 105(19).
324. Marnellos, G., Athanasiou, C., Tsiakaras, P., Stoukides, M. Modelling of solid oxide proton conducting reactor-cells: thermodynamics and kinetics. *Ionics.* 1996, 2(5–6), 412–420.
325. Panagos, E., Voudouris, I., Stoukides, M. Modelling of equilibrium limited hydrogenation reactions carried out in H^+ conducting solid oxide membrane reactors. *Chem. Eng. Sci.* 1996, 51(11), 3175–3180.
326. Marnellos, G., Sanopoulou, O., Rizou, A., Stoukides, M. The use of proton conducting solid electrolytes for improved performance of hydro- and dehydrogenation reactors. *Solid State Ion.* 1997, 97(1–4), 375–383.
327. Giddey, S., Badwal, S. P. S., Kulkarni, A. Review of electrochemical ammonia production technologies and materials. *Int. J. Hydrog. Energy* 2013, 38(34), 14576–14594.
328. Hongsirikarn, K., Goodwin, J. G., Greenway, S., Creager, S. Influence of ammonia on the conductivity of Nafion membranes. *J. Power Sources* 2010, 195(1), 30–38.
329. Văsaru, G. *Tritium Isotope Separation.* CRC Press: Florida, 1993.
330. Chen, F. F. Introduction. In *Introduction to Plasma Physics and Controlled Fusion.* Springer International Publishing: Cham, 2016, pp. 1–18.
331. Howard K. Rae. *Separation of Hydrogen Isotopes.* American Chemical Society, 1978.
332. Ambashta, R. D., Sillanpää, M. E. T. Membrane purification in radioactive waste management: A short review. *J. Environ. Radioact.* 2012, 105, 76–84.
333. Glugla, M., Murdoch, D. K., Antipenkov, A., Beloglazov, S., Cristescu, I., Cristescu, I.-R., Day, C., Laesser, R., Mack, A. ITER fuel cycle R&D: Consequences for the design. *Fusion Eng. Des.* 2006, 81(1–7), 733–744.
334. Tosti, S., Bettinali, L., Rizzello, C., Violante, V. Membrane technologies for tritium recovering in the fusion fuel cycle, 2001.
335. Demange, D., Glugla, M., Günther, K., Le, T. L., Simon, K. H., Wagner, R., Welte, S. Counter-current isotope swamping in a membrane reactor: The PERMCAT process and its applications in fusion technology. *Catal. Today* 2010, 156(3–4), 140–145.
336. Bükki-Deme, A., Demange, D., Le, T.-L., Fanghänel, E., Simon, K.-H. PERMCAT experiments with tritium at high helium flow rates relevant for the tritium extraction systems using the CAPER facility at TLK. *Fusion Eng. Des.* 2015, Online.
337. Albrecht, H., Hutter, E. Tritium recovery from an ITER ceramic test blanket module — process options and critical R&D issues. *Fusion Eng. Des.* 2000, 49–50, 769–773.

338. Demange, D., Stämmler, S., Kind, M. A new combination of membranes and membrane reactors for improved tritium management in breeder blanket of fusion machines. *Fusion Eng. Des.* 2011, 86(9–11), 2312–2316.
339. Demange, D., Borisevich, O., Gramlich, N., Wagner, R., Welte, S. Zeolite membranes and palladium membrane reactor for tritium extraction from the breeder blankets of ITER and DEMO. *Fusion Eng. Des.* 2013, 88(9–10), 2396–2399.